26/6

26/6

Building Design
and Human
Performance

Building Design and Human Performance

Edited by Nancy C. Ruck, Ph.D.

Van Nostrand Reinhold
_____ New York

Copyright © 1989 by Van Nostrand Reinhold

Library of Congress Catalog Card Number 88-20809

ISBN 0-442-27847-0

Printed in the United States of America

Designed by Caliber Design Planning, Inc.

Van Nostrand Reinhold
115 Fifth Avenue
New York, New York 10003

Van Nostrand Reinhold International Company Limited
11 New Fetter Lane
London EC4P 4EE, England

Van Nostrand Reinhold
480 La Trobe Street
Melbourne, Victoria 3000, Australia

Macmillan of Canada
Division of Canada Publishing Corporation
164 Commander Boulevard
Agincourt, Ontario M1S 3C7, Canada

16 15 14 13 12 11 10 9 8 7 6 5 4 3 2 1

Library of Congress Cataloging in Publication Data

Building design and human performance / edited by Nancy C. Ruck.
    p.   cm.
  Bibliography: p.
  Includes index.
  ISBN 0-442-27847-0
  1. Architecture—Human factors. 2. Architecture—Environmental aspects.
I. Ruck, Nancy C.
NA2542.4.B85 1989
720—dc19   88-20809
                                      CIP

Foreword

Prior to the eighteenth century, building services were very simple: buildings were daylit, artificial lighting was by candles and oil lamps, and heating was by open fires and solid-fuel stoves. There was no sound amplification, no cooling, no artificial ventilation, and no air filtration. During the late nineteenth and early twentieth centuries, new methods and new equipment were developed, and it seemed possible that a perfect interior environment could be provided at a price, but a price that an affluent society could well afford.

Then came the energy crisis of the 1970s. It was perhaps politically embarrassing that some major city buildings used more energy than the entire requirement of one of the smaller countries in Central Africa, and it was worrisome that much of this energy had to be imported. The decisive new factor, however, was that creation of the "perfect environment" was no longer cheap.

This initiated a great deal of research and rethinking, and it became clear that much energy and money could be saved by some quite simple measures, such as greater attention to insulation and sunshading and better use of daylight to reduce the amount of artificial lighting.

The 1970s and 1980s also brought complaints indicating that many of the people who worked in these buildings did not perceive the environment as "perfect." It may be that repetitive strain injury (RSI) is an imaginary disease, or it may be that it is caused by a combination of poor illumination design, uncomfortable furniture and equipment, and boring work. People do complain, however, and the least we can do is attempt to remove any possible causes of complaint.

The critical comments on air-conditioning are at least partly due to the current theory that perfect comfort can be achieved by finding the temperature at which the largest number of people are most comfortable. This produces a monotonously uniform environment that might be acceptable if everybody traveled from an air-conditioned home in an air-conditioned car to an air-conditioned office and never set foot outside. Apart from the fact that most of us cannot afford to live this way, it is doubtful whether many of those who can would wish to do so. The indoor temperature should therefore be responsive to the outdoor conditions.

Interactions between and among the various aspects of environmental design are of great importance. Evidently illumination design has a profound effect on the thermal environment, because electric lamps create a heat load that can be the most important single factor in the design of the thermal environment. Air-conditioning ducts, unless given correct acoustic treatment, can transmit sound, so that the people in the waiting room, for example, can listen in on the conversation between a patient and his or her psychiatrist. Air-tight construction and recycling of air save energy for heating and cooling, but the reduced ventilation can cause public health problems, such as increasing the levels of radon gas in buildings where it is present because of certain materials.

The human factor is paramount in the design of buildings and building services. It is what people feel, not what instruments tell us they ought to feel, that matters. This book stresses the human factors; it also deals with acoustic, illumination, and thermal design, the design of building services, and the interactions of these types of designs in one volume. Dr. Ruck has broken new ground, and she has assembled a strong team of experts to discuss these diverse topics in a unified manner.

Henry J. Cowan
Professor Emeritus of
Architectural Science
University of Sydney

Preface

Interest in recent years has been directed toward the efficient use of energy in the built environment. This was a result of the energy crisis of the 1970s and of today's fluctuating energy costs. Although energy conservation measures in buildings and their cost-effectiveness are of major interest, the costs associated with the productivity of the occupants of buildings are invariably the largest component of a building's total operating budget, and small negative impacts as from the presence of glare in a building's interior will reduce the value of any energy savings. Consequently, human well-being and hence performance are of primary importance in the design of buildings from the energy viewpoint, in addition to the human need for an appropriate and stimulating environment.

It can be said that buildings do not consume energy, people do. At the same time, heating, cooling, lighting, and ventilation adjustments need to be made in response to people's needs and desires. Human well-being and energy savings do not necessarily achieve their optimums simultaneously, however, and in any design, basic human needs should have first priority.

In the last 30 years there has generally been one basic standard environmental control solution in which light, air temperature, humidity, and ventilation in buildings are managed internally by equipment such as electric lighting and heating, ventilation, and air-conditioning systems that produce static indoor conditions which ignore the variable stimuli of the outdoors. If the major objective in work situations is to increase the productivity of the occupants, then considerable thought should be given to a more appropriate environment for the required tasks; i.e., the stimulus of the environment must be sufficient to induce energetic action, but not so excessive that it dominates workers' activities. Subjective comfort is closely related to the immediate environment, but it also varies greatly with an individual's age, sex, activity, state of acclimatization, and past experience. Occupant response programs have determined that there can be considerable dissatisfaction with the control of summer and winter temperatures and ventilation. Physical measurements conducted during the winter in various locations have found that there are many instances of overheating. On the other hand, habitual exposure to overheated or overcooled conditions may well lead to a preference for warmer temperatures in the middle of the winter despite an increase in clothing over that worn in summer.

The specification of optimum human comfort in buildings is today becoming a controversial issue. The limitations of some of the traditional theoretical assumptions used in regulating the built environment are now being queried. The question is being asked, for example, whether a uniformity of temperature throughout the day produces the best possible conditions for either work or leisure. With the adoption of such an invariable temperature, independent of outside weather conditions, it is assumed that given a certain type of clothing and a relative humidity at which the largest number of people feel comfortable, fewer people will feel comfortable at any other temperature, whether higher or lower. The validity of this assumption is now being researched. In contrast, nature provides a daily temperature cycle, and it has been demonstrated that some people miss that diurnal change and complain about the boring uniformity of air-conditioning.

Although procedures for data analysis have reached a sufficiently high level of sophistication, more attention needs to be paid to variability rather than constancy in indoor environmental conditions and to the subjective reactions of people in actual environments. Technological advancements have provided the tools to simulate variability. Microprocessor developments have enabled the already versatile thermostat and the photovoltaic cell to undertake sophisticated control roles. In an effort to conserve energy, the thermostat has been modified to anticipate changes in heat flow to and from outdoors and to achieve an increased precision in its operation. These gyrations per se, however, are no more than a means to achieving a temperature constancy on the basis of some predetermined comfort criterion. Lighting controls are also programmed to relate to a given level of illuminance in order to achieve a recommended uniform lighting level regardless of the changing levels outdoors.

These traditional concepts of constancy in temperature and illuminance are the result of hypotheses based largely on laboratory conditions. It is evident that data on conditions where natural adaptation is allowed to operate freely can be obtained only in actual real environments. Research results are limited in this area.

Workspace conditions are complex and interrelated. Changing one feature is likely to alter other interrelated features, and consequently, attempts to improve one workspace condition may or may not improve other conditions. It is impossible to consider the effects of heat, light, and sound as separate, isolated elements in the design of buildings. In order to design for a comfortable visual environment, a detailed understanding of the interaction of the thermal and acoustic properties is also required, for example, the interaction of daylighting with heat gain and loss and noise insulation. A variety of other related stimuli and tradeoffs is involved. Their reactions with one another and their impact on human comfort (and energy conservation measures) need to be considered in total.

However, it is also necessary to study these physical variables separately for the purpose of decision making or to find a realistic solution. The architect or mechanical engineer, for example, must decide how much money should be spent on an air-conditioning system. This book has been planned, therefore, as a comprehensive series of expert individual viewpoints. These cover a wide range of perspectives, for it is in the nature of our individual reactions to the built environment that many perspectives are of value. The selected contributors also have explored human responses to heat, light, sound, and air quality in a number of different ways.

Although the book consists of independent viewpoints, coherence is achieved by dividing the text into four major parts with a concluding integrating chapter. The first part is concerned with human responses to heat, light, sound, and ventilation in the built environment. The second part deals with human requirements and hence criteria for good design, particularly in interior workspaces. The third part addresses the building envelope and the manipulation of its components to provide these requirements. And finally, the fourth part provides an analysis and evaluation of energy-conserving measures and their possible effects on human comfort.

This book can be used in a number of ways. Although its intent is to provide an overall view of current studies on human response to the built environment and the interaction of components associated with overall human comfort in buildings, certain chapters will have greater interest to some readers while for others the book will provide a stimulus for further research in areas that have not as yet been studied. *Building Design and Human Performance* will therefore appeal to a broad technical and professional audience of researchers, architects, engineers, and educators.

Nancy C. Ruck
Associate Professor
University of
New South Wales

List of Contributors

A. Auliciems
Associate Professor, Department of Geographical Sciences, University of Queensland, St. Lucia, Queensland 4067, Australia.

H. J. Cowan
Emeritus Professor, Department of Architectural Science, University of Sydney, Sydney 2006, Australia.

G. Cunningham
Principal Designer, Building Management Systems, Civil and Civic, Australia Square, Sydney 2000, Australia.

A. B. Lawrence
Associate Professor of the Graduate School of the Built Environment, University of New South Wales, Kensington, New South Wales 2033, Australia.

N. C. Ruck
Associate Professor of Architecture, University of New South Wales, Kensington, New South Wales 2033, Australia.

S. Selkowitz
Group Leader, Windows and Daylighting, Lawrence Berkeley Laboratory, University of California, Berkeley, California 94720, U.S.A.

P. R. Smith
Associate Professor of Architectural Science, University of Sydney, Sydney 2006, Australia.

Contents

PART I

Human Response to the Environment

In Part I, on human responses to heat, light, sound, and air quality, the contributors summarize the results of recent studies in these areas and the findings of current research on causes of thermal stress, air pollution, visual fatigue, and reactions to noise. Possible solutions to these response problems are outlined. Although these responses cannot altogether be considered in isolation, they are examined here according to their individual effects on human beings.

In Chapter 1, on the thermal environment, Andris Auliciems evaluates human stress, identifies preferred levels of warmth in various climates, and presents the results of data analyses on human preferences. The thermal constancy hypothesis is questioned, and an alternative hypothesis that takes into account human adaptability is presented.

In Chapter 2, on air quality, George Cunningham details sources of air pollution in and around buildings, how contaminants are brought into buildings, and the effects of pollutants and viruses on the health and well-being of building occupants.

In Chapter 3, on the luminous environment, Nancy Ruck describes the function and limits of performance of our visual system, our response to both the visual and nonvisual effects of light, and the effects of light sources on our social behavior.

In Chapter 4, on the acoustic environment, Anita Lawrence describes human sound perception, defines audio ranges and thresholds, our responses to vibration in buildings, and also describes the physiological and psychological effects of noise.

Thermal Comfort

1

Andris Auliciems

Indoor thermal comfort has been hailed as one of the finest achievements of modern civilization (Benzinger 1978). In this, due recognition also must be given to the part played by the technology of the twentieth century. Rarely would today's affluent citizen need to share the concerns of celebrated nineteen-century English diarist Gilbert White, who frequently noted the unwelcome incursions of winter weather into his country residence. Only the wine cellar, he observed with accuracy and, it seems, with some consolation, offers a near-acceptable abode (temperatures in degrees Fahrenheit): "1877 Jan 9. Frost comes within doors. Thermometer within 28, in the wine vault 43½, abroad 24" (Johnson 1931).

While from the earliest of times heat and cold discomfort has indeed been the normal condition for most people, this same thermal depredation also has provided a major stimulus to human development. In large part, the course of human evolution and distribution on the earth's surface may be charted by the organism's responses to prevailing thermal environments. Physiological adaptation and technological development have gone hand in hand to permit human settlement in climate zones well beyond the likely original warm human birthplace (Burton and Edholm 1955; Dubos 1965; Mayr 1956; Newman 1953, 1955; Sargent 1963; Scholander 1955, 1956).

In the earlier part of the present century, some authors also emphasized that excessive heat constituted a major constraint to human well-being in the low latitudes (Mills 1946; Taylor 1959). The technology of microclimate control, in particular, was seen as the only key to the success of human beings in these environments (Markham 1947). Some authors even suggested that the very rise and fall of certain civilizations was determined by climate (Huntington 1924, 1926; Missenard 1957; Toynbee 1945). For additional reappraisal of this complex and at times highly controversial field, the interested reader is particularly urged to consult works by Chappell (1975), Lamb (1982), and Sargent and Tromp (1964).

Whatever the broader implications, despite the advances in building design and the relative abundance of fuels, even today's affluent Western society cannot be considered to have eradicated the miseries of excessive thermal stress. Surveys in Great Britain (Fox et al. 1973), Belgium, and France (Collins 1979) still find the less privileged groups of people, such as the elderly, suffering from cold in poorly insulated and unheated houses. Recently, even "affluent" Americans have been reminded of the fragility of the thermal fabric of modern society by the shortages of fuel oils during the "crisis" of the winter of 1973-1974. No doubt the documented cases of thermal strain represent little more than the tip of a vast iceberg existing even in today's global society.

Thus, while thermally comfortable indoor environments are taken for granted by many urban dwellers, the present undue dependence on equable and mild indoor warmth, as generated by the conversion of energies external to the human body, must be viewed with some concern. Logically, for the long-term well-being of human beings, the striving toward optimal environments should examine all available options. This, by necessity, requires the questioning of some very entrenched approaches and a reexamination of the concept of what is often referred to as "thermal comfort."

QUANTIFICATION OF THERMAL STRESS

Since human beings are homeotherms, with a need to maintain a constant core temperature near 37°C (98.6°F), they require continuous physiological and behavioral adjustments to balance energy exchanges between the body and the environment. Thus all modern studies of thermoregulatory responses are based on an energy-budget model, most frequently some variant of the original model proposed by Gagge (1936):

$$M \pm R \pm C - E = \pm S \, (\text{W/m}^2) \tag{1.1}$$

where M is the net metabolic rate; R, C, and E are radiation, convection, and evaporation, respectively; S is storage within body tissues.

Based on the well-established physical laws of energy transfer, the determination of particular algorithms has been the preoccupation of much thermophysiological work. Subjects have been repeatedly tested in laboratories under wide ranges of ambient conditions and with varying postures, metabolic rates, states of acclimatization, and clothing insulation. Useful reviews may be found in numerous monographs and journal articles, including those in Hardy (1963), Cena and Clark (1981), and McIntyre (1980). It is sufficient here to summarize that in the maintenance of homeothermy, two personal and four atmospheric parameters need to be considered:

1. Metabolic rate
2. Clothing insulation
3. Air temperature
4. Radiant temperature of surroundings
5. Rate of air movement
6. Atmospheric humidity

Each of these parameters has been quantified, and atmospheric coefficients selected as best by McIntyre (1980) appear in Table 1.1.

TABLE 1.1 Summary of Heat-Loss Equations

Radiation	$R = \epsilon f_{\text{eff}} f_{\text{cl}} h_r (T_{\text{cl}} - T_r)$	W/m²
	$h_r = 4\cdot6(1 + 0\cdot01 T_r)$	W/m²K
	$f_{\text{eff}} = 0\cdot72$	
	$f_{\text{cl}} = 1 + 0\cdot15 I_{\text{clo}}$	
	$\epsilon = 0\cdot95$	
Convection	$C = h_c(T_{\text{cl}} - T_a)$	W/m²
	$h_c = 8\cdot3 \sqrt{v} \qquad v > 0\cdot2\text{m/s}$	W/m²K
	$h_c = 4\cdot0 \qquad v < 0\cdot2\text{m/s}$	W/m²K
Evaporation		
Regulatory	$E_{\text{max}} = h_c(p_{\text{ssk}} - p_a)$	W/m²
	$h_e = 1\cdot65 h_c$	W/m²mb
	$E = w E_{\text{max}}$	W/m²
Insensible:		
Skin diffusion	$E_{\text{is}} = 4 + 0\cdot12(p_{\text{ssk}} - p_a)$	W/m²
Respiration (latent)	$E_{\text{res}} = 0\cdot0017 M(59 - p_a)$	W/m²
Respiration (dry)	$C_{\text{res}} = 0\cdot0014 M(34 - T_a)$	W/m²
Conduction through		W/m²
clothing	$R + C = K = (T_{\text{sk}} - T_{\text{cl}})/(0\cdot155 I_{\text{clo}})$	
Evaporation		W/m²
through clothing	$E_{\text{max}} = f_{\text{pcl}} h_c(p_{\text{ssk}} - p_a)$	

Where ϵ = emissivity, f_{eff} = effective radiation area, f_{cl} = clothing area factor, h_r = radiative transfer coefficient, T_{cl} = surface temperature of clothed body (in °C), T_r = radiant temperature (in °C), I_{clo} = clothing insulation in clo units, h_c = connective transfer coefficient, T_a = air temperature (in °C), v = air speed (in m/s), E_{max} = maximum evaporative loss from wet skin per unit area, p_{ssk} = saturated vapor pressure (mb), p_a = partial vapor pressure (mb), h_e = evaporative transfer coefficient, w = skin wettedness, E_{is} = evaporative loss by diffusion through skin, E_{res} = evaporative loss through respiration, C_{res} = dry respiration loss, T_{sk} = skin temperature (in °C), and f_{pcl} = permeation efficiency factor.

Source: From McIntyre (1980).

The basic heat-balance equation has been further analyzed to produce single-unit indices of heat strain generated by conditions when $S = 0$. These indices have merit in attempts to translate the storage term into objective and meaningful physiological effect. In cold conditions, estimation of human thermoregulatory response appears to be a relatively simple matter of linear compensation for excessive dry heat loss. For example, mathematically, $S = -10$ W/m² can be balanced by a corresponding increase in metabolic rate, whereas the resistance to heat transfer through clothing is in direct proportion to its thickness. With heat, however, account also has to be made of the inefficiencies involved in sweating, since not all sweat produced is effective in body cooling, and the excess production must be incorporated into strain estimates.

Thus most of the analytical indices have concentrated on responses under heat stress. Some of these are the predicted 4-hour sweat rate (McArdle et al. 1947; Smith 1955), the heat-strain index (Belding and Hatch 1955), the relative strain index (Lee and Henschel 1963), and index of thermal stress (Givoni 1963).

In a recent review, Szokolay (1985) has differentiated between the preceding indices and those which are empirically established using social survey methods, such as the comfort-vote technique outlined below. The second group includes effective temperature (Houghten and Yaglou 1923a, b), equivalent warmth (Bedford 1936), operative temperature (Winslow, Herrington, and Gagge, 1937), resultant temperature (Missenard 1935), and the equatorial comfort index (Webb 1959). Table 1.2 shows thermal scales used in comfort research.

Effective temperature (*ET*) and corrected effective temperature (*CET*) are probably the best known of these scales, and the latest variant, derived by further analysis of heat flows, the new effective temperature *ET** (Gagge, Stolwick, and Nishi 1971) is used in the ASHRAE Comfort Standard 55-74 (1981). Simply described, *ET** is the temperature of an isothermal environment at moderate humidity in which a lightly clothed individual would exchange the same total heat as in the atmospheres in question. For example, a person sitting in light clothing in low air movement at a temperature of 25°C (77°F) and 50 percent relative humidity (RH) would experience $ET^* = 25°C$. Some suggested physiological, sensation, and comfort responses for several ranges of effective temperatures, based on laboratory studies, are shown in Table 1.3 (see also Fig. 1.11).

The well-publicized comfort equation and predicted mean vote (PMV) of Fanger (1967, 1970) are also hybrids of both techniques. The equations essentially are empirical translations of the heat exchanges described in Eq. (1.1), but in addition to the achievement of homeothermy, conditions of comfort need to satisfy the criteria of comfort in skin temperature and an appropriate rate of sweating according to given metabolic rates.

The PMV equation (indicating comfort for sedentary populations wearing light clothing, at $-0.5 < PMV < +0.5$) is as follows:

$$PMV = (0.303e^{-0.036M} + 0.028)\{(M - W) - 3.05 \times 10^{-3} \tag{1.2}$$
$$\times [5733 - 6.99(M - W) - p_a] - 0.42$$
$$\times [(M - W) - 58.15] - 1.7 \times 10^{-5}M(5867 - p_a)$$
$$- 0.0014M(34 - T_a) - 3.96 \times 10^{-8}f_{cl}$$
$$\times [(T_{cl} + 273)^4 - (T_r + 273)^4] - f_{cl}h_c(T_{cl} - T_a)\}$$

TABLE 1.2 Thermal Scales for Use in Comfort Research

ASHRAE Code (7-Point)	Sensation/Comfort		Preference[+]	Bedford (1936)
	Extended	*Hybrid**		
+4	Very hot	Hot		
+3 (hot)	Hot	Hot/warm	Much cooler	Much too warm
+2 (warm)	Warm	Warm	Cooler	Too warm
+1 (slightly warm)	Slightly warm	Warm/comfort	Slightly cooler	Comfortably warm
0 (neutral[‡])	Neutral	Comfort	No change	Comfortable[§]
−1 (slightly cool)	Slightly cool	Comfort/cool	Slightly warmer	Comfortably cool
−2 (cool)	Cool	Cool	Warmer	Too cool
−3 (cold)	Cold	Cold/cool	Much warmer	Much too cool
−4	Very cold	Cold		

*Developed by Rohles and Millikin (1981).

[+]Developed by de Dear (1985).

[‡]Original 0 category was "comfortable," e.g., as used by Nevins et al. (1966).

[§]Further explained as "neither warm nor cool" and "just right."

TABLE 1.3 Thermal Relationships Observed in Laboratory Studies of Lightly Clothed, Sedentary College Students in the United States

Environmental Warmth		Physiological Response	Assumed Comfort Level
ASHRAE Code	SET (°C)*		
+3	34.5	Profuse sweating	Unacceptable
+2	30.1–34.5	Onset of sweating	Uncomfortable
+1	25.6–30.0	Vasodilatation	Acceptable
0	22.1–22.5	Minimal	Maximum
−1	17.6–22.0	Vasoconstriction	Acceptable
−2	14.6–17.5	Onset of body cooling	Uncomfortable
−3	14.6	Shivering	Unacceptable

*Standard effective temperature (SET), based on McIntyre (1980).

where PMV = predicted mean vote

    M = metabolic rate, in W/m^2 of body surface area

    W = external work, in W/m^2, equal to zero for most activities,

    p_a = partial water vapor pressure, in Pa

    T_a = air temperature, in °C

    f_{cl} = ratio of human surface area while clothed, to human surface area while nude

    T_{cl} = surface temperature of clothing, in °C

    T_r = mean radiant temperature, in °C

    I_{cl} = thermal resistance of clothing, in m^2/°C·W

    h_c = convective heat transfer coefficient, in W/m^2·°C

    v = relative air velocity, in m/s

and where

$$I_{cl} = 35.7 - 0.028(M - W) - I_{cl}\{3.96 \times 10^{-8} f_{cl}$$
$$\times [(T_{cl} + 273)^4 - (T_r + 273)^4] + f_{cl}h_c(T_{cl} - T_a)\}$$

$$h_c = \begin{cases} 2.38(T_{cl} - T_a)^{0.25} & \text{for } 2.38(T_{cl} - T_a)^{0.25} > 12.1\sqrt{v} \\ 12.1\sqrt{v} & \text{for } 2.38(T_{cl} - T_a)^{0.25} < 12.1\sqrt{v} \end{cases}$$

$$f_{cl} = \begin{cases} 1.00 + 1.290 I_{cl} & \text{for } I_{cl} < 0.078 \text{ m}^2 \cdot °\text{C/W} \\ 1.05 + 0.645 I_{cl} & \text{for } I_{cl} > 0.078 \text{ m}^2 \cdot °\text{C/W} \end{cases}$$

To solve these equations, standard metabolic rates, as in Table 1.4, and clothing insulation values, as in Table 1.5, are used in computer programs (e.g.,

TABLE 1.4 Metabolic Rates for Different Activities

Activity	Metabolic Rate (W/m^2)
Reclining	46
Seated, relaxed	58
Standing, relaxed	70
Sedentary activity (office, dwelling, school, laboratory)	93
Standing activity (shop assistant, domestic and machine work)	116
Medium activity	165

Source: From International Standard (1984).

TABLE 1.5 Insulation Values of Clothing

Garment	I_{cl} m^2·C/W	I_{cl} clo
Pantyhose	0.002	0.01
Socks: light	0.005	0.03
heavy	0.006	0.04
Underwear: bras and panties	0.008	0.05
half slip	0.020	0.13
full slip	0.029	0.19
briefs	0.008	0.05
undershirt	0.009	0.05

Note: 1 clo = 0.155 m^2·°C/W. Total resistance of ensembles is calculated by $I_{cl} = 0.82 \, \Sigma I_{cli}$, where I_{cli} is insulation of individual garments.

Source: Adapted from International Standard (1984).

(continued)

TABLE 1.5 Insulation Values of Clothing—*Continued*

Garment	I_{cl}	
	$m^2 \cdot C/W$	*clo*
Shirt: T-shirt	0.014	0.09
light, short sleeved	0.031	0.20
light, long sleeved	0.043	0.28
heavy, short sleeved	0.039	0.25
Skirt: warm	0.034	0.22
Dress: light	0.026	0.17
heavy	0.098	0.63
Sweater: light, short sleeved	0.026	0.17
heavy, long sleeved	0.057	0.37
Jacket: heavy	0.076	0.49
Trousers: light	0.040	0.26
medium	0.050	0.32
heavy	0.063	0.44
Shoes: light	0.006	0.04
Clothing ensemble:		
Shorts	0.015	0.1
Typical tropical clothing ensemble:	0.045	0.3
Briefs, shorts, open-neck shirt with short sleeves, light socks, and sandals		
Light summer clothing:	0.08	0.5
Briefs, long light-weight trousers, open-neck shirt with short sleeves, light socks, and shoes		
Light working ensemble:	0.11	0.7
Light underwear, cotton work shirt with long sleeves, work trousers, woollen socks, and shoes		
Typical indoor winter clothing ensemble:	0.16	1.0
Underwear, shirt with long sleeves, trousers, jacket or sweater with long sleeves, heavy socks, and shoes		
Heavy traditional European business suit:	0.23	1.5
Cotton underwear with long legs and sleeves, shirt, suit including trousers, jacket, and waistcoat, woollen socks, and heavy shoes		

TABLE 1.6 Comparison of Comfort Indices

Index	Variables*	Range	Comments
Effective temperature (*ET*)	T_a, RH, v 2 levels clo	$0 < ET < 45°C$ $v < 2.5$ m/s	Major index, now superseded. Overestimated effect of humidity at low temperatures.
Corrected effective temperature (*CET*)	T_a, T_r, RH, v		As above. Used in British armed forces.
Resultant temperature (T_{res})	T_a, T_{wb}, v	$20 < T_{res} < 40°C$ $v < 3$ m/s	Similar to *ET*, but more accurate. Separate charts for nude-sedentary and clothed-active.
Equivalent temperature (T_{eq})	T_a, T_r, v	$8 < T_{eq} < 24$ $v < 0.5$	Originally defined as eupatheoscope reading. Later defined by Bedford's regression equation.
Fanger's comfort equation	All	Comfort condition	
Predicted mean vote (*PMV*)	All	$2.5 < PMV < 5.5$	No range stated, but *PMV* becomes less certain away from thermal comfort.
Standard effective temperature (*SET*)	All	Shivering to upper limit of prescription zone	Based on model of physiological response. Most general index.
New effective temperature (*ET**)	T_a, RH	Sedentary, light clothing only	Used by ASHRAE for indoor comfort zone.
Subjective temperature (T_{sub})	T_a, T_r, v, H, I_{clo}	Near comfort	Simple index used to predict comfort conditions, based on Fanger's equation.

*T_{wb} = wet-bulb temperature; other terms as defined for Eq. (1.2) and Table 1.1.

Source: From McIntyre (1980).

see Fortran IV in Annexe D of International Standard, 1984). In the field, it is most convenient to arrive directly at the solution by a thermal comfort meter (Bruel and Kjaer Type 1212), which allows direct measurement of the ambient atmospheric parameters and their integration for specific metabolic rates and clothing values (see Madsen 1979).

Together with the standard effective temperature (*SET*), which allows for variations in *ET** as well as in metabolic rate, clothing amounts, and air movement (Nishi and Gagge 1977), PMV and SET indices represent the most sophisticated attempts to mathematically model human physiological responses to particular environments. Table 1.6 shows McIntyre's (1980) comparison of the main characteristics of selected indices.

Whatever the mathematics involved, homeothermy demands that [within narrow limits in Eq. (1.1), $S = 0$] metabolic heat production, which varies according to activity, must be balanced by its dissipation to the environment over time. During this balancing process, all the terms on the left side of Eq. (1.1) may be varied by either involuntary physiological processes or by behavioral responses as listed in Table 1.7. As regards the body's active adjustments of the individual atmospheric parameters themselves, there is little evidence to suggest that there is preference for adjustment of any particular element in the equation.

Perhaps surprisingly, in field studies the complex thermal indices have not proved to be particularly useful predictors of comfort levels. This is shown in Table 1.8 by the magnitudes of correlation of common indices with comfort

TABLE 1.7 Main Human Thermoregulatory Responses

Physiological	Behavioral
Vasomotor activity	Voluntary postural adjustment
Vasodilatation	Alteration of activity
Vasoconstriction	Relocation
Involuntary postural adjustment	Interposition of clothing
Cardiovascular change	Change of diet
Thermogenesis	Migration
Muscular tension	Construction of housing
Shivering	Heating
Sweating	Cooling
Habituation	
Acclimatization	

TABLE 1.8 Correlation of Thermal Votes with Selected Indices During Normally Encountered Conditions in Field Studies

Author	Number of Observations	Pearson's Correlation with					
		T_a	T_g	T_{eq}	ET	HSI	ITS
Bedford (1936)	2571	0.48	0.51	0.52	0.48		
Hickish (1955)	1537	0.38	0.37	0.37	0.34		
Auliciems (1972)	2624	0.37	0.32	0.36	0.36		
Humphreys and Nicol (1970)	1284	0.67	0.66			0.62	0.59
Woolard (1979)*	1965	0.56	0.57		0.55	0.57	0.56

*Graphic scale as in Fig 1.3; all others Bedford; T_a = dry-bulb thermometer temperature, T_g = globe thermometer temperature, T_{eq} = equivalent temperature, ET = effective or corrected effective temperature, HSI = heat-stress index, ITS = index of thermal stress.

votes by people in climates ranging from the cool of Europe to the equable warmth of the tropics on sensation scales as subsequently described. It would seem that under the ranges of conditions normally experienced by sedentary populations, and characteristically without excessive air movement, radiation, and humidity, there really is little reason to use other estimates of warmth beyond simple air temperature.

EVALUATION OF THERMAL CONDITIONS

If indoor environments are to be optimized, determination of human preferences for particular levels of warmth (i.e., the thermopreferendum) becomes an essential task for microclimatic design. Critical to this are the methods employed in data gathering, data interpretation in the light of known psychophysiological processes, and the logic of applying these methods and interpretations to the highly adaptive human organism.

Over a large range of thermal stimuli, the intensity of sensations may be expected to follow the sigmoidal curve of the power function of Weber-Fechner (Chrenko 1955; Stevens and Stevens 1963). The relationship is given by

$$\psi - \psi_0 = k'(\varphi - \varphi_0)^n \tag{1.3}$$

where ψ is sensation and ψ_0 is absence of sensation, φ is thermal stimulus and φ_0 is the point of neutrality, k' is a constant, and n is an exponent that varies according to different sense modalities (Stevens and Stevens 1963). In general, such a relationship also may be expected to occur with the stimuli of light (Chap. 3) and sound (Chap. 4).

Figure 1.1 shows a computer-fitted relationship as observed by de Freitas (1979) between the response by beach-goers on a thermal sensation scale and the index of thermal stress (Givoni 1963).

Since the early experiments by Houghten and Yaglou (1923a, b), the

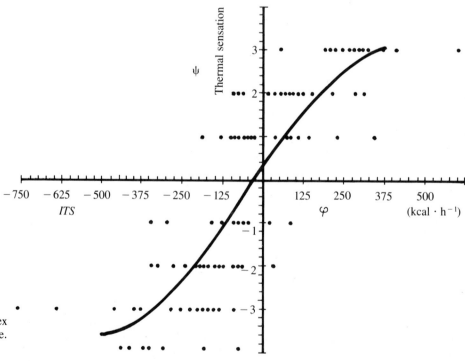

FIGURE 1.1.
Thermal sensations against Giovoni's (1963) index of thermal stress (ITS) in a beach microclimate. (After de Freitas 1979.)

evaluation of thermal conditions has been based mostly on response on verbal scales that report sensory evaluation of the level of warmth of an environment, as in Fig. 1.1. Alternatively, the expression has been one of the degree of subjective discomfort experienced. These responses have been rather loosely assumed to be "ratio" scales, in which discrete thermal votes have been assigned convenient numerical values. As noted by Humphreys (1975), the number of steps have ranged from three to twenty-five, and most have been symmetrical around a minimum response category.

The most frequently applied scales have been the ASHRAE (1966) in laboratories and the Bedford (1936) in field studies (see Table 1.2). At the extremes, and especially with changing temperatures (Gagge, Stolwijk, and Hardy 1967), the particular semantics employed appear to lead to some divergence in interpretation of the categories employed. However, within the central categories, no significant differences in interpretation have become evident, and ASHRAE and Bedford scales tend to produce identical results, especially with English-speaking samples (de Dear 1985, Table 5.5).

Other techniques have included the use of direct assessments of thermal preference by respondents' adjustments of indoor temperatures at Fanger's Laboratory of Heating and Air-Conditioning at the Technical University of Denmark. More recent developments include the normalized certainty scale and application of the semantic differential (Winakor 1982) and the latter's extension to the differential attributes scale (Rohles and Laviana 1985), as in Fig. 1.2. In the semantic differential format, the responses are treated on a continuum and loadings are attributed to each adjective pair to produce an index of thermal comfort. In the differential attribute format, correlation and principal component analysis are used to identify thermal satisfaction and thermal dissatisfaction. Unfortunately, insufficient attention has been paid to validation of some scales, and there is a need to develop culture- and language-free methods of quantifying warmth perceptions.

One seemingly successful attempt to remedy this shortcoming has been Woolard's (1979) transcription of the seven-point verbal ASHRAE scale into both pidgin English and a graphic representation for use with indigenous people in the Solomon Islands (Fig. 1.3). His scales, when administered to 10 percent of the Honiara population, showed correlations with the ASHRAE scale at $r = 0.98$ and $r = 0.54$ with air temperature.

The robustness of the graphic representation is particularly encouraging. Suitable adjustments of the cartoon character to conform with cultural types may facilitate cross-cultural comparisons and provide a tool for research into the requirements of people with communications problems. With young children, for example, scales may be like those shown in Fig. 1.4. With the advent of these techniques it has become apparent that some questions need to be raised as to the precise meaning of the recorded phenomena. "Preference" in particular appears to display cultural bias. In Canada (Auliciems and Parlow 1975) and the United States (McIntyre and Gonzalez 1976), college-age subjects indicated average winter preferences for the 0 to −1 categories of the Bedford and ASHRAE scales, respectively, while in England, Humphreys (1975) and McIntyre and Griffiths (1975) found preferred sensations to lie within the 0 to +1 categories. Rohles and Johnson (1972) observed that while similar in ASHRAE votes, in general elderly people preferred temperatures 1°C (2°F) warmer than middle-aged people who, in turn, showed similar elevations over college-age subjects.

In recognition of these differences, and unless specifically stated otherwise, all subsequent discussion refers to data analyzed on the ASHRAE and Bedford scales. Although the particular semantics employed may have theoretical significance (Auliciems 1981, 1983), for our purposes the central categories "neutral" and "comfortable" are treated as synonymous.

Thermal comfort scale in semantic differential format

	:	:	:	:	:	:	:	:	:	
Comfortable	:	:	:	:	:	:	:	:	:	Uncomfortable
Bad temperature	:	:	:	:	:	:	:	:	:	Good temperature
Pleasant	:	:	:	:	:	:	:	:	:	Unpleasant
Unacceptable	:	:	:	:	:	:	:	:	:	Acceptable
Satisfied	:	:	:	:	:	:	:	:	:	Dissatisfied
Uncomfortable temperature	:	:	:	:	:	:	:	:	:	Comfortable temperature

Thermal comfort scale in differential attribute format

Thermal Environment Ballot

Instructions: Below is a list of words that can be used to describe the thermal environment. We would like you to rate how accurately the words below describe the **Thermal Environment** of this place. Use the 1–7 scale for your answer for each word.

7 = Very accurate
6 = Accurate
5 = Slightly accurate
4 = Neutral, neither accurate nor inaccurate
3 = Slightly inaccurate
2 = Inaccurate
1 = Very inaccurate

The Thermal Environment

1. Uncomfortable	12. Good	23. Intolerable
2. Content with	13. Unacceptable	24. Disagreeable
3. Agreeable	14. Enjoyable	25. Adequate
4. Tolerable	15. Great	26. Desirable
5. Unpleasant	16. Distressful	27. Unsatisfactory
6. Inadequate	17. Bad	28. Gratifying
7. Annoying	18. Acceptable	29. Pleasing
8. Undesirable	19. Discontent with	30. Poor
9. Satisfactory	20. Pleasant	31. Appealing
10. Miserable	21. Dissatisfied with	32. Delightful
11. Satisfied with	22. Comfortable	

Normalized certainty scale

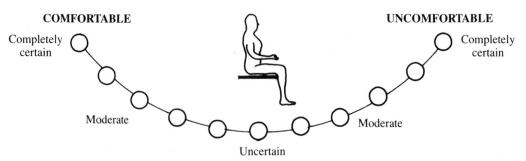

FIGURE 1.2.
Scales in thermal comfort research. (Adapted from Rohles and Lavinia 1985.)

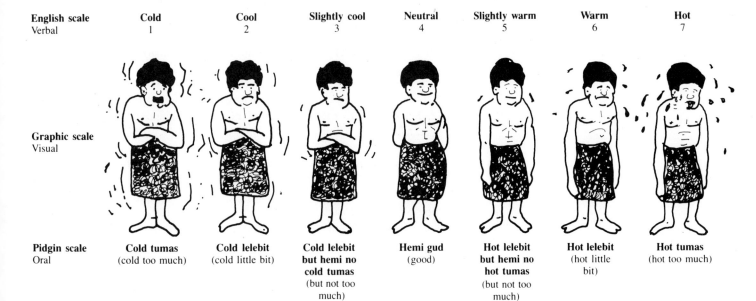

English scale Verbal	Cold 1	Cool 2	Slightly cool 3	Neutral 4	Slightly warm 5	Warm 6	Hot 7
Graphic scale Visual							
Pidgin scale Oral	Cold tumas (cold too much)	Cold lelebit (cold little bit)	Cold lelebit but hemi no cold tumas (but not too much)	Hemi gud (good)	Hot lelebit but hemi no hot tumas (but not too much)	Hot lelebit (hot little bit)	Hot tumas (hot too much)

FIGURE 1.3.
Relationship of thermal sensation scales as used
by Woolard (1979) with Solomon Islanders.

−3	−2	−1	0	+1	+2	+3

FIGURE 1.4.
Graphic thermal sensation scales for use with
children. Verbal instruction: "How warm/cool
do you feel now? Point to the picture that best
describes how you feel."

Thermal Comfort **13**

DATA GATHERING AND ANALYSIS

Basically, thermal scales have been administered either to laboratory samples or to people in the field. Notable among the former are the extensive chamber studies at Kansas State University, at the J. B. Pierce Laboratories in New Haven, Connecticut, at the Electricity Research Council Research Centre in Chester, England, and at the Technical University of Denmark. In such laboratories, factors likely to influence thermal sensations have been reduced to a minimum, especially the influence of clothing.

Field studies are far more numerous, and since personal factors have been left uncontrolled, the results represent more closely real-life conditions. The vast majority of field studies have used groups of people in public buildings, including offices, factories, and schools, and all have employed metabolic rates below 100 W/m². At times, carefully designed comparative studies have been carried out between different locations, perhaps most notably by de Dear (1985) in several climatically different zones.

Methods of recording subjective thermal responses have included individual interrogation, voting on slips of paper, directly punching information into computers, and adjusting dials to regulate ambient warmth. Sample sizes have ranged from one person to groups of several dozen individuals in the same room, with atmospheric parameters being recorded at a central location. The data have been analyzed to identify the location on thermal scales of the central neutral category and the positions of those judged to be stressful. Apart from percentage frequency counts per category, the most usual form of data treatment has involved linear regression analysis, on the assumption that the transformation of thermal responses into real numbers is valid and that, within the central portion of the general power curve in Fig. 1.1, equal increments of warmth produce equal increments of response.

A typical distribution of individual thermal responses against air tempera-

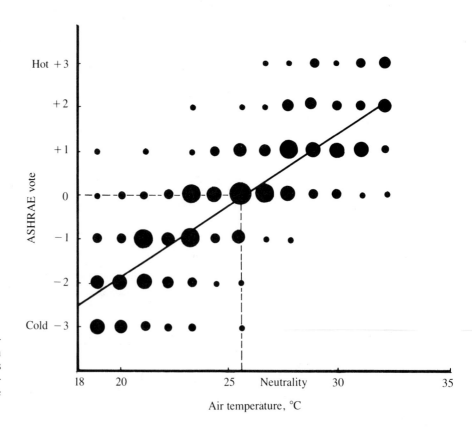

FIGURE 1.5.
The warmth sensation of 1296 subjects as a function of temperature after a 3-hour exposure in light clothing. The area of each circle represents the number of responses. The best-fitting regression is superimposed. (Adapted from McIntyre 1980.)

ture is shown in Fig. 1.5. Neutrality for these particular laboratory results, as determined by the regression, is identified here at 25.5°C (77.9°F). This assumption of linearity is probably accurate enough for the practical purpose of establishing a minimum response under most conditions. It does appear, however, that there may be a generally increased sensitivity to cooler environments (Stevens, Marks, and Gagge 1967) by perhaps up to 15 percent in regression slopes on either side of 25.5°C (77.9F) air temperature (McIntyre 1978).

   Since available data suffer from truncations at thermal extremes, and since most scales use whole numbers, linear regression has recently given way to probit regression following transformation of data to smoothed cumulative frequency curves for particular category boundaries (i.e., the proportions of individuals voting above or below specific values). The latter approach more closely generates the sigmoidal curve distribution, as shown in Fig. 1.1, and one that can be used to define category boundaries, as shown in the almost ideal case in Fig. 1.6. This technique was developed initially in toxicology by Finney (1952), and first applied to thermal sensation studies by Chrenko (1955), but particularly useful and brief descriptions of the probit transformation procedures in thermal comfort work may be found in Ballantyne, Hill, and Spencer (1977) and McIntyre (1980).

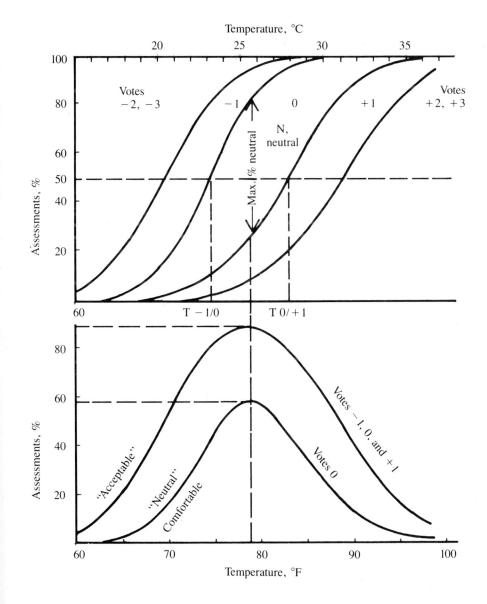

FIGURE 1.6.
(*Above*) Percentages of votes falling within the central vote categories. (*Below*) Accumulative probit boundaries between ASHRAE vote categories. (Adapted from Ballantyne, Hill, and Spencer 1977.)

COMFORT ZONES

Typical standard deviations of seven-point scale votes, both between and among subjects in laboratory studies, have been 0.8 to 1.0 (McIntyre 1980). In the field, standard deviations approach closer to 2 to 3 and double that for school children (Humphreys 1975). Figure 1.6 also shows the percentages of votes falling within the three central categories of the ASHRAE scale, as observed with a laboratory sample. Here neutrality is seen to occur in a maximum of 60 percent of the sample at any single temperature, but apart from air-conditioned places, in the field, the neutral vote is unlikely to occur in more than 40 percent of any sample. Neither, as illustrated by the cumulative percentage curves that enclose the three central categories, is it certain that everyone will experience one of the "comfort" sensations.

Apart from identifying idealized design temperatures, single value recommendations obviously are not as useful as the definition of suitable ranges, or "comfort zones." Given the variability noted above, the choice of particular temperature limits becomes a pragmatic one of what is regarded as constituting acceptable levels of discomfort. Most suitable appear to be the temperatures falling between the median votes of +0.5 (i.e., at $T-1/0$ and $T0/+1$ in Fig. 1.6) or those within which at least some 80 percent of laboratory subjects are likely to vote in either of the three central categories. With school children, however, it has been necessary to resort to no more than 60 percent (Auliciems 1969a).

The relative widths of categories thus defined may vary, but in general, the central comfort zone exceeds the others by some 25 percent. Humphreys (1975) calculated for field data that the mean range for each of the three central categories is approximately 4 K. With a temperature shift of this magnitude, a group of respondents is likely to indicate a whole vote change.

THE THERMAL CONSTANCY HYPOTHESIS

In the first half of this century there was a general acceptance of different comfort-zone requirements for winter and summer. In the United States, these seasonal differences had been identified by Houghten and Yaglou (1923a, b) and Yaglou and Drinker (1928), and in England, they were identified by Bedford (1936) and Hickish (1955). During a period of intensive laboratory experiments, especially at the ASHRAE climate chamber at Kansas State University in the 1960s, however, several large samples of college-age subjects wearing standard clothing and having normal metabolic rates recorded neutralities at the same temperature during all seasons (Nevins et al. 1966; McNall et al. 1968). ASHRAE (1966 and 1974) Standard 55 no longer differentiates between the seasons.

This constancy hypothesis has been most strongly supported by Fanger. Largely on the basis of two experiments in Copenhagen on a small group of "tropical travelers," winter swimmers, and meat packers, the hypothesis has been extrapolated as equally applicable to human beings throughout the world regardless of race, culture, or climatic experience (Fanger 1973a, b). Certainly the hypothesis is still being fostered by the International Standard Organization (1984), equipment manufacturers' handbooks, and the prestigious ASHRAE (1981) Handbook.

Table 1.9 shows the International Standard Organization's ISO 7730 comfort recommendations (1984) for sedentary populations engaged in light activities. In effect, these are identical to the latest ASHRAE (1981) recommendations. The seasonal differences in both are regarded merely as reflecting changes in clothing insulation.

TABLE 1.9 International Standard (1984) Thermal Comfort Recommendation for Light, Mainly Sedentary Activities

	Clothing Insulation	
	1.0 clo (Winter)	*0.5 clo (Summer)*
Operative temperature	20-24°C	23-26°C
Vertical temperature adifferrence 0.1 to 1.1 m	3°C	3°C
Mean air velocity	0.15 m/s	0.25 m/s
Normal temperature of floor*	19-26°C	0.25 m/s

*Floor heating systems may be designed for 29°C radiant temperature asymmetry relative to a plane 0.6 m above floor: from windows and cold surfaces, 10°C; from heated ceilings, 5°C.

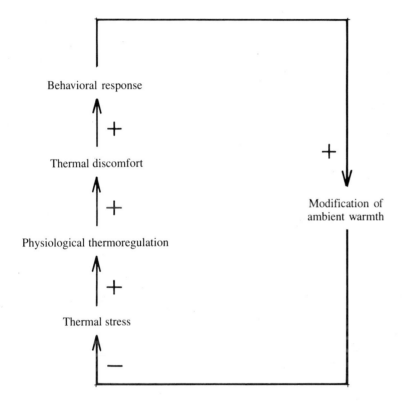

FIGURE 1.7.
Simple path of cause and effect assumed in constancy hypothesis.

For the practical purpose of determining people's behavioral responses, and thus the prescription of optimal levels of ambient warmth, the constancy hypothesis assumes a simple path of cause and effect, as shown in Fig. 1.7. If valid, this simple model seemingly resolves many difficulties in building design.

Applied globally, the established standards seem to provide a ready-made tool for predicting the amount of required technology for the construction of controllable microclimates and the active energy usage for heating and cooling. Thermal design, therefore, might be regarded as no more than a question of engineering design on the readily available criteria for Western people. Logically, also, if there exists a universality in temperature preference, can it not be claimed that at last the intervening cultural variable has been eliminated and that an economically advantageous universal design for a "world house" with standard specifications can be established?

To those readers familiar with the environmental determinist versus possibilist debate (Chappell 1975), the thermal constancy hypothesis, as described above, may well seem to be a revival of the old and rejected climatic determinist theme of the universal temperature optimum (e.g., see Huntington 1924). The acceptance of selective and limited evidence as proof of large issues and the simple extrapolations to other geographic areas and cultures may possibly come as a surprise to those familiar with the problems of social survey sampling.

VARIABILITY IN THERMAL PERCEPTION

It can be said that the thermal constancy hypothesis is being increasingly questioned as to its validity and sweeping generalizations. It is not often realized that the claims of its universal applicability were based on remarkably limited and rather incompletely reported preference studies of only 16 travelers from Copenhagen and 32 Danes (Fanger 1973b; Fanger, Hojbjerre, and Thomsen 1977). Even so, as pointed out by de Dear (1985), Fanger's results, if anything, tend to support adaptation: the tropical travelers showed somewhat elevated temperature preferences over those of winter swimmers and meat packers (Table 1.10). Moreover, the assumptions underlying the thermal constancy hypothesis also appear to be invalidated by research into variations in thermal perception as related to personal factors that affect metabolic heat production.

1. As shown in Figs. 1.5 and 1.6, no single thermal condition can be equally judged as comfortable or acceptable by all. Although some important metabolic parameters, such as body build and menstrual cycle (Fanger 1970), and even the time of day and the circadian cycle (Fanger 1970, 1973a; Fishman and Pimbert 1978; Nevins et al. 1966), could not be found to significantly affect sensations of warmth, there is also no clear evidence to identify the expected six differences in warmth perception (Ballantyne, Hill, and Spencer 1977; Fishman and Pimbert 1978; Yaglou and Messer, 1941) nor differences in thermal neutrality between age groups (Fanger 1970; Langkilde 1977; Rohles and Johnson 1972), as assessed on the ASHRAE scale.

2. Over a period of several weeks, sensation variability in individuals is increased to an average standard deviation of one whole vote category. This variability is reflected in the progressively decreasing magnitudes of regression coefficients observed in field studies, as shown in Table 1.11. Other puzzling phenomena include the drift of sensations during continuous observations. The chamber studies of Griffiths and McIntyre show that the variability within a subject during a single session of several hours is as large as that between individuals in general (Griffiths and McIntyre 1974; McIntyre and Griffiths 1973, 1975).

Sensation votes also appear to drift from warmer to cooler over time (Fanger 1973a; Griffiths and McIntyre 1974; Rohles and Johnson 1972), and the general fluctuations in temperature preference around a specific thermal

TABLE 1.10 Preferred Temperatures Observed in Denmark (Fanger 1973b, Fanger et al. 1977)

Subjects			Preferred Temperature °C	
Sample	*Sex*	*Number*	*Mean*	*Standard Deviation*
Meat packers	F	8	24.5	1.0
	M	8	24.9	0.6
Winter swimmers	F	8	24.3	1.3
	M	8	25.7	1.1
College students	F	32	25.1	1.2
	M	32	25.0	1.2
"Tropical travelers"		16	26.2	?

Source: From Fanger (1973b) and Fanger et al. (1977).

level suggest that short-term instability in thermal perception may be a far more complex phenomenon than merely one of metabolic adjustment (McIntyre 1978).

3. Seasonality in the thermal environment also has been examined as a possible cause of the observed variability. Acclimatization to heat, and to some extent to cold, is a well documented phenomenon (Edholm 1966, 1968; Ellis 1976; Glasser 1966; Sargent 1963). Whether such short-term processes also can alter the relative position of the neutral vote or preference for particular levels of warmth has not been sufficiently researched under controlled laboratory conditions, but there were indications of changed temperature preferences in an experiment by Gonzalez (1979) in a group of subjects acclimatizing through exercise over a period of just 1 month.

Unfortunately, few large-sample studies are available for both summer and winter with adequate information on clothing and a wide enough indoor temperature range to enable adequate comparisons. Those by Bedford (1936) and Hickish (1955) on light industrial workers in England were separated by two decades and very much changed social conditions. Between the 1930s and the mid-1970s, there had been an overall shift in thermal comfort requirements upward, as shown in Table 1.12. These changes probably represented alterations in clothing customs, the increasing availability of lightweight materials, and considerable microclimatic adjustment through active energy usage. However, in the case of office workers in London (Fishman and Pimbert 1979), the subjects managed to maintain neutralities between 21 and 22°C (70 and 71.5°F) throughout the year, but only by alterations in clothing. That is, in this instance, the clothing compensations in themselves indicate that the needs of the subjects had changed with seasons.

4. A recent large sample survey (Auliciems and de Dear 1986) showed that longer-term residents in tropical Darwin, North Australia, have lower neutralities than more recent arrivals. It was found that a person who had lived in Darwin for more than 45 years would have in the same environment (on the average) thermal sensations one full ASHRAE or Bedford unit cooler than the new arrival. On the other hand, in an office building study in New York, Gagge and Nevins (1975) found that despite increased clothing, indoor winter temperature preferences by the same sample actually exceeded those of summer. By contrast to the London and Darwin studies, here the offices were overheated in winter and excessively cooled in summer, to the extent that indoor temperatures were consistently warmer in the cold season. Explanations for this seemingly bizarre shift in neutralities suggest evidence of habituation.

Other direct evidence of the neutrality shift under real-life conditions comes from a comparison of two classes of English secondary school children (Auliciems 1969b) of the same age and socioeconomic background, engaged

TABLE 1.11 Dependence of Regression Slopes on Sampling Over Time

Sampling	Regression Coefficient/°C
Climate chamber, standard clothing and metabolism	0.32
Field observations:	
Extending over days/weeks	0.23
Once per day over 1 year	0.16
Once per week over 1 year	0.10
Monthly means over 1.5 years	0.08
All field studies over decades	0.05

Source: From Humphreys (1975).

TABLE 1.12 Trends in Neutralities at Light Activities

Laboratory, United States			
Year:	1923	1941	1966
Authors:	Houghten and Yagloglou (1923b)	ASHRAE (1968)	Nevins et al. (1966)
ET (°C):	19	20	22
T_a (°C)	18	21	22.5
Field, United Kingdom			
Year:	1936	1954	1978
Sample:	Light industry	Office workers	Office workers
Authors:	Bedford (1936)	Black (1954)	Fishman and Pimbert (1979)
T_a (°C):	18	20	22

in identical activities, wearing similar school uniforms, but habitually exposed to different temperatures. As shown in Fig. 1.8, the positions of their neutralities were significantly different: The group residing in the colder environment showed a 6°C (11°F) reduction in thermal comfort requirements.

5. Perhaps the most surprising source of variability in thermal perceptions is comprised of those associations related to nonthermal parameters [or ones that in themselves do not play a direct part in the energy exchanges as defined in Eq. (1.1)]. These include nonthermal dimensions of humidity (Koch, Jennings, and Humphreys 1960; Roberts 1959), color of surroundings (Fanger, Breum,

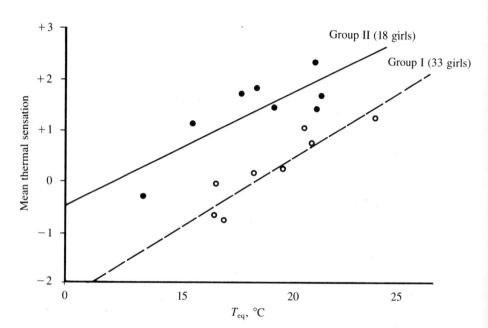

FIGURE 1.8.
Mean thermal assessments on the Bedford scale against equivalent temperature in two classes of English schoolgirls. (From Auliciems 1969*b*.)

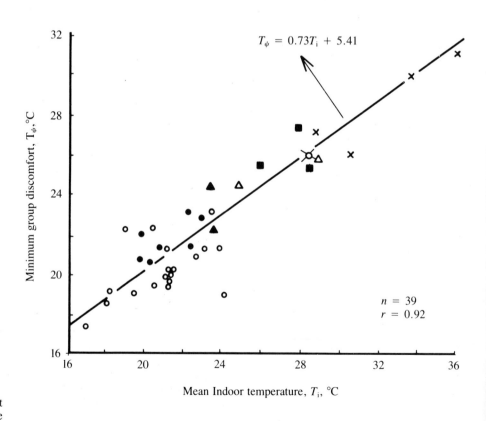

FIGURE 1.9.
Field study of mean thermal neutralities against (*a*) average indoor temperature, and (*b*) average monthly outdoor temperature.

Key: ● Australia   ■ Melanesia   ✕ Asia

(*a*)

and Jerking 1977), physical arrangements of rooms (Rohles and Wells 1977), and personality (Auliciems and Parlow 1975; Carlton-Foss and Rohles 1982).

CLIMATE, ADAPTATION, AND GEOGRAPHIC DIFFERENCES

Whereas the thermal constancy hypothesis should be rejected on the basis of the preceding "anomalies," there also exists evidence from large-sample studies around the world that clearly contradicts the validity of the universal optimum in ambient warmth. From an analysis of the results of field studies, Humphreys (1975) found that the position of Bedford's "comfortable" vote varied by some 13 to 15°C (23 to 27°F), a range that is not explicable by either clothing or metabolic rate. Some form of adaptation has taken place beyond simple adjustment to ambient conditions. This leads to a corollary that since indoor conditions are also a function of outdoor conditions, significant associations between indoor warmth perception and outdoor temperatures in general also could be expected.

For the longer term, Humphreys (1976) calculated the dependence of comfort (T_ψ) for the globally available data on parameters of outdoor warmth. In buildings without climate control, the following equation was observed:

$$T_\psi = 0.53\, T_m + 11.9 \qquad (r = 0.97) \tag{1.4}$$

where r is the simple correlation coefficient, T_ψ is sensation as predicted for groups, and T_m is monthly mean temperature.

A reanalysis of Humphrey's data to remove some incompatible information, to include the results of more recent field studies, and to combine data for both buildings with active and passive climate control produced results as shown in Fig. 1.9. The absence of thermal discomfort again is remarkably well predicted

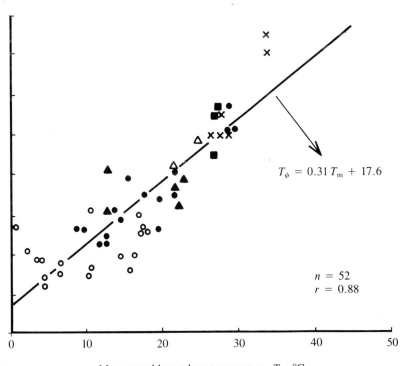

$$T_\psi = 0.31\, T_m + 17.6$$

$n = 52$
$r = 0.88$

Mean monthly outdoor temperature, T_m, °C

△ S. America   ▲ N. America   o Europe

(b)

by simple equations in terms of the mean indoor (T_i) and outdoor monthly temperature (T_m):

$$T_\psi = 0.73T_i + 5.41 \qquad (r = 0.92) \qquad \textbf{(1.5)}$$

$$T_\psi = 0.31T_m + 17.6 \qquad (r = 0.88) \qquad \textbf{(1.6)}$$

$$T_\psi = 0.48T_i + 0.14\,T_m + 9.22 \qquad (R = 0.95) \qquad \textbf{(1.7)}$$

where R is the multiple correlation coefficient.

A direct comparison of the *PMV* technique and the preceding equations was recently conducted with both air-conditioned and free-running buildings in Australia. Large samples were drawn from the office workers during summer in three cities: Darwin, Melbourne, and Brisbane (de Dear 1985; de Dear and Auliciems 1985*a, b*). Data were gathered on individual metabolic rates, clothing worn, and all ambient parameters of warmth. Probit analysis was performed on both ASHRAE and Bedford scale votes, and the neutralities observed were compared to those predicted. In brief, even with the great differences in climatic means of the locations, (1) the best predictor of neutrality within air-conditioned buildings was Eq. (1.7), while (2) for free-running buildings Eqs. (1.4) and (1.6) yielded the best results (Table 1.13).

Therefore, thermal comfort for groups of people living in diverse climatic regions and geographic locations is not a constant, but varies with time and place with adaptation to given environments.

AN ADAPTATIONAL MODEL OF THERMOREGULATION

These observations have led to the creation of a model of human thermoregulation that goes beyond one based merely on immediate physiological responses to ambient warmth as observed in laboratory experiments to the existence of a thermal expectation feedback parameter, as suggested in Fig. 1.10. This enables an evaluation of thermal environments in terms of past experiences of warmth with cultural modifications, i.e., composite levels of satisfaction.

Allowance for such adaptability to environmental stimulus is in line with modern adaptation theories (Wohwill 1974; Bell, Fisher, and Loomis 1978) and serves to explain the 14°C (57°F) range of thermally neutral temperatures evident in Figs. 1.9 and 1.10 and the earlier discussed shifts in thermal neutralities through acclimatization.

TABLE 1.13 Field Comparison of Methods Predicting Thermal Neutralities (°C) for office (de Dear and Auliciems 1985*b*)

	Darwin AC		Brisbane		Melbourne	
	Dry	*Wet*	*AC*	*non-AC*	*AC*	*non-AC*
Observed neutrality (Bedford scale & probit)	24.2	23.9	23.9	25.6	22.7	21.8
PMV prediction (Equation 1.2)	24.7	25.3	25.1	26.2	24.8	25.0
Mean indoor T_i (Equation 1.5)	22.4	22.7	22.8	25.8	22.5	23.1
Mean outdoor T_m (Equation 1.4)	n.a	n.a	n.a	25.2	n.a	21.8
(Equation 1.6)	25.4	26.5	25.2	25.3	23.4	23.4
Indoor + indoor $T_i + T_m$ (Equation 1.7)	23.9	24.6	24.0	26.2	23.1	23.5

Darwin dry = "winter," wet = monsoonal build-up period, AC = air-conditioned buildings, n.a. = not applicable

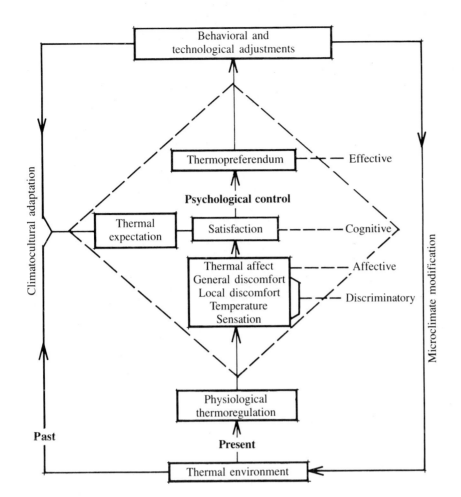

FIGURE 1.10.
An adaptional model of thermoregulation. (From Auliciems 1983.)

VARIABLE *PMV* AND COMFORT CONDITIONS

To allow for the climate factor, as discussed in the preceding section, global human comfort can be described in shifting 4°C bands around the neutral temperature as defined by Eq. (1.6) and delimited by maximum and minimum possible vapor pressures of skin, set at 12 and 4 g/kg absolute humidity, respectively (Szokolay 1987). The positions of the extreme and central zones are shown in Fig. 1.11.

In the absence of local field data, monthly comfort zones in terms of dry-bulb temperatures can be similarly identified for particular places by locating the monthly mean outdoor temperature on scale T_m and converting to T_ψ. These then become the central points in the comfort zones, which can be extended $2ET^*$ to either side.

Similarly, adjustments can be made to levels of indoor warmth in terms of the comfort equation and *PMV*. Given the seasonal effect as estimated in Eq. (1.6), and given that a shift of 4°C (7°F) in environmental warmth is required to change one whole ASHRAE, Bedford, or *PMV* unit, an appropriate indoor temperature adjustment to achieve the central point of the comfort zone (T_ψ) would be numerically equivalent to

$$T_\psi = 0.31 T_m + 17.6 - 4PMV \qquad \textbf{(1.8)}$$

Alternatively, such calculations also enable a ready redefinition of *PMV* to account for climatically mobile optima. Outdoor monthly mean temperature appears to equal that preferred indoors—near 25.5°C (78°F) dry-bulb temperature (see Fig. 1.9). Since the humidities found in field studies were mostly below 50 percent RH, this value is very close to the $24ET^*$ identified earlier as the optimum for U.S. laboratory samples. Given that the rate of change in the

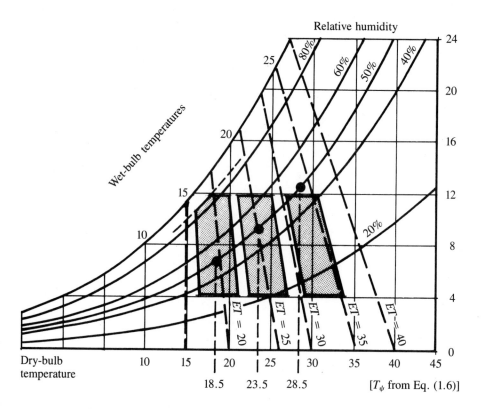

FIGURE 1.11.
Lowest, average, and highest positions of the comfort zone in terms of air temperature, humidity, and *ET**. Ready estimation of *ET** can be made by extension of *ET** lines passing through the intercepts of temperature and humidity. For example, a person in a warm and humid environment at 27°C (79°F) and 70 percent humidity would experience near 34*ET**. (Adapted from Szokolay 1987.)

perception of warmth according to Eq. (1.6) is $0.31 T_m$, there is need for adjustment of indoor temperatures by 0.31 of every degree T_m falls above or below this value. Again, taking account of the whole *PMV* unit shift by a change of 4°C, an adjusted predicted mean vote (*PMV**) should be calculated using the regression coefficient from Eq. (1.6) as follows:

$$PMV^* = PMV - \frac{0.31(T_m - 25)}{4} \tag{1.9}$$

that is,

$$PMV^* = PMV - 0.08(T_m - 25) \tag{1.10}$$

where lower and upper limits to T_m are set at 0 and 40°C (32 and 104°F), respectively.

By the same logic, since every *PMV** unit calculated in particular environments above and below 0 equates stress equivalent to 4 K, an appropriate adjustment has to be made to ambient warmth so that

$$T = 25 - 4PMV^* \tag{1.11}$$

SUMMARY

1. People cope with thermal stress by physiological and behavioral responses. These responses are dynamic and may be immediate adjustments or longer-term adaptations.
2. Thermal comfort is a complex psychophysiological state of satisfaction that occurs within a narrow range of these responses.
3. Assessment of thermal comfort depends on the conditions of testing: Extrapolations of results from laboratory studies to real-life situations may not be valid. Therefore, specification of optimal indoor levels of warmth from generalizations may require validation in the field.
4. Under most real-life circumstances, levels of ambient warmth may be adequately assessed by simple measurements of dry-bulb temperatures.

5. Levels of achieved indoor thermal comfort may be predicted for a given population from metabolic rate, clothing, ambient indoor thermal parameters, and levels of outdoor warmth as varying with weather and season.

6. To ensure minimum discomfort for any population, comfort zones may be defined as air temperatures falling within the equivalent of 2°C (3.5°F) of the maximal comfort temperature for that population.

7. For some European and American samples, thermal sensation neutrality and preference appear to be near 25°C (77°F) air temperature. In general, however, comfort levels will vary with prevailing levels of warmth both indoors and outdoors. Therefore, prescriptions for optimal indoor conditions need to rely on the expressed thermopreferendum for any population, depending in part on its own past climatocultural experiences and expectations.

REFERENCES

American Society of Heating, Refrigerating, and Air Conditioning Engineers (ASHRAE). 1966. Standard 55-56: Thermal Environmental Conditions for Human Occupancy. ASHRAE, Atlanta, Ga.

ASHRAE. 1981. *Handbook of Fundamentals.* ASHRAE, Atlanta, Ga.

Auliciems, A. 1969*a.* Thermal requirements of secondary school children in winter. *J. Hyg. (Lond.)* 67: 59-67.

Auliciems, A. 1969*b.* Some group differences in thermal comfort. *Heat. Vent. Engr* 42: 562-564.

Auliciems, A. 1972. *The Atmospheric Environment.* University of Toronto Press, Toronto.

Auliciems, A. 1983. Psycho-physiological criteria for global thermal zones of building design. *Int. J. Biometeorol.* 26(Suppl.): 69-86.

Auliciems, A., and de Dear, J. R. 1986. Air-conditioning in a tropical climate: Impacts upon European residents in Darwin, Australia. *Int. J. Biometeorol.* 30: 259-282.

Auliciems, A., and Parlow, J. 1975. Thermal comfort and personality. *Build. Serv. Eng.* 43: 94-97.

Ballantyne, E. R., Hill, R. H., and Spencer, J. W. 1977. Probit analysis of thermal sensation assessments. *Int. J. Biometeorol.* 21: 29-43.

Bedford, T. 1936. The Warmth Factor in Comfort at Work. Report Industrial Health Research Board 76. H.M.S.O. London.

Belding, H. S., and Hatch, T. F. 1955. Index for evaluating heat stress in terms of resulting physiological strain. *Heat. Pip. Air Condit.* 27: 129-136.

Bell, P. A., Fisher, J. D., and Loomis, R. J. 1978. *Environmental Psychology.* Saunders, Philadelphia.

Benzinger, T. H. 1978. In introducing part III of the First International Indoor Climate Symposium, at Copenhagen, Aug. 30 to Sept. 1.

Black, F. W. 1954. Desirable temperatures in offices. *J. Instn. Heat. Vent. Engrs.* 22: 319-328.

Burton, A. C., and Edholm, O. G. 1955. *Man in a Cold Environment.* Edward Arnold, London.

Carlton-Foss, J. A., and Rohles, R. H. 1982. Personality factors in thermal acceptability and comfort. *ASHRAE Trans.* 88: 776-788.

Cena, K., and Clark, J. A. 1981. *Bioengineering, Thermal Physiology and Comfort.* Elsevier, Amsterdam.

Chappell, J. E. 1975. The ecological dimension: Russian and American views. *Ann. Assoc. Am. Geogr.* 65: 144-162.

Chrenko, F. A. 1955. The assessment of subjective reactions in heating and ventilation research. *J. Instn. Heat. Vent. Engrs.* 23: 281-295.

Collins, K. J. 1979. Hypothermia and Thermal Responsiveness in the Elderly. In Fanger, P. O., and Valbjrn, O. (Eds.), *Indoor Climate: Proceedings of the First International Indoor Climate Symposium.* Danish Building Research Establishment, Copenhagen.

de Dear, R. J. 1985. Perceptual and Adaptational Basis for the Management of Indoor Climate. Ph.D thesis, University of Queensland.

de Dear, R. J., and Auliciems, A. 1985a. Validation of the predicted mean vote model of thermal comfort in six Australian field studies. *ASHRAE Trans.* 91: 452-468.

de Dear, R. J., and Auliciems, A. 1985b. Thermal neutrality and acceptability in six Australian field studies. In Fanger, P. O. (Ed.), *Clima 2000,* Vol. 4, pp. 103-108. VVS Kongres-VVS Messe, Copenhagen.

de Freitas, C. R. 1979. Beach Climate and Recreation. Ph.D. thesis, University of Queensland.

Dubos, R. 1965. *Man Adapting.* Yale University Press, New Haven.

Edholm, O. G. 1966. Problems of acclimatization in man. *Weather* 21: 340-350.

Edholm, O. G. 1968. *Man Hot and Cold.* Edward Arnold, London.

Ellis, F. P. 1976. Heat illness: Acclimatisation. *Trans. R. Soc. Trop. Med. Hyg.* 70: 401-411.

Fanger, P. O. 1967. Calculation of thermal comfort: Introduction of a basic comfort equation. *ASHRAE Trans.* 73: III.4.I-III.4.20.

Fanger, P. O. 1970. *Thermal Comfort.* Danish Technical Press, Copenhagen.

Fanger, P. O. 1973a. Assessment of man's thermal comfort in practice. *Br. J. Ind. Med.* 30: 313-324.

Fanger, P. O. 1973b. The influence of age, sex, adaptation, season and circadian rhythm on thermal comfort criterion for man. *Bull. Inst. Int. Froid Meeting IIR* (Suppl.): 91-97.

Fanger, P. O., Breum, N. D., and Jerking, E. 1977. Can colours and noise influence man's thermal comfort? *Ergonomics* 20: 11-18.

Fanger, P. O., Hojbjerre, J., and Thomsen, J. O. B. 1977. Can winter swimming cause people to prefer lower room temperatures? *Int. J. Biometeorol.* 21: 44-50.

Finney, D. J. 1952. *Probit Analysis.* Cambridge University Press, Cambridge.

Fishman, D. S., and Pimbert, S. L. 1978. Survey of the Subjective Responses to the Thermal Environment in Offices. In Fanger, P. O., and Valbjorn, O. (Eds.), *Indoor Climate: Proceedings of the First International Indoor Climate Symposium.* Danish Building Research Establishment, Copenhagen.

Fox R. H., Woodward, P. M., Exon-Smith, A. N., Green, M. F., Donnison, D. V., and Wicks, M. H. 1973. Body temperature in the elderly: A national study of physiological, social and environmental conditions. *Br. Med. J.* 1: 200-206.

Gagge, A. P. 1936. The linearity criterion as applied to partitional calorimetry. *Am. J. Physiol.* 116: 656-668.

Gagge, A. P., and Nevins, R. G. 1975. Effect of Energy Conservation Guidelines on Comfort, Acceptability, and Health (Report to Federal Energy Administration). J. B. Pierce Foundation Laboratory, New Haven.

Gagge, A. P., Stolwijk, J. A. J., and Hardy, J. D. 1967. Comfort and thermal sensation and associated physiological responses at various ambient temperatures. *Environ. Res.* 1: 1-20.

Gagge, A. P., Stolwijk, J. A. J., and Nishi, Y. 1971. An effective temperature scale based on a simple model of human physiological regulatory response. *ASHRAE Trans.* 75: 108-125.

Givoni, B. 1963. *Man, Climate and Architecture.* Elsevier, Amsterdam.

Glasser, E. M. 1966. *The Physiological Basis of Habituation.* Oxford University Press, London.

Gonzalez, R. R. 1979. Role of Natural Acclimatization (Cold and Heat) and Temperature: Effect of Health and Acceptability in a Built Environment. In Fanger, P. O., and Valbirn, O. (Eds.), *Indoor Climate: Proceedings of the First International Indoor Climate Symposium.* Danish Building Research Establishment, Copenhagen.

Griffiths, I. D., and McIntyre, D. A. 1974. Sensitivity to temporal variation in thermal conditions. *Ergonomics* 17: 499-507.

Hardy, J. D., Ed. 1963. *Temperature, Its Measurement and Control in Science and Industry,* Vol. 3. Reinhold, New York.

Hickish, D. E. 1955. Thermal sensations of workers in light industry in summer: A field study in southern England. *J. Hyg. (Lond.)* 53: 112-123.

Houghten, F. C., and Yaglou, C. P. 1923a. Determining lines of equal comfort. *ASHVE Trans.* 29: 163-175.

Houghten, F. C., and Yaglou, C. P. 1923b. Determination of the comfort zone. *ASHVE Trans.* 29: 361-379.

Humphreys, M. A. 1975. Field Studies of Thermal Comfort Compared and Applied. Building Research Establishment, Current Paper (76/75), U.K. Dept. of Environment.

Humphreys, M. A. 1976. Comfortable indoor temperatures related to the outdoor air temperature. Building Research Establishment (Note PD117/76), U.K. Dept. of Environment.

Humphreys, M. A., and Nicol, J. F. 1970. An investigation into the thermal comfort of office workers. *J. Instn. Heat. Vent. Engrs.* 38: 181-189.

Huntington, E. 1924. *Civilization and Climate.* Yale University Press, New Haven.

Huntington, E. 1926. *The Pulse of Progress.* Simpson, New York.

International Standard. 1984. ISO 7730: Moderate Thermal Environments. International Standards Organization, Switzerland.

Johnson, W., Ed. 1931. *The Journals of Gilbert White.* Routledge & Sons, London.

Koch, W., Jennings, B. H., and Humphreys, C. M. 1960. Environment study II: Sensation and responses to temperature and humidity under still air conditions in the comfort range. *ASHRAE Trans.* 66: 264-282.

Lamb, H. H. 1982. *Climate History and the Modern World.* Methuen, London.

Langkilde, G. 1977. Thermal comfort for people of high age. Institut National de la Santé et de la Recherche Medicale 75: 187-194.

Lee, D. H. K., and Henschel, A. 1963. Evaluation of Thermal Environment in Shelters. Division of Occupational Health Technical Report 8, U.S. Dept. of Health Education, and Welfare, Cincinnati.

Madsen, T. L. 1979. Measurement of Thermal Comfort and Discomfort. In Fanger, P. O., and Valbjorn, O. (Eds.), *Indoor Climate: Proceedings of the First International Indoor Climate Symposium.* Danish Building Research Establishment, Copenhagen.

Markham, S. F. 1947. *Climate and the Energy of Nations.* Oxford University Press, London.

Mayr, E. 1956. Geographical character gradients and climatic adaptation. *Evolution* 10: 105-108.

McArdle, B., Dunham, W., Holling, H. E., Ladell, W. S., Scott, J. W., Thomson, M. L., and Weiner, J. S. 1947. The Predication of Physiological Effect of Warm and Hot Environments. Medical Research Council Report 47-391. H.M.S.O., London.

McIntyre, D. A. 1975. Determination of individual preferred temperatures. *ASHRAE Trans.* 81: 131-139.

McIntyre, D. A. 1978. Seven point scales of warmth. *Build. Serv. Eng.* 45: 215-226.

McIntyre, D. A. 1980. *Indoor Climate.* Applied Science Publishers, London.

McIntyre, D. A., and Griffiths, I. D. 1973. Subjective response to radiant and convective environments. *Environ. Res.* 5: 471-482.

McIntyre, D. A., Griffiths, I. D. 1975. Subjective responses to atmospheric humidity. *Environ. Res.* 9: 66-75.

McIntyre, D. A., and Gonzalez, R. R. 1976. Man's thermal sensitivity during temperature changes at two levels of clothing insulation and activity. *ASHRAE Trans.* 82: 219-233.

McNall, P. E., Ryan, P. W., Rohles, F. H., Nevins, R. G., and Springer, W. E. 1968. Metabolic rates of four activity levels and their relationship to thermal comfort. *ASHRAE Trans.* 74: IV.3.1-IV.3.12.

Mills, C. A., 1946. *Climate Makes the Man.* Gollan, London.

Missenard, A. 1935. Theorie simplifié du thermometre resultant. *Chauf. Vent.* 12: 347-352.

Missenard, A. 1957. *In Search of Man.* Hawthorn Books, New York.

Monteith, J. L., and Mount, L. E. 1974. *Heat Losses from Animals and Man: Assessment and Control.* Butterworth, London.

Newman, M. T. 1953. The application of ecological rules to racial anthropology of the aboriginal world. *Am. Anthropol.* 55: 311-327.

Newman, M. T. 1955. Adaptation of man to cold environments. *Evolution* 9: 101-105.

Nevins, R. G., Rohles, F. H., Springer, W., and Feyerherm, A. M. 1966. A temperature-humidity chart for thermal comfort of seated persons. *ASHRAE Trans.* 72: 283-291.

Nishi, Y., and Gagge, A. P. 1977. Effective temperature scale useful for hypo- and hyperbaric environments. *Aviat. Space Environ. Med.* 48: 97-107.

Pepler, R. D. 1972. The thermal comfort of students in controlled and non-climate-controlled schools. *ASHRAE Trans.* 78: 97-109.

Roberts, B. M. 1959. Environmental testing. *J. Instn. Heat. Vent. Engrs.* 27: 238-250.

Rohles, F. H., and Johnson, M. A. 1972. Thermal comfort in the elderly. *ASHRAE Trans.* 78: 131-135.

Rohles, F. H., and Laviana, J. E. 1985. Indoor Climate: New Approaches to Measuring How You Feel. In Fanger, P. O. (Ed.), *Clima 2000,* Vol. 4, pp. 1-6. VVS Kongres-VVS Messe, Copenhagen.

Rohles, F. H., and Milliken, G. A. 1981. A Scaling Procedure for Environmental Research. In *Proceedings of the 25th Annual Meeting of the Human Factors Society,* pp. 472-476. Human Factors Society, Rochester, N.Y.

Rohles, F. H., and Wells, M. V. 1977. The role of environmental antecedents on subsequent thermal comfort. *ASHRAE Trans.* 83: 21-29.

Sargent, F., II. 1963. Tropical neurasthenia: Giant or windmill? In *Arid Zone Research, Environmental Psychology and Physiology: Proceedings of the Lucknow Symposium.* UNESCO, Paris.

Sargent, F., II, and Tromp, S. W., Eds. 1964. *A Survey of Human Biometeorology.* WMO Technical Note No. 65, Geneva.

Scholander, P. R. 1955. Evolution of climatic adaptation in homeotherms. *Evolution* 9: 15-26.

Scholander, P. F. 1956. Climatic rules. *Evolution* 10: 339-340.

Smith, F. E. 1955. Indices of Heat Stress. Medical Research Council Memo 29. H.M.S.O., London.

Stevens, J. C., Marks, L. E., and Gagge, A. P. 1967. The quantitative assessment of thermal discomfort. *Environ. Res.* 2: 149-165.

Stevens, J. C., and Stevens, S. S. 1963. The Dynamics of Subjective Warmth and Cold. In Hardy, J. D. (Ed.), *Temperature, Its Measurement and Control in Science and Industry,* Vol. 3. Reinhold, New York.

Szokolay, S. V. 1985. Thermal Comfort and Passive Design. In Boer, K. W. (Ed.), *Advances in Solar Energy 2.* American Solar Energy Society, New York.

Szokolay, S. V. 1987. Climate analysis based on the psychometric chart. *Int. J. Ambient Energy* 7: 1-12.

Taylor, G. T. 1959. *Australia: A Study of Warm Environments and Their Effect on British Settlement,* 7th Ed. Methuen, London.

Toynbee, A. J. 1945. *A Study of History,* Vol. 2, 3d Ed. Oxford University Press, London.

Webb, C. G. 1959. An analysis of some observations of thermal comfort in an equatorial climate. *Br. J. Ind. Med.* 16: 297-310.

Winakor, G. 1982. A questionnaire to measure environmental and sensory factory factors associated with personal comfort and acceptability of indoor environments. *ASHRAE Trans.* 88: 470-481.

Winslow, C.-E. A., Herrington, L. P., and Gagge, A. P. 1937. Physiological reactions of the human body to varying environmental temperature. *Am. J. Physiol.* 120: 1-20.

Wohlwill, J. F. 1974. Human adaptation to levels of environmental stimulation. *Hum. Ecol.* 2: 127-147.

Wollard, D. S. 1979. Thermal Habitability of Shelters in the Solomon Islands. Ph.D. thesis. University of Queensland.

Yaglou, C. P., and Drinker, P. 1928. The summer comfort zone, climate and clothing. *J. Ind. Hyg.* 10: 350-363.

Yaglou, C. P., and Messer, A. 1941. The importance of clothing in air conditioning. *J. Am. Med. As.* 117: 1261-1262.

Air Quality

2

George Cunningham

In providing an acceptably comfortable environment in buildings, one of the considerations must be the quality of the air itself. The air must contain sufficient oxygen to support respiration, and both particulate pollutants and harmful or irritant gaseous pollutants, including odors, must be kept to an acceptably low level. To gain an understanding of the problem, one must consider the nature and source of the various pollutants normally encountered in and around buildings.

THE NATURE OF POLLUTANTS

Pollutants can be classified in various ways, for example:

Particulate or gaseous
Organic or inorganic
Visible or invisible
Submicroscopic, microscopic, or macroscopic
Toxic or harmless

A common and useful classification used by ventilation engineers (ASHRAE 1985) is based on the source or method of formation of the pollutant. Thus pollutants can be loosely classified as follows:

Dusts, fumes, and smokes, which are primarily solid particulate matter, although smoke often contains liquid particulates.

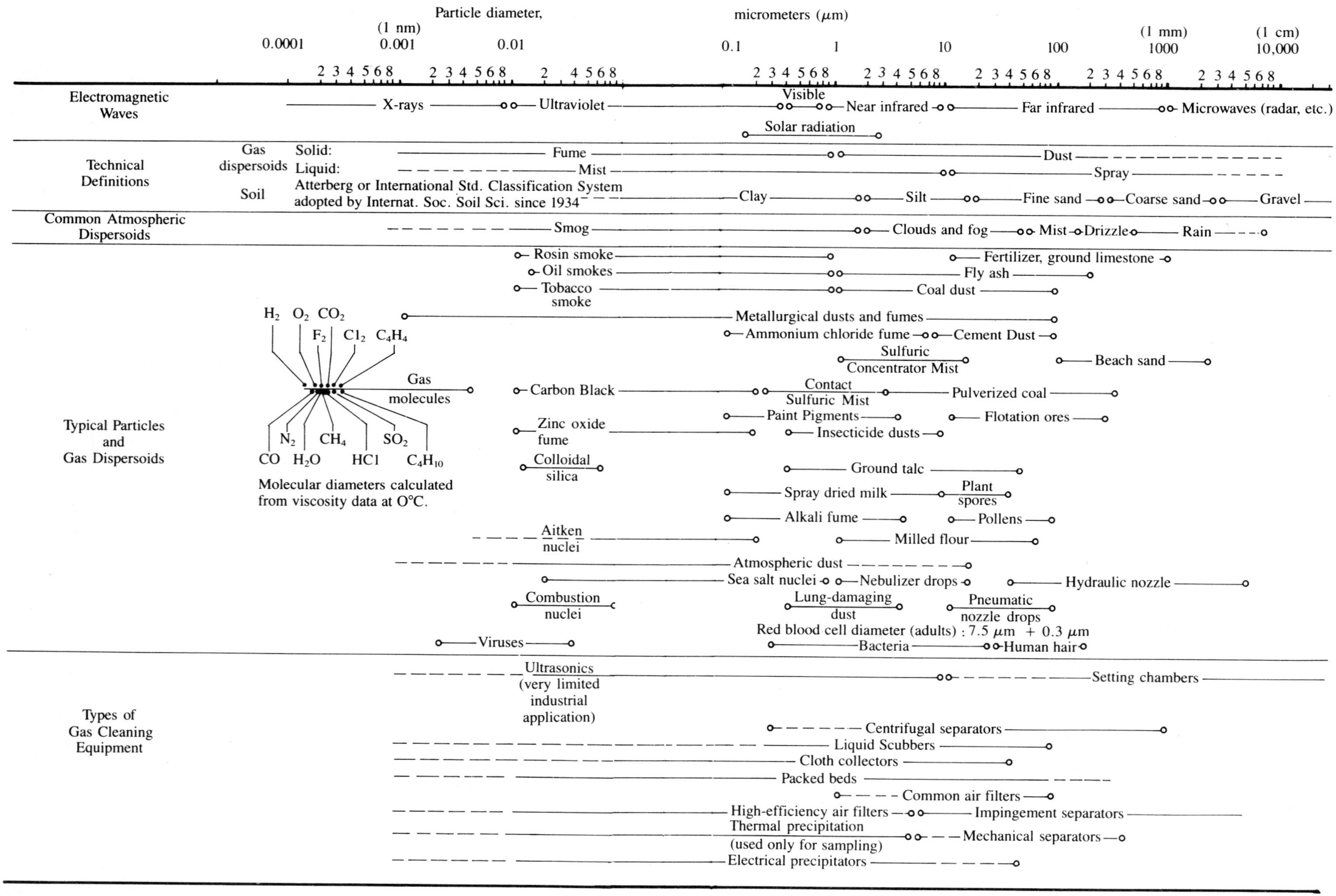

FIGURE 2.1.
Characteristics of airborne particulates.

Mists and fogs, which are liquid particulate matter.

Vapors or gases, which are nonparticulate.

Each of these loose classifications can be examined in detail according to nature and source. It is particularly important if one is attempting to control the environment within a space to recognize if the pollutants are being generated within the space or from a source outside the space.

Generally, particles are not considered to be dust unless they are smaller than 100 μm in size. They can be mineral (such as rock, metal, or clay), vegetable (such as grain, flour, wood, cotton, or pollen), or animal (such as wool, hair, silk, or feathers). Dusts can be generated by natural forces such as wind erosion or by mechanical forces in such activities as demolition and drilling. Inside a building, dusts can be generated by sweeping or even walking across carpets.

Fumes are solid particles commonly formed by the condensation of vapors of solid materials. Typical sources are molten metals and chemical reactions or processes such as distillation or sublimation which create airborne particles predominantly smaller than 1 μm.

Smokes are extremely small solid and/or liquid particles produced by incomplete combustion of organic substances such as tobacco, wood, coal, and oil. The term "smoke" is commonly applied to a mixture of solid, liquid, and gaseous products, although some technical literature distinguishes between individual components. Smoke particles vary considerably in size, the average often being in the range of 0.1 to 0.3 μm. For a frame of reference, viruses range in size from 0.005 to 0.1 μm, although they usually occur in colonies or attached to other particles. Fungus spores are usually from 10 to 30 μm, whereas pollen grains are from 10 to 100 μm.

Mists are very small airborne droplets of materials that are ordinarily liquid in the environment. They can be formed by atomizing, spraying, or chemical reactions. Sneezing expels very fine droplets as a mist which can contain microorganisms.

Fogs are very fine airborne droplets usually formed by condensation of vapor. There is a fine distinction between mists and fogs. The volatile nature of most liquids reduces the size of their airborne droplets from the mist to the fog range and eventually to the vapor phase.

The term "smog" is commonly used to describe air pollution and implies a mixture of smoke particles, mist, and fog droplets of such a concentration and composition as to impair visibility and be irritating or harmful. The term is often applied particularly to haze caused by a sunlight-induced photochemical reaction involving the materials in automobile exhausts.

The terms "gas" and "vapor" are often used interchangeably to describe a common state of a substance. "Gas" is normally used to describe a substance that always exists in the gaseous state at normal temperature and pressure, e.g., oxygen, helium, or nitrogen. "Vapor" is used to describe a substance in the gaseous state that can also exist in the liquid state at normal temperature and pressure, e.g., water, benzene, or carbon tetrachloride.

Figure 2-1, extracted from data prepared by the Stanford Research Institute, is a tabulation of the sizes and characteristics of airborne solids and liquids.

THE SOURCE AND TRANSPORT OF CONTAMINANTS

Contaminants usually are brought from their sources to the occupied space by means of an airstream. From outside a building, the wind can bring into a building road dust, dust from nearby demolition; smoke and fumes from boilers, incinerators, or industrial processes; mists and fogs from sea spray or cooling towers; smog from automobile exhausts; and vapors and gases from automobile exhausts, boilers, and industrial processes.

Inside a building contaminants are distributed from their sources by means of natural convection or the ventilation system itself. These contaminants include dust from occupants' clothing (particularly shoes); lint from carpets, clothing and furnishings; smoke and fumes from cigarette smoking; mists and fogs from coffee urns and cooking and washing processes; and vapors and gases from human activity (perspiration and body odor), printing processes, cleaning, unvented heaters, and synthetic furnishings and other building products.

Of increasing interest to researchers is the presence of radon and its immediate radioactive decay products because of the indicative relationship between radon exposure and lung cancer (Nero et al. 1984). Radon contamination appears to originate mainly from the soil, with wide variations from one geographic area to another. Currently, research at the Lawrence Berkeley Laboratory in California (Grimsrud et al. 1985) is underway to determine the sources of radon contamination and the factors that influence its transport in buildings.

The radon problem came to the attention of Swedish researchers in the mid-1970s when high levels of radon were found in new housing projects. The source was the alum shale used in the concrete. This substance had a radium content hundreds of times higher than that found in normal soil or rock.

It also was thought that the airtight nature of this energy-efficient housing and the much lower ventilation rates were preventing the dilution of radon and its products. Subsequent research by the Lawrence Berkeley Laboratory has shown that high radon levels exist in ordinary housing as well as housing designed specifically for airtightness. There was found to be little correlation between radon concentration and ventilation rates.

Some radon products that cling to particles can be removed by filtration, but this still leaves unattached products. One means suggested for reducing the entrance of radon into buildings is to drive pipes through the basement floor and use small fans to exhaust air from the underlying soil or gravel to outdoors. For buildings above ground level, subfloor ventilation would be effective. Radon concentration does not appear to be a problem in upper floors of buildings (except where the gas is present in building materials). This research is continuing.

Viruses can be generated inside or outside buildings. They are transportable via airstreams and are of particular interest because of their direct effect on the health of occupants rather than their mere discomfort. Control of airborne distribution of viruses is important in hospitals, particularly in operating theaters, where viruses are both plentiful and have a great opportunity to cause infection. Recently, the most publicized virus associated with air-conditioning systems is *Legionella,* which causes legionnaires' disease. This virus occurs in nature in moist soil or tepid water and has been found in buildings in shower cubicles. It also has been found in water cooling towers associated with air-conditioning plants and in water spray systems which form part of some air-conditioning equipment.

Odors are those substances which can be detected by our olfactory senses. With few exceptions (e.g., hydrogen sulfide, ozone), inorganic substances are odorless. There are, however, thousands of organic substances that provide olfactory stimuli, some so potent that they can be perceived at concentrations too low to be detected by existing instruments.

Most of these are unpleasant even at barely perceptible levels and often well below concentrations that are considered dangerous or toxic. Nevertheless, unpleasant odors at even nontoxic concentrations may cause nausea, headaches, and loss of appetite. Even a room with a recognizable odor of low concentration may cause uneasiness among occupants.

Because of the very low levels of odor concentration required to stimulate the olfactory senses, it is difficult to devise quantitative criteria for odor control. This is further complicated by the range of sensitivities among individuals to different concentrations of odor and the subjective appraisal as to whether an odor is pleasant or unpleasant. Work is proceeding in the evaluation of odor measurement and control by ASHRAE and others.

The source of odors may be from within a building or from the outside. From outside, odors in the air may be from automobile exhausts, furnace effluents, or industrial effluents. Inside a building, odors may be generated from smoking, body odors, cleaning and other processes using organic chemicals, food and cooking, and furnishings and building materials that are made from or include organic materials.

AIR QUALITY CONTROL

The control of air quality within a building is usually achieved by one of two mechanisms:

Removal of contaminants
Dilution of contaminated air with noncontaminated air

Removal of Particulate Contaminants

The removal of contaminants depends on whether the contaminants are particulate or gaseous. Particulate contamination is usually removed by air filters placed in the circulating airstream of an air-conditioned or mechanically ventilated space. In an air-conditioned building, the circulating airstream is made up of two components: the air drawn from outside and the air recirculated from within the space. These two airstreams are usually mixed together immediately upstream of the air-conditioning apparatus.

Air filters are generally located on the upstream side of any heating, cooling, or humidifying equipment and the circulating fan, since the performance of the air-conditioning equipment is impaired by impurities borne by the airstream. Dust settling on the surface of the heating and cooling heat exchangers reduces the heat transfer to or from the conditioned air. If dust continues to collect, it forms a thick layer that can eventually block off the passage of the air through the heat exchanger and is very difficult to remove.

The efficiency of the filtration device is generally selected by the air-conditioning design engineer. For normal comfort applications, there are no standards that define the efficiency of filtration required or indeed the levels of contaminants that are acceptable in the conditioned space. The definition of a standard based on the end result would be difficult and complex. The major difficulty arises from the fact that much of the contamination (such as cigarette smoking or carpet lint) is generated within the space itself or brought in by other than the airstream (such as, for example, dust from shoes and clothing).

A familiar sight in many air-conditioned buildings is the dark "smudge" around the air-supply outlets. Although the first reaction as to the cause may be ineffective filtration, in actual fact the mark is generally formed from particles within the room being induced into the incoming airstream and deposited on the surfaces adjacent to the air-supply outlet.

Air Filtration Efficiency and Costs

The efficiencies of various air filter materials or systems are well documented. There are a number of well-defined standards (ASHRAE 1971; BSA 1971) against which filter performance can be tested. However, the designer still must choose a level of filtration "appropriate" to the particular function of the building. Some work has been done (Gibbs 1985) to establish a quantitative approach to filter selection on the basis that filters chosen should, at the very least, maintain the indoor pollution level at a value no higher than that of the outdoor air.

Generally, as air filtration efficiency increases, either the initial cost of the filter increases or the air pressure drop across the filter increases. An increase in air pressure drop results in an increase in the required fan power and thus an increase in running cost. The cost and frequency of replacing and/or cleaning filters also must be considered in assessing the right filter for a given project.

Examination of a residential room air-conditioner will show that the filter is very low cost, with a very low pressure drop and efficiency. Indeed, the major role of such a filter would appear to be to keep the heat exchange equipment and fan clean. Factory-packaged air-conditioners tend to include higher-cost filters with slightly higher efficiencies.

The selection of filters for "engineered" systems will vary according to initial and maintenance costs, energy costs, and perceived levels of "quality" required by the client. Despite this variation, many of the engineered systems throughout the world use filters of the extended-surface dry media type. However, even within this broad family of filters there is a wide range of performance in terms of efficiency, dust-holding capacity, first cost, maintenance cost, and air pressure drop.

One type of air-cleaning device has both high efficiency and a low pressure drop, although the initial cost is high. This is the electrostatic air cleaner. In this device, air is drawn through an electrostatic field in which particles in the air take on an electrostatic charge and are attracted to electrostatically charged plates where they are collected. An advantage of electrostatic air cleaners is that they are effective even on very small particles, such as cigarette smoke. Unfortunately, these devices generate small quantities of ozone, which some believe is more beneficial than harmful to health, while others believe that proof is still required that there is no potential health hazard.

Despite this, electrostatic air cleaners are provided in many applications where large concentrations of people gather and where cigarette smoking is prevalent, such as in conference rooms, ballrooms, bars, and taverns. The effectiveness of these air cleaners is so well demonstrated that many authorities, under whose jurisdiction ventilation requirements fall, will permit a reduction of the "dilution" outside air quantity required by governing codes if electrostatic air cleaners are employed.

In some special areas, such as, for example, laboratories, clean rooms, and hospital operating theaters, it may be necessary to keep particulate contamination to extremely low levels. Such rooms use very high efficiency filters called "absolute" or "HEPA filters" to achieve the required result. These filters are introduced into the airstream immediately before it enters the room so that any contaminants generated within the system itself are also caught. They also are used in conjunction with "laminar flow" rooms, in which the airflow is introduced through the high-efficiency filters along one wall or sometimes the ceiling and flows like a piston across the room to the opposite wall (or floor).

The cleanest part of such a room is close to the filter. In such rooms it is normal for occupants to wear special clothing and masks to prevent contamination from themselves as they enter the room. The pressure drop across the high-efficiency filters is high, the air quantity required to produce the laminar-flow effect is high, and thus the energy required to provide such clean conditions is high. In order to reduce the high energy usage, some designers provide for the ultraclean space in more localized areas, e.g., at individual workbenches in a laboratory or only over the operating table in a hospital operating theater.

Removal of Gaseous Contaminants

Gaseous contaminant removal is commonly achieved by "wet scrubbing" or adsorption. Wet scrubbing is a process in which the gas-laden air is passed through liquid sprays (most commonly water) that dissolve the gaseous contaminants and remove them in the liquid stream. This process is not commonly used in comfort air-conditioning applications, but more in industrial applications.

In commercial building applications, the most common method of gaseous contaminant removal is by the adsorption process, in which the contaminant gas is adsorbed into the surface of a solid over which it passes. The two most common adsorbents are activated alumina and activated carbon, and of these, the latter is much more common. The gases removed by activated carbon are primarily organic and usually are those gases classified as odors. Other gases that are products of combustion, e.g., the oxides of carbon and nitrogen, are usually dealt with by dilution.

Activated carbon filters are usually made up from trays of activated carbon some 25 mm deep and about 100 mm in the direction of airflow. When the surface of the carbon becomes saturated with adsorbed gases, it must be reactivated. This is usually done by applying steam to the carbon, which removes the adsorbed material.

Once again, there are no codes governing the allowable concentrations of the thousands of nontoxic organic materials that inflict discomfort by means of their odorous nature. It is usally left to the judgment of the air-conditioning designer to decide on the necessity for gaseous contaminant removal.

It is quite uncommon to see gaseous contaminant removal facilities in comfort air-conditioning applications. They are most common where odors from cooking or vapors from chemicals being used in adjacent areas, e.g., laboratories, are likely to migrate into other occupied areas. Another application might be where the outside air is drawn from an area that is contaminated with unpleasant odors. One such application that springs to mind occurred where the office's immediate neighbors were a vegetable oil refinery, a soap factory, an industrial gas distribution center, and a fellmonger.

Dilution of Contaminated Air

A very common means of dealing with contaminated air, particularly where the contaminants are gaseous in nature, is by dilution with outdoor air. This has been a long-standing practice, and in most areas federal, state, or local codes and standards specify the amount of outdoor air required for dilution of contaminants for many applications. In the area of ventilation or air-conditioning for human comfort, the application of dilution goes back to early in this century, and many practices defined in codes up to the 1970s were based on studies done early in this century.

The outdoor air quantity required was based on the number of occupants in the space. Typically, outdoor air quantities of 15 ft^3/min or about 7 liters/s were required for each occupant. This amount is many times greater than that required to make up for oxygen depletion from respiration. In fact, it was originally based on the quantity required for adequate control of body odor.

During the 1970s, as a result of the energy crisis, the subject of outdoor air for contaminant control was seriously reexamined for the first time in decades. It was found that since the time that the first codes were proposed, personal hygiene habits had changed significantly. In particular, it was now common for each household to have its own showering and bathing facilities and that these were used much more regularly than they were decades ago.

After considerable research it was found that the outdoor air quantity for normal office environments could be safely reduced to about 3 liters/s per person. This resulted in a significant reduction in the energy required to heat or cool a building.

It was about the same time that the term "outdoor air" superseded the term "fresh air" that had been more commonly used before that time. Indeed, the increasing pollution in cities from automobiles and industrial sources had made the suitability of outdoor air for contaminant dilution somewhat questionable.

The emphasis on energy conservation was accompanied by a new form of discomfort, the "sick building" syndrome (see page 37). This phenomenon has been particularly evident in cold climates where tight sealing of buildings is customary to minimize winter heating energy. In response to this phenomenon, ASHRAE, in their Technical Committee on Ventilation Requirements and Infiltration, reexamined their recommendations on outdoor air requirements and have now introduced a revised draft ASHRAE Standard 62-1981R: Ventilation for Acceptable Indoor Air Quality. The recommendations have been reported (Janssen and Grimsrud 1986), and the major change is an increase in the minimum ventilation level from 2.5 to 7.5 liters/s per person. It is considered that this level of ventilation is required to achieve acceptably low levels of carbon dioxide.

Many areas associated with air-conditioned spaces are also commonly mechanically ventilated primarily to ensure dilution of contaminants. Toilet and bathroom ventilation is traditionally handled by an exhaust system. Kitchen ventilation is also handled in this way. This system ensures that the contaminated area is at a negative pressure with respect to other occupied areas. Thus airflow will be from the "clean" area to the "dirty" area. The amount of clean air required for dilution of contaminants in these rooms is well defined in local codes and standards. Note that the air in the clean area is usually conditioned and has had significant energy expended on it to cool, heat, humidify, or dehumidify it. This waste of energy could be reduced or reclaimed.

A reduction in energy could be achieved if makeup air is drawn from outside rather than from the conditioned space. This approach is fairly common in kitchen ventilation systems, but its use in restroom ventilation systems is limited by the need to keep these rooms at a significant negative pressure with respect to other spaces.

Another approach is to provide a heat exchanger to transfer heat from the exhaust airstream to, say, the outdoor air being introduced into the air-conditioning plant. It is essential that there be no risk of contaminated air from the exhaust system entering the makeup air to the air-conditioning system. The commercial viability of these heat-exchange systems depends on the cost of energy, the cost of heat exchangers, the ease with which the two airstreams can be brought in contact with the heat-exchange device, and the hours of running. In general, longer operating hours and higher costs of heating and cooling energy will favor the installation of a heat-exchange system. In assessing the energy consumed by the system, it is important not to overlook the additional fan energy required to overcome the pressure drop of the heat exchanger.

Garage or car park ventilation systems also make use of the dilution principle to keep contamination at a safe level. The quantity of air required is defined in local codes. If a garage or car park has sufficient openings in walls and/or the roof to provide effective natural ventilation, then mechanical ventilation may not be required at all. Again, local codes usually define the requirements for adequate natural ventilation.

Should mechanical ventilation be required, then the local code will provide the rules to be used to determine the capacities of supply and/or exhaust ventilation systems required. Such rules are usually prescriptive, e.g., minimum number of liters per second of air per square meter of floor area or per car park bay. Some codes (SAA, AS1668) now include in a limited way an optional performance specification. Under this option, it is up to the designer to demonstrate that the system will operate in such a way as to keep levels of carbon monoxide below a specified value. To achieve this, carbon monoxide sensors are used to activate the ventilation systems only when the carbon monoxide level exceeds a predetermined value. Such systems may provide energy savings in those areas where cars are running for only short periods of time each day, e.g., where the car park serves an office block that has peak activity in the morning and evening as staff arrive and leave, with very little activity in between.

THE HEALTH EFFECTS OF POLLUTANTS

Specific Pollutants

Research into the health effects of pollutants in the indoor environment is sparse. Most of the research on the indoor environment is concerned with the measurement and control of pollutants. However, the health effects of pollutants in outdoor air and the industrial environment are well documented, and these results can be applied to the indoor environment. These effects are summarized by Strauss and Mainwaring (1984). It is of interest that despite much research, the direct relationship between many pollutants and health problems cannot be proven. However, there is sufficient evidence to warrant some action in controlling the levels of many pollutants.

High levels of pollutants are generally associated with respiratory problems, since most pollutants are introduced to the body by means of the respiratory system. The complexity of clearly determining the effect on health of specific pollutants, however, is demonstrated by the effects of "classical" smog. There is a strong correlation between high levels of smog and mortality and illness rates. Nevertheless, no single pollutant has been singled out as being responsible. It is probable that a synergistic association between sulfur dioxide particulates and their reaction product, sulfuric acid, is operative. Studies point to an association between sulfur dioxide particulate pollution and bronchitis, emphysema, and asthmatic attacks. The effect is greatly aggravated in adults by cigarette smoking and is possibly synergistic.

"Photochemical" smog is a complex mixture of products formed from the interaction of sunlight with nitric oxide and hydrocarbons (the two major components of automobile exhaust). The products formed are oxidants, predominately ozone, and peroxyacetyl nitrates. A high level of photochemical smog has not been unequivocally linked with human mortality, but it has been linked with eye irritation, sore throats, and respiratory infection.

There is a demonstrable relationship between carbon monoxide and changes in human physiology. While death occurs at concentrations around 1000 ppm (corresponding to blood levels of 60 percent carboxyhemoglobin), impaired function occurs at blood levels between 10 and 20 percent carboxyhemoglobin. Actual levels for urban dwellers in moderate to highly polluted cities vary from 0.8 to 3.7 percent carboxyhemoglobin for nonsmokers and from 1.2 to 9 percent for smokers. It also has been suggested that high levels of carbon monoxide are a contributing factor to heart disease (Strauss and Mainwaring 1984).

Considerable research has been carried out on carcinogenic compounds, but difficulties result from the delay factor of 20 to 40 years between cause and effect and from factors other than pollution that are associated with an urban lifestyle. It is suspected, however, that polyaromatic hydrocarbons found in urban air are carcinogens.

Radioactivity is also a cause for concern, particularly since it has been demonstrated (see page 32) that there can be a buildup of radon in energy-efficient homes, although many ordinary homes also have been found to exhibit high radon levels (Nero et al. 1984). The average radon level to which the population of the United States is exposed represents a 0.3 percent average cancer risk. This is much higher than the risk of other suspected carcinogens, e.g., about 0.001 percent for benzene, vinyl chloride, and pesticides. However, houses have been identified with radon levels representing a 2 to 5 percent cancer risk.

The "Sick Building" Syndrome

Since the move to energy conservation in the midseventies, a new phenomenon has been observed, that of the "sick building." A great deal of international discussion and research is now focusing on this issue. The difficulty in dealing with this syndrome is summarized in an article on the subject (Field 1984) that states,

> It would be easy to define office building syndrome (OBS) if it were one specific theory—but its manifestations are various and the basic causes still in doubt. Essentially OBS is a dissatisfaction with the environment other than feeling too warm or too cold. It can include lassitude, headaches, dizziness, nausea, irritation of eyes, nose and throat, and feelings of stuffiness, disorientation and oppression. (Page 39)

It is generally considered that an important facet of "sick building" syndrome is particulate and gaseous contamination emanating inside the building from occupants, furnishings, building materials, and other sources. One reaction to this has been to revert to the higher outdoor air ventilation rates that were common before the energy crisis. This results in higher energy consumption and a greater concentration of outdoor pollutants.

Until much more conclusive research is completed, there are a number of other logical steps that should relieve the syndrome. One is that proposed by Gibbs (1985) to improve the efficiency of filtration systems in air-conditioning plants. This will reduce the contamination by particulates and other attached contaminants from both indoor and outdoor sources.

Major causative agents contributing to a large number of "sick buildings" have been reported by Robertson (1985) as being primarily fungal (74 percent) and bacterial (70 percent) contamination. The levels of contamination posed a danger to health in 27 percent of buildings as a result of excessive fungal contamination and in 25 percent as a result of excessive bacterial contamination. The problems were rectified primarily by cleaning accumulations of dusts and dirts from the systems and providing improved filtration to prevent further contamination.

Where very low ventilation rates and very well sealed buildings have been introduced in response to the energy crisis, as in Scandinavia, research suggests that the amount of outdoor air introduced into a ventilation system is only part of the story (Skaret 1981). It has been proposed that the ventilation air should be evenly distributed throughout the occupied areas to be most effective. This requires the careful selection of ventilation supply outlets and exhaust grills and their location. A measure of the effectiveness of this approach has been introduced and is termed "air exchange efficiency."

Bacteria and Viruses

From the time of Pasteur it was believed that the spread of infection from viruses was by person-to-person contact, but in the 1930s it was determined that respiratory infections were spread by airborne means. The transport mechanism for viruses is by means of droplets (e.g., generated by sneezing or coughing) or by clinging to dust particles. It was not generally considered that air-conditioning and ventilation systems posed a significant health threat related to bacteria and viruses until the advent of legionnaires' disease, which resulted in the loss of lives. It has been found that air-conditioning equipment can spread and in some cases can support the growth of the legionnaires' disease bacteria, *Legionella pneumophilia*. Recent research (Gwaltney, Hendley, and Dick 1985) has shown that rhinoviruses, the common cold viruses, also can be spread by airborne means.

The level of risk to occupants is still considered to be small, and further research is proceeding to quantify the risks and identify those parameters which influence the risk. Indeed, ventilation systems themselves hold the key to minimizing the risk. Specially designed ventilation systems with very high filtration levels are used to minimize the risk of viral infection in hospital operating theaters and laboratories.

SUMMARY

Air quality is an important ingredient in the human comfort equation. Indoor air quality is a function of pollutant levels out of doors, over which the building designer has little control, and pollutants being generated within the building. Since most air pollutants affect the human body by means of the respiratory system, there is a close link between air quality and ventilation systems. Some pollutants, e.g., bacteria, viruses, and carcinogens, directly affect the health of the occupants, while others, e.g., odors and dusts, can cause significant discomfort.

Pollutants generated within a building can be controlled by dilution with outdoor air, provided, of course, that the outdoor air is of an acceptable quality. Historically, there have been codes governing the amount of outdoor air required to maintain a comfortable indoor air quality. The energy crisis of the late 1970s resulted in a general reduction of outdoor air ventilation rates. This threw focus on the significance of pollution generated within buildings, and there is now a trend toward increasing the minimum outdoor air quantities required.

So that energy is conserved without compromising human health and comfort, there is a growing emphasis on the effective distribution of outside air to the occupied spaces within a building.

Air quality also can be controlled by the removal of contaminants. Commonly this is done by filtration for particulate contaminants and adsorption for gaseous contaminants.

Research is increasing in the area of indoor air quality, and it is expected that there will be significant developments in measurement and control over the next decade.

REFERENCES

American Society of Heating, Refrigerating, and Air Conditioning Engineers (ASHRAE). 1985. *ASHRAE Handbook 1985: Fundamentals.* ASHRAE, Atlanta, Ga.

ASHRAE. 1971. Filter Test Code. ASHRAE, Atlanta, Ga. British Standards Association (BSA). 1971. BS Filter Test Codes: BS2831:1971 Filter Test Code. BSA, London.

Field, A. 1984. The sick building syndrome. *CIBS Build. Serv. J.* (November 1984): 39-40.

Gibbs, G. 1985. Energy Conservation and Health Inside Buildings with High-Efficiency Filtration. Energy '85 Conference, Sydney Australia.

Grimsrud, D. T., et al. 1985. Building Ventilation and Indoor Air Quality. In *FY1985 Annual Report, Energy Efficient Buildings Progress.* Lawrence Berkeley Laboratory, California.

Gwaltney, J., Hendley, J. D., and Dick, E. 1985. Rhinoviruses Spread by Airborne Means. Conference of Infectious Diseases, Washington, D.C.

Janssen, J. E., and Grimsrud, D. T. 1986. Ventilation standard draft out for review. *ASHRAE* Journal 28: 43-45.

Nero, A. V., Schwehr, M. B., Nogaroff, W. W., and Revgan, K. L. 1984. Distribution of airborne 222 radon concentrations in U.S. homes. Lawrence Berkeley Laboratory.

Skaret, E. 1981. Ventilation Efficiency and Indoor Air Quality. International Symposium on Indoor Air Pollution, Health and Energy Conservation, Amherst, Mass.

Strauss, W., and Mainwaring, S. J. 1984. *Air Pollution.* Edward Arnold, London.

Robertson, G. 1985. Statement of Gray Robertson, of ACVA, Atlantic, Inc., to the Board of Health, Nassau County, New York, August 14, 1985.

3

Luminous Environment

Nancy Ruck

THE FUNCTION OF LIGHTING

People have a tendency to take light very much for granted, and yet approximately 80 percent of the information we receive comes from our eyes and is therefore of a visual nature. Seeing is not a passive response to images of light; it is an active information-seeking process whereby images are focused on the light detectors of the eye's retina. This information is then collected and transferred along the optic nerve to the brain for interpretation. Seeing is therefore dependent on both light and the human visual system, which is a highly developed combination of light detection and image processing.

The quantity and quality of light that our eyes receive have a direct influence on the way we see things. Great architects, including the designers of the Parthenon, the craftsmen of the Gothic cathedrals, and indeed certain twentieth-century architects, have understood the impact of natural light and its importance for putting human beings in touch with their environment. Although both natural and electric light have their own individual characteristics and different qualitative attributes, light, in general, can be thought of in architectural terms as a structural material that can be used in a building to serve various functions. Like other materials such as brick, steel, stone, and concrete, it should therefore not be treated as applied decoration or as equipment that can be added after the building is designed.

In light terms, an environment can be said to be comfortable when an eye task such as reading a book or searching for an object is quickly and easily performed without distraction or the creation of stress. Generally, the attainment of visual comfort relies on such parameters as adequate illuminance, limitation of glare, and such subjective considerations as achieving an appropriate color scheme, and, in the case of daylighting, avoiding a gloomy interior and providing an acceptable size and shape of window to maintain contact with the outside world.

It is possible that meeting the criteria of adequate illuminance and avoidance of glare may result in a failure to meet other requirements, such as those associated with thermal comfort. It remains the task of the designer to consider the impact of all relevant variables, including heat, sound, and air quality, together with their impact on and interactions with one another, and to decide which aspects have priority according to their function in the building design.

Lighting, however, can do more than just make things visible. It can contribute to people's impressions of an interior by giving the space a character or atmosphere. What is regarded as appropriate for a particular interior will depend on the interior's function. What is considered appropriate for a display area will not suffice for a general office.

Light also can be considered on physiological and biological grounds as being essential for the well-being of a building's occupants owing to its nonvisual effects such as brain stimulation and body orientation and balance. A good luminous environment therefore depends not only on environmental and task lighting design, but also on the spectral composition effects of the light on individuals.

This chapter is concerned with describing the impact of both the nonvisual and the visual aspects of light on people. The following comments are applicable to all kinds of working and living environments. Although some discussion is relevant to the context of offices, where a good working environment is expected, it is also applicable to a much larger range of buildings with regard to both the work done and the equipment used. The units of measurement used are generally SI units, but can be converted to Imperial units by multiplying by 10.76.

EFFECTS OF VISIBLE RADIATION

The Nature of Light

The controversy over the nature of light is one of the most interesting in the history of science. Light is now regarded as encompassing a narrow region of the total electromagnetic spectrum, which includes radio waves, infrared light, ultraviolet light, and x-rays. The physical differences lie solely in the wavelength of the radiation, but the effects are very different. The range to which the eye is sensitive is between 380 and 760 nm (a nanometer is equivalent to a millionth of a meter). Within this range, eye sensitivity differs, and different wavelengths are perceived by the eye as different colors. The peak sensitivity for light-adapted (photopic) vision is 555 nm, and for dark-adapted (scotopic) vision it is about 505 nm. The actual human spectral response has been standardized in an internationally agreed-upon form represented by the Commission Internationale de l'Eclairage (CIE) standard observer, as shown in Fig. 3.1.

Present thinking is that light also consists of packets of energy, or "quanta," thus combining the characteristics of corpuscles and waves. Light of short wavelength has more waves in each quantum than light of longer wavelength. The quantal nature of light has an important implication for vision and has inspired some particularly elegant experiments that bridge the physics of light and its detection by the eye and the brain. It has been found, for example, that the quantal nature of light is important in considering the ability of the eye to detect fine detail. Insufficient quanta falling on the retina to build a complete image within the time span required to integrate energy give a lack of conspicuity.

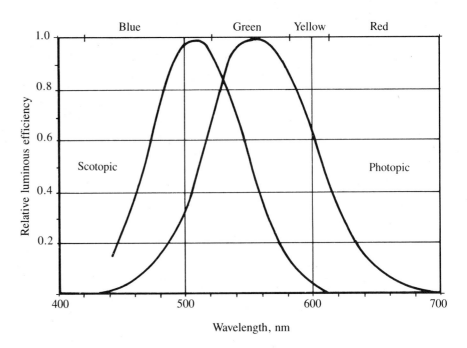

FIGURE 3.1.
The relative spectral sensitivity of the eye for photopic (light-adapted) and scotopic (dark-adapted) vision.

Effects of Natural and Electric Light

The properties and characteristics of natural and electric light sources vary. Natural light is principally characterized by its variability in both quantity and spectral distribution. Electric light sources are more stable, but they differ considerably in their efficiency at converting electricity to light, their life, and their color properties.

Natural light can be divided into two components, sunlight and skylight. All natural light comes from the sun, but it is the sunlight diffused and scattered in the atmosphere that is the principal source of daylight in buildings. Unless properly controlled, sunlight produces glare and thermal problems.

The spectral composition of natural light varies with the nature of the atmosphere through which it passes and the path length. The correlated color temperature can vary from 4000 K for an overcast sky to 40,000 K for a clear blue sky. As the correlated color temperature rises, the blue end of the spectrum becomes more dominant. However, regardless of the variations in spectral composition and hence the rendering of colors under natural light, it is current practice to take an appropriate phase of daylight as a reference source, and therefore, conventionally, the color-rendering ability of natural light is considered almost perfect.

A large number of electric light sources are available, each with different color-rendering properties, and they can be broadly classified into two groups, incandescent lamps and discharge lamps. Both rely on electricity as a source of energy, but each has a different basic mechanism. Incandescent lamps have low correlated color temperatures (2700 to 2900 K), being warm in appearance, and discharge lamps have varying color-rendering abilities depending on the composition of the gases in the discharge tube and/or the fluorescent coating on the inner surface of the tube or lamp.

Continuing research is being carried out in new lamp technologies to increase luminous efficacy and hence energy efficiency as well as color-rendering ability. However, the impact of electric light on human health and productivity has yet to be evaluated. Various research projects have been carried out on the effect of fluorescent lighting on building occupants in terms of visual and nonvisual reactions. Results have shown that there is little difference in terms of fatigue between fluorescent and incandescent lighting.

However, it has been found that there are significant differences in pupil size when people are exposed to indirect high-pressure sodium lamps as compared with indirect incandescent lighting when the light intensities are photopically matched. It has been suggested that these observed differences in pupil size are due to the differences in the spectral power distribution of the two light sources. Since pupil size can affect visual performance, the spectral power distribution of light sources requires consideration in lighting design and application (see page 54).

NONVISUAL EFFECTS OF LIGHT

Natural light is vital to our well-being because it promotes not only a visual response to the world around us, but also nonvisual biological effects that can improve human health and well-being. In recent years, medical research has concentrated on the influence of natural and electric light on human physiological functions, such as, for example, brain stimulation and activity, blood circulation, skin resistance, and body orientation and balance.

It is now known that sunlight can influence the rate of development and growth of certain bacterial and viral illnesses. Research also has shown that skylight through glazing (especially that of a cloudless sky) has a similar effect if the dosage is long enough (3 to 4 hours), for although ultraviolet radiation from the sun cannot totally penetrate clear glass, such glass is very transparent in the visible spectrum of sun and sky radiation. (Clear glass has a visible transmittance of 80 to 90 percent.)

Components of Optical Radiation

"Optical radiation" can be defined as that part of the electromagnetic spectrum which includes ultraviolet (UV), visible, and infrared (IR) radiation. This encompasses a wavelength range of 100 nm to 1 mm. Figure 3.2 shows the spectral distribution of direct solar radiation after scattering by air particles and absorption by ozone, water vapor, and oxygen in the atmosphere.

The ultraviolet and infrared ranges can be further subdivided into smaller wavelength bands on the basis of their various effects or measuring standards. Designation of wavelength bands on the basis of the German standard (DIN 5031 1979) is given in Table 3.1.

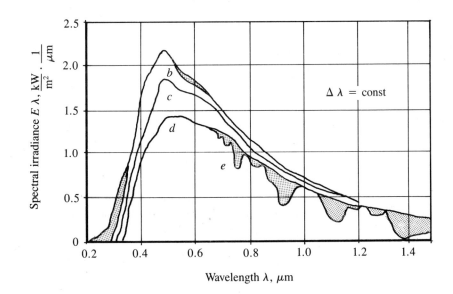

FIGURE 3.2.
Spectral distribution of direct solar radiation (according to CIE Publication No. 20): (a) solar radiation, (b) after absorption by ozone, (c) after Rayleigh scattering, (d) after absorption and scattering by aerosol, and (e) after absorption by water vapor and oxygen, i.e., direct solar radiation on the ground.

TABLE 3.1 Designation of Wavelength Bands

Optical radiation	100 nm to 1 mm
Ultraviolet radiation (UV)	100 to 380 nm
Shortwave UV-C	100 to 280 nm
Middle-wave UV-B	280 to 315 nm
Long-wave UV-A	315 to 380 nm
Visible radiation	380 to 780 nm
Infrared radiation (IR)	780 nm to 1 mm
Shortwave IR-A	780 nm to 1.4 μm
Middle-wave IR-B	1.4 μm to 3 μm
Long-wave IR-C	3.0 μm to 1.9 mm

Effects of infrared and visible radiation

The nonvisual biological effects of infrared radiation are mainly related to the heating of the skin and tissue by radiation absorption. This effect results in an inversion in body temperature that causes a greater blood flow and hence has an effect on bodily functions such as circulation, respiration, and the nervous system.

Visible light not only stimulates vision, but also influences, by means of the pineal gland, all glands of the inner secretory system and hence the whole human organism. The relationship between pineal gland function and the central nervous system and endocrine glands has been reviewed by Lewy et al. (1980). In many countries, medical research is being directed toward the influence of natural light on the function and performance of these glands in terms of internal secretion by means of the hypothalamus and hypophysis. The pineal gland and its major hormone, melatonin, appear to play a central role in regulating the indirect effects of light (Wurtman 1975).

Ultraviolet radiation has the widest and most crucial effect, although its proportion in the radiation of natural and artificial sources is relatively low compared with the visible and infrared wavelength ranges.

Effects of Ultraviolet Radiation

The effects of ultraviolet radiation are both external (on human skin) and internal (via the skin). The most recognizable effect on the skin is a reddening effect known as "ultraviolet erythema." People lose (skin) pigmentation when deprived of sunlight, and subsequent exposure results in a reddening of the skin. The degree of reddening depends on the individual, the geographic location, and the wavelength of the ultraviolet radiation. Shortwave ultraviolet radiation (UV-C) has less effect than long-wave ultraviolet radiation (UV-A). After irradiation of the skin by short- or middle-wave ultraviolet radiation, a secondary pigmentation (tanning) of the skin can occur as a consequence of the acceleration formation of melanin. Continued repetition of exposure of the skin to ultraviolet-B radiation can lead to changes in the epidermis known as the preceding stages of skin cancer. However, this is possible only by exposure to strong sunlight, which can be avoided by appropriate shading.

Sometimes, presumptions are made as to the origin of skin cancer from the light from fluorescent lamps. Such presumptions can be examined by a comparison of the calculated carcinogenic exposure that a person receives under natural global radiation and under artificial illumination in the course of his or her life. For example, one working day (8 hours) under an artificial illumination of 1000 lux from fluorescent lamps (color temperature 4300 K, color rendering index 75) corresponds to an outside exposure in Davos, Switzerland, of only 3 minutes. Figure 3.3 shows the spectral irradiance of global radiation in Davos, Switzerland, on a cloudless day at noon in June compared with a 40-W universal white fluorescent light (color temperature 4000 K, color rendering index 70) at 1000 lux.

Recent investigations by the Commission Internationale de l'Eclairage (CIE) on malignant melanoma and fluorescent lighting indicated that available evidence did not support the existence of any substantial association between melanoma risk and exposure to fluorescent lighting, although further research is warranted because the ultraviolet radiation emitted in some lighting situations can be a considerable proportion of the ultraviolet radiation received from the sun and this radiation is known to have deleterious effects on human skin. Reduction of ultraviolet radiation can be carried out easily by a suitable choice of phosphors or glass envelopes. Complete removal of shortwave ultraviolet radiation is possible by using styrene covers.

The majority of fluorescent lamps in general use do not provide the beneficial radiation levels of sunlight. The continuous spectral distribution

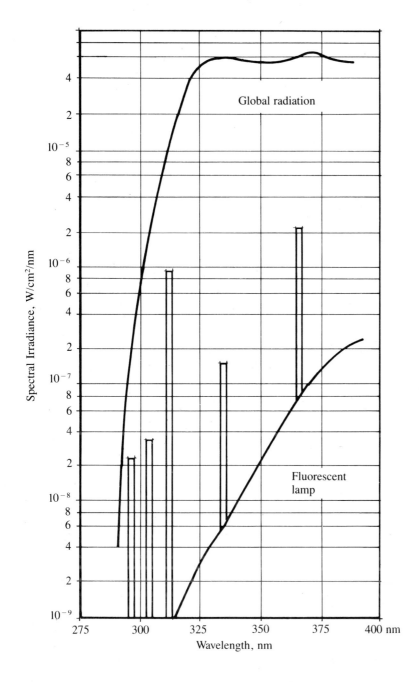

FIGURE 3.3.
Spectral irradiance of global radiation in Davos, Switzerland, compared with a 40-W universal white fluorescent lamp. (After Steck 1973.)

curve of sunlight not only produces the necessary levels of ultraviolet radiation, but also encompasses the whole spectrum so that a balanced form of radiation is received. Fluorescent lighting does not usually emit radiation at wavelengths as short as that of solar radiation, nor are the levels of radiation at the short end of the visible spectrum as high as that of solar radiation. Therefore, constant exposure to these lamps without adequate exposure to natural light may deprive people of necessary forms of light.

Lamps that more accurately reproduce extra daylight, i.e., with the use of "full-spectrum lighting," have been found to improve the health and efficiency of building occupants. Currently, a fluorescent lamp manufactured by Duro-Test meets this specification. However, the lamp is listed with the U.S. Federal Food and Drug Administration as a medical device. The specification of an electric light source for the simulation of a balanced ultraviolet spectrum of solar global radiation for use in indoor illumination is shown in Table 3.2.

It also should be noted that at present many glazing materials do not transmit the useful long-wave ultraviolet radiation, but several new multilayer coating techniques are currently being investigated, and these have been shown

Table 3.2 An Electric Light Source to Simulate Sunlight (according to Hughes 1983)

Correlated color temperature	5500–6500 K
Color rendering index (CRI)	≥ 90
Long-wave ultraviolet radiation	$220 \pm 60\,\mu$ W/lm
Middle-wave ultraviolet radiation	$8 + \frac{12}{0}\,\mu$ W/lm
Long-wave ultraviolet/middle ratio	$28 + \frac{7}{20}\,\mu$ W/lm
Daily indoor dose level	Between ⅛ and ¾ MED* or 8-hour exposure at 538–3228 lx

*Minimum erythema dose.

to give very high solar transmission, even higher than clear, uncoated glass (see Chap. 10).

Another important biological effect of ultraviolet radiation is the formation of vitamin D_3, which controls the calcium-phosphorous exchange and hence has a direct influence on bone development and construction (Holick et al. 1982). For example, the photobiological effect of ultraviolet sun and sky radiation is used to cure rickets and vitamin D deprivation. Ultraviolet radiation also exerts a nourishing and regenerating effect on the muscles and has a normalizing influence on the nervous system. These latter influences have been proven by experimental and clinical results (Seidl 1969).

PSYCHOLOGICAL EFFECTS OF LIGHT

The importance of the relative spectral power distribution of a light source for optimizing visual comfort is evident in recent research (Wake et al. 1977; Hughes and Neer 1981). The information our brain receives from the illuminated environment is an essential element in shaping our moods, reactions, and psychological well-being (Hughes 1983). Plank and Schick (1974) summarize the effect of color on nonvisual processes in human beings. The potential effects include changes in mood and emotional state, muscular activity, rate of breathing, pulse rate, and blood pressure.

In addition, MacLaughlin et al. (1982) have concluded that the spectral character of natural sunlight has a profound effect on the photochemistry of dehydrocholesterol in human skin and induces physiological and biochemical responses in humans. The yellow and orange sunlight, for instance, at sunrise and sunset affects sexual activity and preparation of the whole human organism for stimulation and awareness. Higher levels and dynamic changes in daylight illuminances therefore have an ergonomic importance, whereas a relaxed atmosphere is associated with the low light levels and warmer colors of the interior environment. In this sense, such associated reflexes and stereotypes can purposefully be used in architectural design to stimulate productivity, decrease errors, and increase creativity or initiative action in industrial interiors, offices, and schools.

The work of Flynn and Spencer (1977) on subjective impressions of illuminated spaces supports the conclusion that artificial lighting that simulates natural light is perceived as significantly more pleasant and stimulating than conventional fluorescent lighting. Observers feel more relaxed and have greater eye comfort.

Visual Contact with the Exterior

Parallel to the growing awareness of the potential of daylight for human well-being and hence the growing desire for daylight-oriented architecture for health and aesthetic reasons, there is a growing appreciation of the importance of windows for visual contact with the outside world and for meaningful views. This subjective psychological role of windows implies that windows should be relatively large, whereas the thermal aspects of the building envelope require that windows be relatively small to counteract heat gain and heat loss. For these reasons research on acceptable minimal window dimensions has been carried out in recent years (Keighly 1973; Ne'eman and Hopkinson 1970). In Keighly's studies, using a model of an open-plan office, glazed areas of less than 15 percent window-to-wall area were considered disagreeable, but above 30 percent there was almost complete satisfaction. These results were in reasonable agreement with the studies of Ne'eman and Hopkinson, where the importance of the type of view was of major concern. In this case, 25 percent of the window-to-wall area was the minimum acceptable window size for 50 percent of the observers, but this had to be increased to about 32 percent if 85 percent of the observers were to be satisfied. These experiments were done in temperate climates. Differences in climate may place more stringent restrictions on the admission of natural light and hence on window design. Results have shown that windowless buildings are rarely inhabited with pleasure unless there is a specified reason why no windows are required.

Human Biorhythms

The periodic daily changes of many physiological processes in the human body are stimulated by daytime light and nighttime darkness, and many experiments in underground caves and windowless interiors have shown that the 24-hour biorhythm is essential for human beings. This fact was stressed by Hallberg (1962). For example, the dissinchronization of biorhythms, which can be caused by long-distance travel, influences the whole nervous system as a result of the changing of the natural radiation cycle. Recent research has found that an antidepressant effect can be obtained by artificially extending the number of daylight hours.

Rhythms of human response and activity are also regulated by variations in the quality and quantity of light. As the pattern of light changes in the winter, for example, difficulties in physiological and psychological adjustment in certain people can be observed.

Recent findings that human melatonin secretion is suppressed by sunlight but not by low-intensity light (500 lx) normally found in office environments have led researchers to examine the possibility that human circadian and seasonal rhythms might be modifiable by employing simulated sunlight of the correct intensity. Cases have been reported (Lewy et al. 1980) in which simulated natural light has been used successfully to experimentally manipulate biological rhythms to reduce winter depression. There is need for additional research to investigate light-behavior relationships, however.

VISUAL SYSTEM COMPONENTS

Light is necessary for the visual system to operate, but it is important to realize the meaning of the term "visual system," since it encompasses different functions. In physiological terms, the visual system includes both the eye and the brain, and psychologically, it involves both immediate and past experience. As with the perception of noise, the system's interpretation is controlled not by sensory input alone, but by a belief in what the input is in actuality. It is interesting to note that one of the areas of current research concerns defining the way in which signals reaching the brain from the eye are coded at various stages in the system and how this coding affects perception. Our present state of knowledge in this area is limited.

In this section, the components of the visual system and the parameters that are relevant to visual performance are discussed together with their performance limits.

Pupil Size

Rays of light enter the eye through the cornea and are brought into focus by the lens onto the photoreceptor layer, the retina, as shown in Fig. 3.4. The cornea, which is the transparent section of the sclera, the outer covering of the eye, has a curvature that dictates the refraction at the corneal surface and hence its contribution to the overall power of the eye as an optical system. The light passes through an aperture, or pupil, in a colored diaphragm called the iris. The size of the pupil is determined by several different factors: the retinal feedback, which concerns the quantity or intensity of available light; the distance at which the eye is focused; the spectral power distribution of the light source, as mentioned previously; human emotion (Duke-Elder 1944); and the spectral composition of the light source.

The spectral composition of the light source and retinal feedback are important as far as visual comfort is concerned. If the amount of light reaching the retina is too small, then pupil size is increased. If the amount is too large for the prevailing state of adaptation, then pupil size is reduced in an attempt to restrict the light input. Pupil diameter can vary from about 2 to 8 mm and these diameters produce an effective change in the amount of light reaching the retina of about 10 to 1. In comparison with the eye's operating luminance range of 10^{12} to 1, variations in pupil diameter can be said to be a minor aspect of the visual system's adaptation process. However, it should be noted that continual use of the muscles that control pupil size can produce fatigue and affect visual performance if the quantity of light within the visual field or its spectral distribution varies extensively over short time spans. Extreme variations in brightness of the visual field should be avoided in work situations. (The effect of the spectral composition of the light source on pupil size is dealt with in greater detail in Chap. 6.)

Retinal Response

The retina is a photochemical sensitive surface that has three layers: a layer of photoreceptors, conventionally divided into two types, rods and cones; a layer

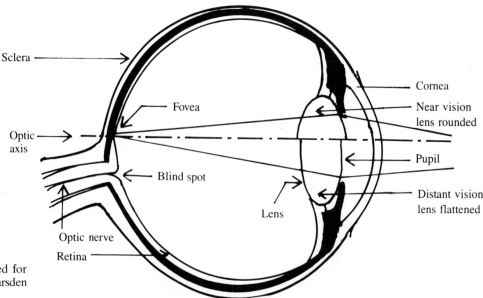

FIGURE 3.4.
A vertical section through an eye adjusted for near and distant vision. (After Cayless and Marsden 1983.)

of bipolar cells; and a layer of ganglion cells which are individually linked to a nerve fiber that transmits signals by means of the optic nerve to the brain, as shown in Fig. 3.5.

It is possible that the bipolar layer of the retina allows some analysis of the information contained in the retinal image and that the ganglion cells convey this information by nerve fibers to the brain.

The response to light varies over the retina, and this has important implications as far as visual comfort is concerned. The rods and cones are not evenly distributed, and have different sensitivities to light. The cones, which perceive detail and color, are concentrated in an area called the "fovea." Outside the fovea there is a rapid increase in the number of rods, which do not perceive color or detail, but are sensitive to movement. In comparison with the cones, the rods are more sensitive and differ in the wavelengths to which they are responsive. The peak sensitivity of cones is about 555 nm, but that of rods is about 505 nm. It can be said generally that the cones operate during the daytime (photopic vision) and the rods at nighttime (scotopic vision) when light levels are lower. The qualitative differences between daytime and night-time vision are well known: In daytime we see colors, but at night no colors are visible.

Fundamentally, therefore, the human being can be considered as having two retinas whose spectral responses differ on moving from scotopic to photopic conditions. For visual field luminances between 10^{-3}, and 3 cd/m^2, an intermediate state exists, called "mesopic," in which both cones and rods operate. This classification of conditions relates to major changes in our visual capabilities. Under scotopic conditions, colors are not visible. As the mesopic condition is reached, color vision starts to appear and strengthens until photopic conditions are reached, at which time complete color vision is available.

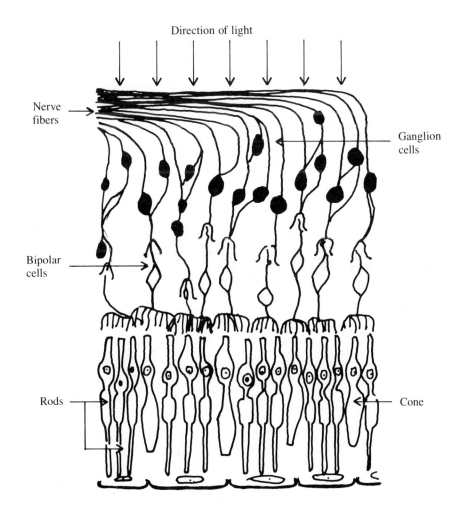

Direction of light

Nerve fibers

Ganglion cells

Bipolar cells

Rods

Cone

FIGURE 3.5.
A schematic representation of the retina. (After Gregory 1965.)

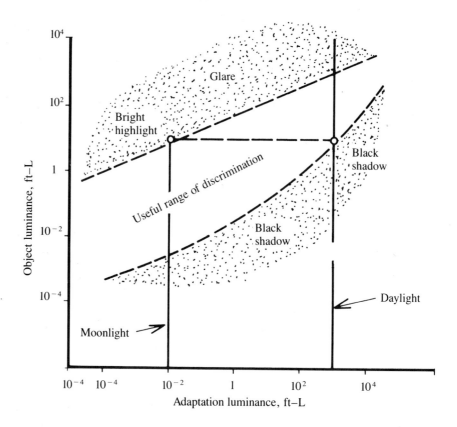

FIGURE 3.6.
The range of discriminable luminances at any given level of adaptation. (After Hopkinson and Collins 1970.)

Discrimination of detail can be achieved only in the periphery of the fovea when cones alone are operating.

When considering the interaction between people and lighting, there is need to consider the pattern of luminances in the visual field, since the eye cannot operate over the range of 10^{12} to 1 simultaneously. The process of adjustment of the visual system, or adaptation, is a primary consideration when designing for visual comfort. When the eye is adapted to a given luminance, much higher luminances are glaring and much lower luminances appear as black shadows. Figure 3.6 gives an indication of the limits within which differences in luminance and hence details can be discriminated for any particular adaptation luminance.

The actual process of adaptation relies on a photochemical reaction that is dependent on time. The actual time taken depends on the starting and final adaptation luminances, since the adaptation processes for rods and cones have different time constants—approximately 7 to 8 minutes for rods and 2 minutes for cones. If both the starting and final adaptation luminances are in the photopic range, adaptation is rapid. If the final adaptation luminance is in the scotopic range, adaptation is much slower.

The seeing of tasks and objects depends not only on the state of adaptation of the eye, but also on the size and contrast of the object with the luminance of the background. The question of what is the most suitable ratio of luminance of the task/object to that of the immediate surroundings (combined effect of illuminance and reflectance) has been investigated by many researchers (Balder 1957; see also page 52). From this research it has been possible to derive criteria for a comfortable and effective balance between the luminance of room surfaces as recommended in country codes.

Visual Perception

Besides being a detector of light in a visual environment and an organizer of information contained in the retinal image, the visual system is also an interpreter of information. Results from studies of the visual system as an interpreter are limited, but some understanding of its basic principles has been achieved.

When considering how we perceive the real world, it is apparent that our overwhelming impression is one of stability regardless of variations in luminance. Despite changes in the retinal image, objects maintain their apparent size, shape, color, and lightness. A chair is still a chair no matter from what direction or how it is seen. This aspect of perception is known as "perceptual constancy." Distortion can occur, however, if there is misleading or insufficient information from the surrounding visual field.

The most relevant of these constancies as far as lighting is concerned is lightness constancy. Two surfaces with varying luminances but the same reflectance will appear the same if viewed in the context of the whole visual field. Color, size, and shape constancy also can occur depending of the setting in which an object is seen. This suggests that perception is very much a matter of mutual support among all the objects and surfaces in the visual world. The more information available, the more likely constancy will be maintained. Lack of information, or constancy failure, can be linked to visual acuity and performance. Conditions under which there can be high levels of performance (e.g., good acuity and high contrast sensitivity) are those most suitable for obtaining information about the visual environment and therefore for achieving or maintaining constancy.

LIGHTING AND VISUAL PERFORMANCE

Early studies of the capabilities of the visual system used "visual acuity" as a measure, which is simply the angle subtended at the eye by detail that can be discriminated on 50 percent of the occasions it is presented according to Boyce (1981). However, other factors also influence recognition in addition to the size of the detail. The illuminance and luminance of the task or object, the luminance of the background against which the task or object is seen, the observer's experience, and his or her expectations all are involved (Overington 1978).

Illuminance

Of all the lighting parameters that have been researched, by far the most thoroughly investigated is illuminance. The effects of illuminance on a wide range of different tasks have been examined by both field and laboratory experiments using simulated tasks. Results indicate that the effect of increasing illuminance on the performance of tasks follows a law of diminishing returns; that is, equal increments of illuminance produce less and less improvement in task performance until performance saturates (Neston 1945). The illuminance at which performance saturates depends on the visual difficulty of the task, which can be classified according to the size and contrast of the critical detail. The smaller the size and contrast, the higher the illuminance at which the task saturates.

Visual performance as related to the task is generally considered to be some combination of speed and accuracy, attributes that determine real productivity for clerical and industrial tasks. IES recommendations have traditionally been made in the form of minimum levels, and this tends to lead to overlighting. With electrical costs rising and the need to curtail the risk of overlighting, maximum as well as minimum levels or perhaps target or maintained levels are more appropriate. The CIE Guide for Electric Lighting of Interiors (1986) now gives a range of illuminances for a particular task that can be modified according to frequency of occurrence, degree of detail involved, and the age factor.

Having specified a design illuminance the actual lighting installation will produce a range of illuminances over the interior. The problem is how large the range should be. Present practice specifies a uniformity ratio of 0.8 (minimum illuminance to average illuminance) or 0.7 (minimum illuminance to maximum illuminance). This applies to an interior in which the design approach adopted is that of uniform lighting. Recently there has been a renewal of interest in local lighting because of its energy saving potential. Uniformity criteria do not apply in this situation. However some recommendation is required to accommodate the eye's adaptive power. Current recommendations are that the background illuminance in a local lighting installation should be not less than one third of the task illuminance.

There are other differences that occur within a lighting installation and these are generally revealed by changes in luminance, that is, by the combined effects of illuminance and reflectance. Increases in illuminance produce increases in luminance and, as a result, generate increases in contrast sensitivity and visual acuity.

In research on visual acuity and the discernment of detail, there have been many different measurements of threshold contrast for different conditions (Blackwell, 1959). However, it is questioned whether research on threshold performance is relevant and realistic in terms of everyday work and whether the activities requested of the observer in laboratory experiments are generally different from real work carried out in a work situation. However, measurements of threshold performance can be useful indicators of the effect of lighting conditions on actual work.

It has been found that equal step changes in size, changes in luminance, or presentation time become steadily less effective in changing threshold contrast. At very low luminances, very small sizes or very short times, a small change in any of these parameters will be effective. At high levels of these variables, however, the same changes will have little effect. This relationship between threshold contrast and luminance holds for a wide range of tasks, such as printed words, gratings, and the like (Guth and McNelis, 1969).

A common feature of all these experiments on factors influencing the threshold values of visual acuity, and contrast sensitivity is that they depend on the retinal location of the stimulus, the state of adaptation of the eye, and the presentation time. Generally, the conditions are such that viewing takes place under optimal conditions (a well-defined target occurring in a known position on a uniform luminance field). Real viewing seldom gives the same results, for targets generally occur off the visual axis, i.e., away from the fovea, where sensitivity is greatest.

Although measurements of threshold performance can be useful indicators of the effect of any lighting condition, the accuracy of any prediction will be strongly influenced by the laboratory conditions under which the data were obtained. In actual interiors, increased effort may be necessary to mask the effect of poor lighting conditions and the indirect effects of lighting, such as distraction caused by glare. In addition, differences between individuals, particularly the worsening of visual performance with age, must be taken into account.

Luminance

Consideration of the contrast of the task with the luminance of the background is also important. It has been found by Foxell and Stevens (1955) that optimum visual acuity occurs when the luminance of the background is slightly less than the luminance of the task. If the luminance of the background is much lower or higher than the luminance of the task, visual acuity is greatly reduced, as shown in Fig. 3.7. The most comprehensive experiment in background luminance ratios is that by Balder (1957). He found that at illuminances recommended for offices (500-750 lx), the optimal luminance ratio is about 0.55 for walls facing the task area and 0.48 for walls alongside the task. He also found that the ceiling-to-task luminance ratio was in the order of 0.5 to 0.8. Ranges of

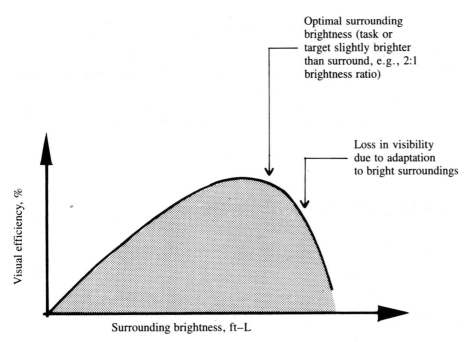

Optimal surrounding brightness (task or target slightly brighter than surround, e.g., 2:1 brightness ratio)

Loss in visibility due to adaptation to bright surroundings

Visual efficiency, %

Surrounding brightness, ft–L

FIGURE 3.7.
Visual efficiency as a function of surrounding luminance.

acceptable luminance ratios have been derived from such data and are given in various codes. In the preceding findings it was assumed that the surfaces had uniform luminances.

LIGHT, HEALTH, AND PRODUCTIVITY

Eyestrain

Light is necessary for the visual system to operate, but as discussed earlier, it can be harmful to health. The most common deleterious effect of light on health is the occurrence of eyestrain. Symptoms of eyestrain occur when the visual system is called on to act at or above the limit of its capabilities as a result of either poor lighting or the inherent features of the task itself. It has also been found that light can be used as a visual aid to the partially sighted only if suitable environmental illuminances are achieved (Julian 1983).

Factors that can cause eyestrain and hence affect visual performance and productivity are due to muscular strain from controlling the pupil size or lens and/or the lack of contrast in visual tasks. These are due to too little or too much light on the task, veiling reflections, disability and/or discomfort glare, and flicker, all of which affect the controlling systems of the eye. These are specified in detail in Chap. 6. Light can also affect health in other ways not related to vision. Human tissue can be damaged and a number of physiological processes can be influenced by light.

The muscle systems that control accommodation, convergence, and pupil size of the eyes can experience strain if a task is difficult to see. There is a tendency for the viewer to get closer, and this requires greater accommodation, a smaller pupil size, and more convergence between the eyes, which call for greater muscular exertion.

The illuminance on a task can be too strong or too weak. If a lighted object has a much higher luminance than the immediate surroundings, pupil control, in an attempt to keep the pupil restricted in size, will require strong muscular action. A case in point is reflected glare, in which high-luminance reflections bounce off a glossy surface adjacent to a task area. The same applies to other forms of high luminance, such as from luminaires or the bright sky through windows.

Flicker also can cause discomfort. For example, with screen-based equipment there may be two types of flicker in the visual field: the phosphor of the CRT screen with its inherent refresh rate and the luminance of fluorescent lighting reflected from the screen. Further research is needed to determine the effects on performance of these phenomena and if subjective discomfort is increased by the combination of CRT display terminals and standard low-frequency lighting.

Pupillary Size Differences under Different Lamp Types

Studies of pupil size have been carried out at three levels of luminance using different light sources (Berman et al. 1987), and the results have indicated that the spectral distribution of a light source can affect pupil size. It has been shown that significant differences in pupil size do occur when subjects are exposed to indirect high-pressure sodium as compared with indirect incandescent lighting even when the lighting intensities are photopically matched. The observed differences have been attributed to differences in the spectral power distribution of the two lighting systems.

Since pupil size can affect visual functioning, these results suggest that control of pupil size should be considered in lighting design. In the past, differences in spectral power distribution associated with different lighting technologies have been presumed not to affect visual performance when the task is achromatic (Bullet and Fairbanks 1980), but pupil size is known to have important effects on depth of field and on the ability of the visual system to resolve fine detail as reflected by visual acuity and contrast sensitivity (Campbell and Green 1965). Therefore, depending on the specific nature of the visual task, improvements in visual performance can result from spectrum control of pupil size.

Discomfort from Glare

The presence of high luminances or an unsuitable range of luminances in the field of view can cause discomfort from glare and in some special cases reduced visibility, or what can be termed "disability glare." The disability from glare is caused by the intensity of the glare source, which scatters light in the eye, producing a luminous veil across the retinal image of the observed object. This raises the adaptation level and hence reduces visibility. A common example is approaching car headlights at night on a poorly lighted street. The resulting increased adaptation level reduces visibility. At the other end of the glare spectrum there is no reduction in visibility, but there is discomfort from the excessive contrast produced by variations in luminance across the visual field. This may be caused principally by the contrast of such light sources as lamps, luminaires, and windows with background luminances of the ceiling or walls. Distraction or disability also can be caused by reflection of light sources from high-reflectance or glossy surfaces.

Disability Glare

Disability glare is rarely experienced in artificially lighted work environments, since light sources are generally mounted well above the line of sight and the luminance differences across a space are relatively small. This type of glare can, however, be experienced as a result of very bright skies or direct sunlight seen through windows if consideration is not given to shading controls. The luminance of the sky visible through windows can in some locations rise to 10.000

cd/m² even on overcast days and may be several times greater than this if brightly sunlit clouds are present.

Disability glare rapidly diminishes with deviation from the line of sight. It is possible to quantify disability glare in terms of the reduction of contrast produced by the glare source due to a given illuminance at the eye from the source and at a given displacement from the line of sight (Holloday 1926).

Holloday has shown that the glare source, in terms of an "equivalent veiling luminance" can be expressed by the following formula:

$$L_v = kE/\theta^2$$

where:

E = the illumination produced by the glare source at the eye

θ = the angular distance from the line of sight to the direction of the glare source

Fischer (1970) has studied these relationships in respect of the view toward windows and van Ierland (van Ierland and Jansen 1973) has related these observations to the theoretical relationships of disability glare to derive provisional criteria.

Discomfort Glare

Discomfort glare, which is related to the distribution of luminance across an interior, is a more common experience in working interiors. The results of various investigators into glare discomfort all agree that the magnitude of the glare sensation is related directly to the luminance of the glare source and its apparent size as seen by the observer and that the discomfort is reduced if the source is seen in surroundings of high luminance.

There now exist a number of formulae that relate characteristics of a lighting system to subjective judgments of the degree of discomfort experienced; and these provide a useful guide to the causes of discomfort. For a single light source most of these equations have the following form:

Glare sensation = (luminance of the glare source)^m ×

$$\frac{\text{(angular subtense of glare source at eye)}^n}{\text{(luminance of background)}^x \times \text{(deviation of glare source from line of sight)}^Y} \qquad \textbf{(3.1)}$$

where m, n, x, and y are exponents that vary in different countries and organizations (IES of North America 1973; IES of London 1967). Equations (3.2) through (3.4) give the values of the different exponents. For IES of North America, the formula by Guth (1963) is

$$M = \frac{0.50 L_s Q}{PF^{0.44}} \qquad \textbf{(3.2)}$$

where M is the glare sensation index, L_s is the glare source luminance, Q is equal to $20.4 W_s + 1.52 W_s^{0.2} - 0.075$ (where W_s is the solid angle subtended at the eye by the glare source, in steradians), P is an index of the position of the glare source with respect to the line of sight, and F is average luminance of the entire field, including the glare source (in candelas per square meter).

The British glare index is determined as follows:

$$G = \frac{0.48 L_s^{1.6} W_s^{0.8}}{L_b P^{1.6}} \qquad \textbf{(3.3)}$$

where L_s is the luminance of the glare source (in candelas per square meter), W_s is the solid angle subtended by the glare source at the eye (in steradians), L_b is the average luminance of the field of view, excluding the glare source (in candelas per square meter), and P is an index of the position of the glare source with respect to the line of sight. For multiple glare sources, the combined effect is given by addition of the glare sensation (G) values. For convenience, the total glare sensation for any particular installation is converted into a quantity called the "glare index" (GI) by the following formula:

$$GI = 10 \log_{10} (\Sigma G) \tag{3.4}$$

The European glare-limiting method links the luminance distribution of the luminaire to a rated experience of glare (Fischer 1972). The Fischer method is based on luminance limits. Australia has developed a similar method specifying a series of shielding angles applicable to open reflector type cut-off luminaires (Standards Association of Australia 1976). Despite the differences in methods that have resulted in different values for the exponents in the equations, there is reasonable agreement for the predicted degree of discomfort glare. All methods enable the lighting designer to calculate the extent to which the electric lighting installation will produce discomfort glare.

The equations show that increasing the luminance of the glare source and increasing its angular size or decreasing its deviation from the line of sight increases the sensation of glare experienced. Increasing the luminance of the background would decrease glare. In practice, it is not quite so simple, since the variables can rarely be varied independently. Using a more intense light source will increase the luminance of both the source and the background.

Occupants of an interior may also experience discomfort glare from windows and large luminous areas as compared with the small source areas experienced in electric lighting installations. Only comparatively recently has it been possible to link laboratory studies of discomfort glare from large sources with the discomfort due to glare from windows. As a result of experiments at the Building Research Station (BRS) in England and at Cornell University in the United States, it was proposed that the most practical way to deal with problems of adaptation and position in the field of view was to modify the original BRS glare formula (IES of London 1967) so that the prediction of glare from large sources could be aligned with the prediction of glare from small sources (Chauvel et al. 1982). The modifications produced a formula subsequently known as the "Cornell formula." This gave the glare constant (G) as follows:

$$G = K \frac{L_s^{1.6} W^{0.8}}{L_b + 0.07 w^{0.5} L_s} \tag{3.5}$$

where K is a constant depending on the units, L_s, L_b, and W are, respectively, the source luminance, the surrounding luminance, and the solid angular subtense of the source at the eye in exactly the same form as in the BRS formula for small sources, and w is the solid angular subtense of the source modified for the effect of the position of its elements in different parts of the field of view, as defined by Petherbridge and Longmore (1954).

From the studies on discomfort glare by Luckiesh and Guth (1949) and Petherbridge and Hopkinson (1950), there appears to be some evidence that as the source increases above a size subtending a solid angle of 0.01 sr, the glare does not increase to the extent predicted. Predictions of the worst case (i.e., a direct view of a window) using the Cornell formula indicate that the discomfort glare index in interiors from windows is reasonably constant for all room sizes in which the area of windows is greater than 2 percent of the floor area. This means that a room or space has a relatively constant glare character for all practical daylighting situations. The glare index values given in Tables 3.3, 3.4 and 3.5 are the results of studies in England. In Table 3.3 the daylight glare index is compared with the British Glare Index in terms of the degree of glare discomfort; Table 3.4 demonstrates the change in the daylight glare index with change of surface reflectance; and Table 3.5 shows, for example, that in a room with an average surface reflectance of 0.6, a daylight glare index of 22.5 (just

TABLE 3.3 Comparison of IES Glare Index (Electric Lighting) with Daylight Glare Index for Similar Discomfort Criteria (from CIE 1989)

Glare Criterion Corresponding to Mean Relation	IES Glare Index	Daylight Glare Index
Just imperceptible	10	16
	13	18
Just acceptable	16	20
	19	22
Just uncomfortable	22	24
	25	26
Just intolerable	28	28

TABLE 3.4 Daylight Glare Indices of Average Sky Luminance of 8900 cd/m² (from CIE 1989)

Average Reflectance of Room Surfaces	Daylight Glare Index
0.4	26
0.6	24

TABLE 3.5 Percentage of Working Hours for which Different Values of Daylight Glare Index Will be Exceeded in a Temperate Climate (from CIE 1989)

Horizontal Illuminance from Whole Sky (lx)	Average Sky Luminance (cd/m²)	Proportion of Annual Working (percent)	Reduction in Daylight Glare Index
28,000	8,900	25	0
20,000	6,400	38	0.5
15,000	4,800	54	1.5
10,000	3,200	68	2.5
5,000	1,600	87	4.0

acceptable) will be exceeded for 54 percent of annual working hours in a temperate climate such as experienced in England.

Basically, glare control for electric light sources involves control of the source luminance in the direction of the eyes of the occupant. Usually it is assumed in a work situation that the viewing direction is horizontal and that the field of view is limited upward at an angle of 45 degrees. With natural light, the higher the window in a given room, the larger will be the source luminance and, for a given observer position, the higher will be the adapting effect on the eye. At a point further from the window, the angular subtense is less, and hence the adapting effect of the source is also less.

The reflectance of the surrounding surfaces has a significant influence on the glare experienced, but by far the greatest effect is produced by the sky luminance as seen through the windows. Thus the principal means of limiting daylight glare is by limiting the luminance or visibility of the sky as seen through the windows. Practical methods to control glare from windows and electric light sources are given in Chap. 6.

SUMMARY

Seeing is an interactive process involving light, the eye, and the brain. Although information is still limited on various aspects of seeing, a comfortable luminous environment can be generally defined as an environment that facilitates seeing by providing adequate illuminance, ensuring the appropriate distribution of luminances and directional lighting, and eliminating distraction and stress, in the accomplishment of visual tasks, generally caused by glare.

The eye can adapt to a wide range of luminances and, once adapted, is capable of detail discrimination. Visual performance, and hence productivity, are influenced by illuminance level, the size of the details of the task, and the relative luminances of the task and its background.

The presence of strong luminances in the field of view can cause discomfort or reduction in visibility as a result of glare. Glare may occur directly from light sources such as lamps, luminaires, and windows or by reflection from glossy surfaces or surfaces that have a high reflectance.

Light also has nonvisual effects on human health and well-being that can be either positive or negative. Dosages of some forms of radiation need to be restricted to prevent such afflictions as keratoconjunctivitis (inflammation of the cornea) and erythema or reddening of the skin. In the case of erythema, protection should be provided from electric light sources used for special scientific, industrial, or social purposes that emit considerable quantities of ultraviolet radiation.

Natural light also influences human well-being in that it enables visual contact with the outside world. Studies of acceptable minimum window dimensions have shown that for temperate climates, window areas should be between 20 and 40 percent of the window-to-wall area.

REFERENCES

Balder, J. J. 1957. Erwunschte leuchtdichten in Buroraumen. *Lichttechnik* 9: 455.

Berman, S. M., Jewett, D. L., Bingham, L. R., Nahass, M., Perry, F., and Fein, G. 1987. Pupillary Size Differences under Incandescent and High-Pressure Sodium Lamps. *Journal of the Illuminating Engineering Society.* 16, 1, 3-20.

Blackwell, H. R. 1959. Specification of interior illuminance levels. *Illum. Eng.* 54: 317.

Boyce, P. R. 1981. *Human factors in lighting.* Applied Science Publishers, London.

Bullet, D., and Fairbanks, K. 1980. An ambient/task high-intensity source office lighting system. *Light. Design Appl.* 10: 41-49.

Campbell, F. W., and Green, D. G. 1965. Optical and retinal factors affecting visual resolution. *J. Physiol.* 181: 576-593.

Cayless, M. A., and Marsden, A. M. 1983. *Lamps and lighting.* Edward Arnold, London p. 23.

Chauvel, P., Collins, J. B., Dogniaux, R., and Longmore, J. 1982. Glare from windows: Current views of the problem. *Lighting Research and Technol.* 14, 1: 31-47.

Commission Internationale de L'Eclairage (CIE). 1986. Guide on interior lighting. CIE Publication No. 29. 2.

Commission Internationale de L'Eclairage (CIE). 1989. Guide on daylight of building interiors. Final draft.

Duke-Elder, W. S. 1944. *Textbook of Ophthalmology,* Vol. 1. Mosby, St. Louis.

Fischer, D. 1970. Lichttechnische Konditionierung von Arbeitsraumen. *Lux* 57, 206.

Fischer, D. 1972. *The European Glare Limiting Method. Lighting Research and Technol.* 4, 97-100.

Flynn, J. E. and Spencer, T. J. 1977. The effect of light source colour on user impression and satisfaction, *J.I.E.S.* 6, 167.

Foxell, C. A. P., and Stevens, W. R. 1955. Measurements of visual acuity. *Br. J. Ophthalmol.* 39: 513-533.

German Standards. 1979. DIN 5031, Vol. 10: Pre-Standard; 1986. Guide on Interior Lighting, Publication CIE No 29/2 (TC4.1).

Gregory, R. L. 1966. *Eye and Brain,* Weidenfeld and Nicholson, London.

Guth, A. K., and McNelis, J. F. 1969. Threshold contrast as a function of target complexity. *Am. J. Optom. Physiol. Opt.* 46: 491-498.

Hallberg, F. 1962. Physiologic 24-Hour Rhythm: A Determinant of Response to Environmental Agents. In Schaefer (Ed.), *Man's Dependence on the Earthly Atmosphere.* Macmillan, New York.

Holick, F., et al. 1982. The Photochemistry and Photobiology of Vitamin D_3. In Regan, J. D., and Parrish, J. S. (Eds.), *Photomedicine.* Plenum, New York.

Holladay, L. L. 1926. The fundamentals of glare and visibility. *J. Opt. Soc. Amer.,* 12, 271-319.

Hopkinson, R. G., and Collins, J. B. 1970. *The ergonomics of lighting.* MacDonald, London.

Hughes, P. C. 1983. *An examination of the beneficial action of natural light on the psychobiological system of man.* Proceedings CIE 20th Session, Amsterdam D603/1-D603/4.

Hughes, P. C., and Neer, R. M. 1981. Lighting for the elderly: A psychobiological approach to light. *Hum. Factors* 23: 65-85.

Illuminating Engineering Society. 1967. *Evaluation of discomfort glare: The Glare Index system for artificial lighting installations,* Technical Report 10, London.

Julian, W. G., 1983. *The use of light to improve the visual performance of people with low vision.* Proceedings of the 20th Session of the CIE, Amsterdam. Publication CIE No. 56 pp. D116/1-D116/4.

Keighly, E. C. 1973. Visual requirements and reduced fenestration in offices—A study of multiple apertures and window area. *Build. Sci.* 8: 321-331.

Lewy, A. J., Wehr, T. A., Goodwin, F. K., Newsome, D. A., and Markey, S. P. 1980. Light suppresses melatonin secretion in humans, *Science* 210: 1267-1269.

Luckiesh, M., and Guth, S. K. 1949. Brightness in visual field at borderline between comfort and discomfort (BCD). *Illum. Eng.* 44: 650-670.

MacLaughlin, J. A., et al. 1982. Spectral character of sunlight modulates photosynthesis of previtamin D_3 and its photoisomers in human skin. *Science* 216: 1001-1003.

Ne'eman, E., and Hopkinson, R. G. 1970. Critical minimum acceptable window size: A study of window design and provision of view. *Light. Res. Technol.* 2: 17-27.

Overington, I. 1978. *Vision and Acquisition.* Pentech, London.

Petherbridge, P., and Hopkinson, R. G. 1950. Discomfort glare and the lighting of buildings. *Trans. Illum. Engineering Soc. (Lond.)* 15: 39-79.

Petherbridge, P., and Longmore, J. 1954. Solid angles applied to visual comfort problems. *Light Lighting* 47: 173-177.

Plank, J. J., and Schick, J. 1974. The effects of color on human behavior. *J. Assoc. Study for Perception* 9: pp. 4-16.

Seidl, E. 1969. The influence of ultraviolet radiation on the healthy adult (Urbach, E.). The biological effects of ultraviolet radiation. Pergamon Press Standards Association of Australia. 1976 Australia Standard 1680 code of practice for interior lighting and the visual environment.

Steck, B. 1973. Die Einwirkung der opticshen Strahlung (Licht, UV, IR) auf den Organisms des Menschen. Techn. wiss. Abhandl.

van Ierland and Jansen, E. 1973. Limitation of disability glare from windows. In *Proceedings CIE TC 4.2 Symposium, Istanbul.* CNBE, Brussels.

Wake, T., et al. 1977. The effects of illuminance, color temperature, and color rendering index of light sources on comfortable visual environments—Case of an office. *J. Light Visual Environ.* 1: 31-39.

Weston, H. C. 1945. The relation between illuminance and visual performance. Industrial Health Research Board Report No. 87, HMSO, London.

Wurtman, R. J. 1975. The effect of light on man and other mammals. *Ann. Rev. Physiol.* 37: 467-483.

4 Acoustic Environment

Anita Lawrence

It has been suggested that deafness is a more severe handicap than blindness. This would be strongly denied by many at first thought; however, sound signals can give far more information about the world around us than can light signals, since the latter are restricted to our line of sight. For example, we can hear a neighboring dog barking, which might alert us to the presence of visitors, but we cannot see the visitors if they are on the other side of a brick wall.

The relative merits of sight and sound need not concern us here, but it is true to say that the importance of the acoustic environment in buildings is sometimes overlooked. The study of "acoustics" is thought to be associated with concert halls, auditoriums, and similar buildings, but acoustic comfort is as necessary as visual and thermal comfort if any building is to provide the best possible environment for human occupation. It also must be emphasized that the decision whether to use air-conditioning or natural ventilation has considerable acoustical implications.

People react more or less passively to their visual and thermal environments—they may adjust lighting and temperature to suit their individual preferences in some cases—but they interact directly with their acoustic environment whenever they speak or sing or use appliances or machinery. Excessive noise can prevent or degrade speech communication, and it also can prevent wanted sound signals from being heard or understood. Excessive noise may not be controllable by the building's occupants, and acoustic discomfort will be experienced.

In addition to sound that is heard, there may be vibration, either associated with audible sound or independent of it. Perceived vibration in a building is difficult, if not impossible, to control.

HUMAN HEARING

The human audio range is conventionally considered to extend between 20 and 20,000 Hz (formerly called cycles per second). In practice, as people age, they tend to lose the ability to hear high-frequency sounds. It also has been discovered, with the advent of powerful, large sources, that low-frequency sounds below 20 Hz may be present at sufficiently high levels to be audible. Such sources are not common in buildings, being mainly associated with rockets, etc., but there have been some cases reported in which strong low-frequency sounds associated with high-velocity air-conditioning or ventilation systems cause discomfort (Tempest 1976).

In building acoustics, the frequency range usually considered is limited to 100 to 5000 Hz. This limitation exists partly because people at any age are much less sensitive to sound at the extremes of the frequency range and partly because it is very difficult to obtain acoustical data for materials and for design purposes outside this range.

In order to take into account people's varying sensitivities to sounds of different frequencies, it is necessary to use a weighting system. The most common weighting used internationally is known as the "A weighting network." If measurements of overall sound levels are made using this network, it can be assumed that the frequency sensitivity of the human ear has been taken into account. Since sound sources have different amounts of energy at different frequencies and building materials and systems react differently also according to frequency, it is usually necessary, for design purposes, to subdivide the audio range into octave or one-third-octave bands. As in music, an "octave" is the interval between one frequency and its double. The bands are described by their geometric center frequencies; that is, the octave band centered on 1000 Hz extends from 707 to 1414 Hz (Fig. 4.1).

Another aspect of hearing to be considered is the range of intensities perceived from the very quietest (threshold of hearing) to the very loudest

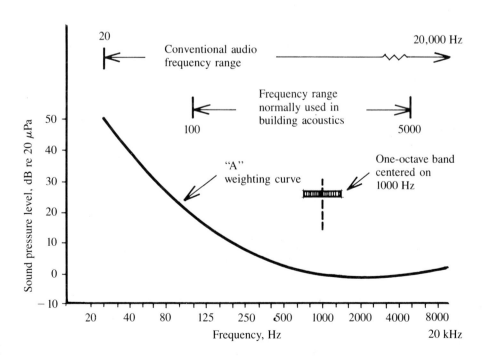

FIGURE 4.1.
Human hearing thresholds and audio range.

(threshold of pain). As with other areas of perception, people react to changes in intensity of a stimulus according to the Weber-Fechner law; that is, equal increments of intensity are perceived if the ratio between successive stimuli is constant. This response characteristic, together with the very large range of intensities perceived, leads to the use of a logarithmic scale, the "decibel scale," to measure this quantity.

Until recently, it had not been possible to measure sound intensity directly; this is changing with the advent of new instrumentation. However, intensity measurement is not yet widely available, and the equipment tends to be expensive. Instead of sound intensity, the changes in stimulus magnitude are measured in terms of the resulting changes in pressure in the medium (usually air) as the sound waves pass through. Sound level meters are calibrated to measure sound-pressure levels in decibels with reference to a standard rms sound pressure of $20 \mu Pa$. If the A weighting network is included, as it usually is, the results are referred to as "A-weighted decibels dB(A)." The audio sound-pressure level range is approximately 0 dB (corresponding to the threshold of hearing at a frequency of 1000 Hz) to about 120 dB (the threshold of pain) (Fig. 4.2).

Up to this point, human perception of sound follows closely that of human perception of light—there is a limited range of frequency (or wavelength, as it is usually described in lighting) and intensity to which people respond. There is a third aspect that also must be considered with sound, i.e., the temporal character of the stimulus. Sound waves are *not* part of the electromagnetic spectrum; they are purely mechanical in origin and travel relatively slowly, particularly in a gaseous medium such as air (velocity of sound in air at room temperature and pressure is approximately 344 m/s). The importance of this is that a finite time elapses between the emission of a sound signal from a source and its reception by a listener some meters away. This may produce the common phenomenon of an echo, which occurs when a reflected sound from some surface has traveled much further than the direct sound from the source to the receiver and thus arrives at a perceptibly later time. Echoes may be perceived if the difference in path length between the direct and reflected sound is 17 m or more, corresponding to an arrival time difference on the order of 50/1000 of a second (or 50 ms) delay (Fig. 4.3). Another illustration is the well-known phenomenon of the flash of lightning occurring several seconds ahead of the roll of thunder, although both sources emit concurrently.

Echoes, as such, are usually considered only in special buildings such as auditoriums. In fact, detailed analyses of the reflection patterns are required when designing concert halls and other rooms for music. These studies may be done by geometric analysis of the shape of the room at the design stage (manually or using computers) or by investigating acoustical scale models. For rooms intended for drama or lectures, it is important that echoes and other long-delayed reflections be suppressed so that speech intelligibility is not degraded. If sound reinforcement systems are to be used, it is important that the first signal received by the audience be from the real source, since this signal defines the aural location of the source. In some cases it becomes necessary to incorporate time delays into the loudspeaker circuits so that there is no conflict between the visual and aural locations of the source.

In all buildings, the temporal aspects of sound are of interest, since many real sources do not emit constant signals; for example, road traffic noise constantly varies as different vehicles pass by and an aircraft flyover has a distinct temporal pattern. It is known that people are more annoyed by intermittent signals, e.g., a compressor cycling on and off, than by steady sounds such as an air-conditioning register. This has led to a number of methods of describing time-varying noises with respect to their possible annoying effect, and these will be discussed later. One of the most important sound signals, the human voice, is itself constantly changing in amplitude and frequency.

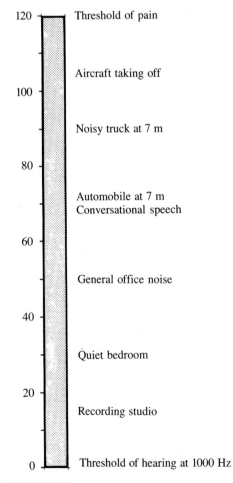

FIGURE 4.2.
Audio range of sound-pressure levels in decibels.

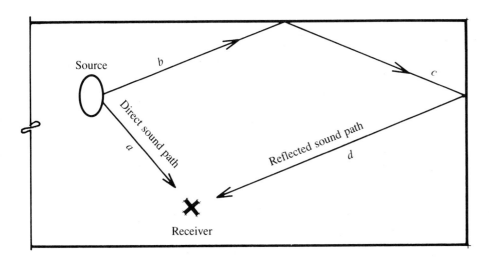

FIGURE 4.3.
Echo perception. If $(b + c + d) - a > 17$ m and time delay > 50 ms, an echo may be heard.

HUMAN RESPONSE TO VIBRATION

The ear is the principal receptor of audible sound, but with vibration, all parts of the body may be affected; that is, various receptors may transmit response to vibration to the central nervous system. Much less is known about human response to vibration than to audible sound. It is thought that very low frequency vibration is detected by the nonauditory labyrinth, situated adjacent to the cochlea (frequency range about 0 to 15 Hz); higher-frequency response is mainly through the skin. Although humans may respond to vibration frequencies up to 100 kHz, the main frequency range of interest is from 0.5 to 100 Hz.

Vibration may be measured in several ways: as displacement, velocity, or acceleration (the latter is normally used when determining human acceptability). The threshold of perception of vibration (and discomfort) depends not only on the frequency and intensity of the stimulus, but also on the posture of the recipient (standing, sitting, or lying down) and on the direction of motion (up and down, sideways, fore and aft, rotation, etc.) (Fig. 4.4). Much of the research into human response has related to transportation (smooth riding in automobiles, for example), and some of the more recent work has investigated the effect of strong vibration on performance (the maximum vibration that will allow pilots to function properly, tolerances for astronauts, etc.). More research is desirable, but some guidelines are available from the International Standards Organization (ISO 2631 1978) as well as from national standards authorities.

Building vibrations are usually of a much lower magnitude than those in transportation vehicles. However, they may cause psychological discomfort. It is well known that wind velocities tend to increase with height above the ground, and with the advent of very tall commercial and residential buildings, problems have occurred. Wind blowing against a tall building will tend to cause the building to lean over to a mean position about which it will sway under the dynamic forces of wind gusts, turbulence, and vortices. Apart from any structural damage that may result, occupants may be disturbed by swinging chandeliers and water moving from one end to the other of a bathtub, for example. This may exacerbate their latent fear of heights. Such problems have been accentuated with the development of very tall, relatively lightweight buildings (Brown and Maryon 1975).

Vibration also may be transmitted to a building through the ground. Blasting is a well-known source of ground vibration and can cause disturbances through startle effects, rattling of windows and objects, and so on. In severe cases, building damage may occur, but most prudent miners and construction organizations will avoid this by limiting explosive charges. (In many places,

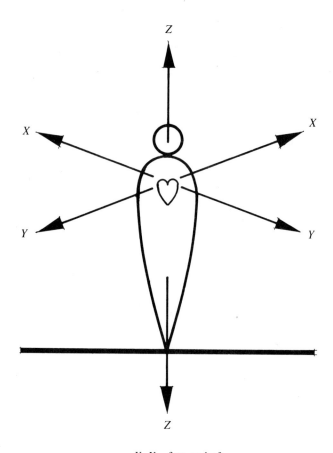

FIGURE 4.4.
Directions of vibratory movement (relating to a standing person).

$X–X$	fore and aft
$Y–Y$	side to side
$Z–Z$	up and down

state authorities limit air blasts and ground vibrations from blasting to levels well below those known to cause any possible building damage.)

More constant sources of vibration in buildings arise from transportation—underground railways and heavy road transportation vehicles in particular. Much more research is needed regarding the transmission of sound through different types of ground materials into building foundations. In a number of instances it has been necessary to isolate underground railway tracks on resilient supports as well as to isolate the building footings themselves. It must be remembered that once vibration has entered the structure of a building, it will travel throughout the building with little attenuation; it is usually extremely difficult, if not impossible, to rectify the situation after construction.

Other sources of vibration occur within a building and may be exacerbated by modern structural designs that are less massive and involve less redundancy than older buildings. Lightweight, long-span floors that fulfill code requirements for structural sufficiency and deflection may be lightly damped and easily set into vibration. This may cause an unpleasant sensation for occupants as they themselves or other people walk across the floor (particularly since many building elements have natural resonant frequencies around 5 to 8 Hz, which coincide with maximum human sensitivity). Floor vibrations also may cause difficulties with sensitive instruments and machinery, and it may be necessary to isolate special laboratories, etc., from the main structure of such a building (Northwood 1973).

Ground vibrations that cause walls and floors to vibrate can cause unacceptable levels of low-frequency airborne sound in studios. It is possible to build studios in areas with high levels of ground vibration, but special construction techniques are required and specialist advice should be sought (Smith, Walkers, and Mathers 1986).

PHYSIOLOGICAL AND PSYCHOLOGICAL EFFECTS OF NOISE

Hearing Damage

It is common to find that as people age their hearing sensitivity is reduced. This may be so because of a general slowing down of neural response, or it may be the result of physiological deterioration of muscle tone and/or other alterations to the middle ear. The general term given to this loss of hearing due to aging is "presbycusis." In addition, many people suffer loss of hearing due to exposure to excessive noise (Noise Induced Hearing Loss, or NIHL). This may arise in the course of their employment or from acoustic trauma from hobbies such as shooting. Unfortunately, this type of hearing loss most severely affects the frequencies of most importance for speech intelligibility (1000–4000 Hz) (Fig. 4.5). The physiological effect is damage to the hair cells in the cochlea which transduce mechanical energy resulting from sound waves in the cochlea to the neural impulses interpreted by the brain as sound. This damage to the hair cells is irreversible. It is called "sensorineural hearing loss," and it is difficult to design hearing aids that provide much assistance. More recently, research has led to the development of the "bionic ear," or cochlear implant. This provides an alternative pathway to the neural system, although as yet only limited signal transmission is available.

There are considerable individual differences in susceptibility to noise-induced hearing loss, but the extent of damage is related to the level of noise and the duration of exposure. The acceptability of damage is a politicoeconomic decision; it is usually related to the loss of ability of people to communicate aurally. In several countries, the "allowable safe exposure" has been set at an equivalent energy level of 90 dB(A) for 8 hours in any 24-hour period. (The equivalent energy level $L_{Aeq,T}$ is the level of a steady sound signal with the same total energy over the relevant time period T given in A-weighted decibels.) A 3-dB tradeoff for doubling of time duration is used in most cases (although in

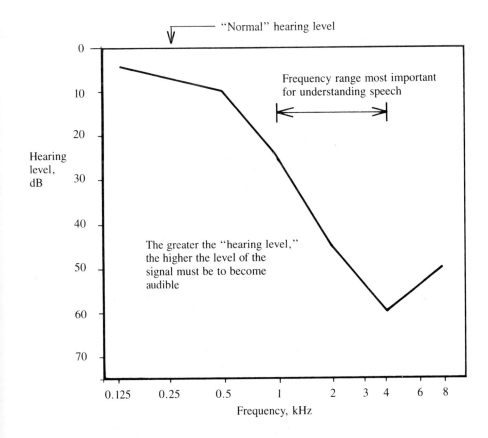

FIGURE 4.5.
Typical audiogram of a worker suffering noise-induced hearing loss.

the United States a 5-dB tradeoff is normal); that is, if the equivalent energy level is 93 dB(A), then the duration of exposure must be halved, to 4 hours. There is considerable evidence to suggest that these so-called safe exposure limits will cause a significant proportion of the working population to incur a severe hearing handicap over a working life of 40 years. Many organizations prefer to limit the exposure to 85 or even 80 dB(A) for an 8-hour period.

It should be emphasized that such sound levels have nothing to do with human comfort. Unfortunately, however, they are prevalent in many industrial situations.

Nonauditory Physiological Effects

Nonauditory physiological effects are much more subject to debate, and although some claims have been made that noise may cause physiological stress and eventual health disorders, they have not been proved satisfactorily. Since sound can act as a warning signal, a sudden, loud sound can elicit a startle-type response, and many laboratory experiments have indeed shown significant physiological reactions in people exposed to loud sounds. For example, as summarized by Kryter (1985), measurements have shown vasoconstriction of the peripheral blood vessels and other blood pressure adjustments throughout the body, a reduced rate of breathing, a change in the electrical resistance of the skin (galvanic skin response), and a brief change in skeletal muscle tension. However, repeated stimuli generally elicit less response; in other words, there is rapid adaptation or habituation to noise.

Two reflex-type responses of the autonomic system to noise have been suggested by Sokolov (1963). One is called the "orientation reflex" (OR), and it readies the organism to receive and respond to the stimulus. This orientation reflex becomes stronger with weaker stimuli, since more effort is required to react. The second is the "defensive response" (DR), and it prepares the organism for flight or defense. This becomes stronger with increasing loudness. For both responses, after repetition of the stimulus, the meaning of the noise becomes recognized and habituation occurs (Kryter 1985).

Psychological Effects of Noise

There have been many studies of the psychological effects of noise, ranging from laboratory experiments to widescale community surveys. The results show considerable variations in response between individuals and also that other nonacoustic factors may be important. For example, in one community noise survey in Paris it was found that people would tolerate sound levels 5 dB(A) higher if they were satisfied with the general character of their neighborhood, as assessed by ease of access to transport, shops, and schools and general appearance (Alexandre et al. 1975). Conversely, if people are dissatisfied with some aspects of their environment or employment, they may use "excessive noise" as an excuse for complaint.

Laboratory studies of the effect of noise on performance are inconclusive. For example, when a task is monotonous, a higher noise level may increase an individual's arousal level and thus improve performance; alternatively, if mental effort is required, noise may be distracting and have the opposite effect. It also has been reported that making some changes, and thus taking notice of people's complaints, may result in a perceived improvement, even though the noise level is unchanged.

Nevertheless, annoyance reactions to noise are common. They may result from interference with communication, in which case the degree of annoyance will depend on the extent of the interference and the importance placed on the communication. Since the human hearing and speaking systems presumably evolved concurrently, it is not surprising to find that the frequency and dynamic ranges of speech fall neatly within the human hearing range. In any language there is a relatively small number of unique speech sounds, called "phonemes," from which all words are constructed (in English the number of phonemes is about 48). Most of the energy in speech is in the lower frequencies, which contain the vowel components of phonemes; however, the consonant components (of which there are many more than vowels) have most of their energy in the higher frequencies. Compared to vowels, consonants have much less energy and are of short duration. In order to understand speech, it is most important that the consonants can be heard, although, fortunately, there is considerable redundancy in connected speech, and in addition, familiarity with a language and a knowledge of the general subject matter of the speech also assist in comprehension. Conversely, nonnative speakers or unfamiliar topics require superior acoustic conditions for good communication. Whether or not speech communication (either two-way, face-to-face, or one-way, as when listening to radio or TV) will be possible depends on the loudness and clarity of the speech, on the transmission path characteristics, on the general acoustic environment, and on the hearing ability of the listener.

A major source of acoustic discomfort in buildings is difficulty with speech communication. This may be caused by excessive noise from other sources either inside or outside the building and/or by excessive sound reflections in the room that cause echoes and long reverberations. Alternatively (and frequently concurrently), there may be a problem with speech privacy, whereby conversation can be overheard by others. Calculation methods are available to estimate speech intelligibility and speech privacy at the design stage of a building, and existing buildings also can be assessed for their performance in these respects (SAA 2822, 1985). However, as is the case with many acoustic problems, it is frequently difficult and expensive, if not impossible, to improve the situation once a building is completed.

Annoyance also may result if an unwanted sound can be identified above the general background noise in an area. For example, the hum of machines from a nearby factory or a neighbor's radio may not be very loud compared to other noises in an area, but some people may become highly annoyed if they can hear them and psychological stress may occur. Again, there have been many studies concerning the acceptability of different noises; not only the absolute sound level, but the character of the noise (tonal, impulsive), whether or not it is continuous or intermittent, whether it conveys information, and the time of day at which it occurs all have been found to affect annoyance responses. There are now several methods available to assess the effect of aircraft, road traffic, industry, etc. on nearby communities; however, these methods are usually intended to protect the majority of people from undue annoyance and it is still likely that the more noise-sensitive people will not be satisfied.

Sleep disturbance by noise has been studied both in laboratories and in people's homes. People normally have fairly complicated sleep patterns (divided into four stages); one stage, in which dreaming occurs, is known as the "REM stage," characterized by rapid eye movements. Stage of sleep can be determined by fixing electrodes to the head and recording "brain waves." Even after considerable periods of habituation, it appears that noise continues to disturb the normal sleep pattern; in particular, the period spent in the REM stage is reduced. People may not be aware that their sleep has been disturbed, although their performance in reaction-time tests may be affected. Griefahn (1986) tentatively suggested that for fairly steady noise, such as that generated by dense road traffic, an L_{Aeq} level of 40 dB(A) should not be exceeded indoors.

Most studies are carried out on people known to be reasonably healthy and without psychological problems. There are some suggestions that people who are unwell or who have some psychological disturbance may be more sensitive to noise than others. Since such people will generally form a significant proportion of any community, their needs should not be neglected.

SUMMARY

Auditory signals are extremely important. Wanted signals are frequently contaminated with noise, and this can reduce efficiency and cause annoyance. Quite low noise levels can be annoying when people are relaxing or concentrating on mental tasks, and the character of the sound affects the response. The acoustic environment is often neglected in the design of a building, but it will affect people's responses to their total environment. If a building does not sound pleasant, it will produce an adverse reaction.

REFERENCES

Alexandre, A., Barde, J.-Ph., Lamure, C., and Langdon, F. J. 1975. *Road Traffic Noise.* Applied Science, London,

Brown, H. and Maryon, J. 1975. The tall building experience: Perception of wind movements. *Build. Mater. Equip.* (Aug./Sept.): 36-39.

Griefahn, B. 1986. Road Traffic Noise and Sleep: Long-Term Effects, Critical Load. In *Proceedings of the 12th International Congress on Acoustics, Toronto,* p. C4-3 Canadian Acoust. Assoc., Toronto.

International Standards Organization. 1978. ISO 2631: Guide for Evaluation of Human Response to Whole-Body Vibration. ISO,

Kryter, K. D. 1985. *The Effects of Noise on Man,* 2d Ed. Academic, New York.

Northwood, T. 1973. Isolation of Building Structures from Ground Vibration. In *Proceedings of the American Society of Mechanical Engineers Seminar on Isolation of Mechanical Vibration, Impact, and Noise,* pp. 87-191. Am. Soc. Mech. Eng., New York.

Smith, T. J. B., Walkers, R., and Mathers, C. D. 1986. Studies on a Vibration-Isolated Radio Studio Mock-up. In *Proceedings of the 12th International Congress on Acoustics, Toronto,* p. D1-5. Canadian Acoust. Assoc., Toronto.

Sokolov, E. N. (S. W. Waydenfeld, transl.) 1963. *Perception and the Conditioned Reflex.* Pergamon.

Standards Association of Australia. 1985. AS 2822: Acoustics: Methods of Assessing and Predicting Speech Privacy and Speech Intelligibility. SAA, Sydney.

Tempest, W., Ed. 1976. *Infrasound and Low Frequency Vibration.* Academic, London.

PART II

Design of the Interior Environment

Part II is concerned with human requirements in buildings and includes problems dealing with air-conditioning and design criteria for thermal, visual, and acoustic comfort. These criteria are based on studies and evaluations of physical conditions, particularly in work environments. However, it can be inferred that the principles applied to produce a good working environment can also be appropriate for other environments.

In Chapter 5, on the human dimensions of air-conditioning interiors, Andris Auliciems refers to the simplifications that have been introduced as a result of the thermal constancy hypothesis and possible reasons for human dissatisfaction. From the results of studies in Australia and Japan, he makes a case for new approaches to indoor climate management.

In Chapter 6, on lighting design, Nancy Ruck introduces the basic principles for good viewing conditions. Problems arising from the use of screen-based equipment and their remedies are discussed, as well as issues arising from the integration of daylight with electric light. The importance of subjective appraisals of luminous environments is shown to pave the way to further research in this area.

The selection and construction of interior elements and surfacing materials have a pronounced effect on acoustic comfort in buildings. In Chapter 7, on acoustic design, Anita Lawrence discusses the use of appropriate materials and acoustic design criteria with regard to ambient sound levels and reverberation time. Noise prediction and assessment are noted.

Human Dimensions of Air-Conditioning

5

Andris Auliciems

OVERSIMPLIFICATION IN THE MANAGEMENT OF INDOOR CLIMATES

Whereas the technologies of reducing cold stress are inseparable from human development for the extent of civilization and before, it is only within the past few generations that there has existed a potential for active air cooling. Perhaps, not surprisingly, mankind has not yet learned to cope with such rapid technological change, and both in environmental and human terms, the price of this learning has been high.

During the past decade, as a result of increasing energy costs, many managers of large, air-conditioned buildings have found it necessary to reduce the intake of relatively warm outdoor air in favor of filtering and recirculating air that has already been cooled. This has necessitated an increased central control of operations to prevent individuals from tampering with openings to the outdoors. Moreover, within shared spaces, an individual's choice and ability to adjust microclimates have been reduced. Not unexpectedly, this has tended to lead to complaints by large numbers of individuals.

Since air cooling at times has been assumed to replace the need for other climate design features, there also has been a tendency to neglect thermal insulation and other passive control measures. Consequently, many air-conditioned spaces are of poor thermal design and create, apart from high demands for energy, uneven conditions indoors and, during mechanical breakdowns or interruptions in energy supplies, very little relief from debilitating thermal stress.

At this time also, air-conditioning is being used for purposes other than the reduction of thermal stress. In certain commercial centers, hotels, and even some office buildings, cooled areas have been extended beyond the proverbial doorstep. A deliberate spillage of cooled air onto uncovered walkways and streets has been carried out in an effort to add prestige to particular buildings and to entice people to enter shops and arcades.

Whatever the location, such an excessive generation of cold air is not consistent with either environmentally sound philosophies or energy-saving measures. The principles of thermodynamics determine that unless the transformed high-entropy energy produced in cooling processes is vented outside the urban atmospheric "dome," the creation of such cool pools of air can do no more than impart a sizable net heat load to the already hot urban fabric. One suggested solution has been to place covers over whole cities and expel the generated heat beyond, although in economic terms this would only tend to generate even further externalities.

Microclimatic modifications in today's world have been dominated by simple engineering solutions, especially as developed during and beyond the seemingly "abundant" energy era of World War II. Architects have delegated the responsibility for indoor climate design to heating and refrigerating engineers, whose concern, perhaps understandably, has been more with machines and the mechanics of energy transformation than with human comfort. Unfortunately, also, this approach has been characterized by an unquestioned acceptance of air-quality criteria based on the convenient, but unsupportable claim of universality in thermal preference, as discussed in Chap. 1.

Rejection of this constancy hypothesis questions the validity of certain present-day oversimplified, yet entrenched management practices. Clearly, if people's thermal perceptions are to some extent in accord with the naturally occurring levels of warmth, there can be no universally applicable standard to set the quantity of heating and cooling required. At most, standards could apply only to specific climatocultural zones as determined by the preferences of local inhabitants and the existing climatic milieu.

To achieve such regionalization of microclimate control strategies, three basic questions need to be resolved:

1. What cultural parameters need be considered, and to what extent is cross-cultural generalization possible?
2. Given the climatic factor in variability of comfort concepts, can zones of heating and cooling be defined in terms of readily available atmospheric data?
3. Can human adaptability be effectively incorporated into management practices of microclimate control?

Unfortunately, the past concentration of both field and laboratory studies on highly selected European populations does not enable, at present, a definitive resolution of the first of these questions. The handful of studies of thermal perceptions on indigenous peoples in India and Melanesia, such as those by Mookerjee and Murgai (1952), Rao (1952), Ballantyne et al. (1979), Budd et al. (1974), and Woolard (1979), are far too few to represent the variety of other racial characteristics and non-Western cultural experiences of the global society. This quite remarkable gap in knowledge provides considerable potential for field research.

REGIONALIZATION OF INDOOR CLIMATE-CONTROL REQUIREMENTS

The Neutral Zone

Assuming that future research with different cultural groups supports the remarkably high correlation between perception and outdoor temperatures as observed with largely European samples, it is possible to make at least first-order global estimates of variable thermal comfort requirements. As had been concluded earlier and shown in Table 1.8, dry-bulb temperature is the simplest and perhaps best measure of ambient warmth for conditions approaching thermal neutrality under real-life conditions. For the remaining sections of this chapter, therefore, all definitions of thermal neutrality refer to ambient air temperature only.

The magnitudes of adaptations in thermal perceptions that have been observed in past research are indicated by Eqs. (1.4) to (1.7). With reference to prevailing outdoor warmth, the rate of change in thermal satisfaction is approximately 0.31°C per 1°C change in monthly mean temperature (T_m), with the indoor-outdoor gradient approaching 0 near 25.5°C. This particular neutrality will be hereafter designated as the "design zero," or $T_\psi = 25.5$. Other neutrality values are also appropriately designated by those predicted by Eq. (1.6) and are globally distributed as shown in Figs. 5.1 and 5.2.

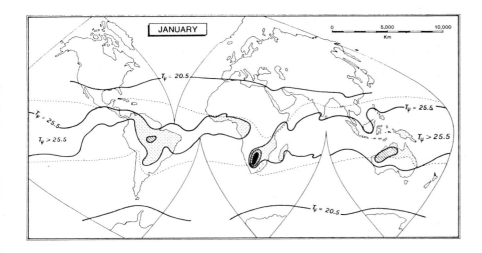

FIGURE 5.1.
Air cooling requirements in January as determined by variable neutralities and humidity. Cooling is required in shaded areas. (Based on Auliciems 1983.)

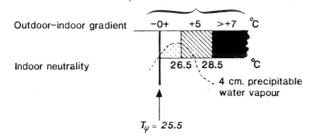

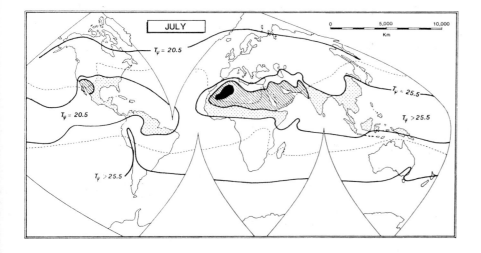

FIGURE 5.2.
Air cooling requirements in July as determined by variable neutralities and humidity. Cooling is required in shaded areas. (Based on Auliciems 1983.)

Thermal neutrality as discussed so far in this section refers to a person subjected to the outdoor environment. Its relevance to buildings is discussed in the following sections, and the idea of thermal neutrality for a building is developed in Chap. 8.

If other climatic parameters, especially solar radiation, could be held constant, maximum duration of $T_\psi = 25.5$ would presumably represent the optimal location of human habitation in terms of comfort and energy. Apart from shading from direct solar radiation, places at design zero should not require any particular provision for either the reduction or conservation of heat. Design zero does occur in some parts of the world, such as, for example, along the Tropic of Cancer and especially near Hawaii, where in equable environments seemingly salubrious shelter is provided by the much romanticized traditional grass huts of native inhabitants.

Since the rates of heat transfer and, therefore, for practical purposes, the insulation properties of a building's envelope are a linear function of differences in temperatures, the thermal gradients between estimated indoor comfort temperatures and existing outdoor means provide an estimate of the amount of technological response required to maintain subjective thermal comfort for adapted individuals. The zonation in the maps in Figs. 5.1 and 5.2 is a pragmatic one. Neglecting for the moment regions of excessive humidity, the areas enclosed by isotherms $T_\psi = 24.5$ and $T = 26.5$ were taken to represent zones of negligible thermal gradients between predicted T_ψ and T_m and were further divided into 5-degree zones. Between $T_\psi = 18.5$ and $T_\psi = 28.5$, taken conservatively for the present to represent acceptable extreme values of group neutralities, this yields an additional four zones of thermal design. Some of these relationships between outdoor warmth, maximum indoor comfort levels, the suggested gradient zones, and potentials for control options are shown graphically in Fig. 5.3. Thus the global patterns for the coldest and warmest months of the year in Figs. 5.1 and 5.2 are defined both in terms of indoor thermal comfort as determined by mean monthly temperatures and by indoor-outdoor gradients.

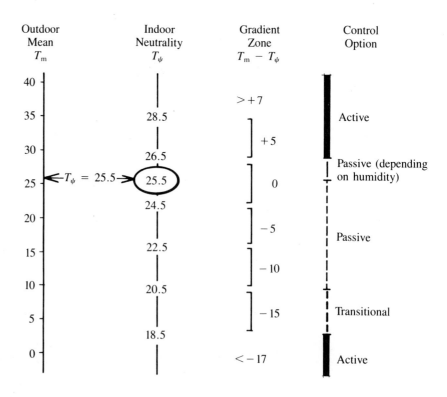

FIGURE 5.3.
Relationships between indoor neutralities and outdoor temperatures.

The Cooling Zone

In general, to the warm side of design zero, i.e., when $T_\psi = 25.5°C$, the achievement of a sizable negative outdoor-indoor temperature gradient is difficult with purely passive systems of microclimate control. In equable climates, i.e., climates within which nighttime losses of long-wave radiation are retarded by moisture within the atmosphere, dry-bulb temperatures do not vary greatly over the diurnal cycle and no convenient heat sink is readily available. Such climates have been described at times by their high measures of relative humidity, such as, for example, minimum values in excess of 55 percent or those on average 75 to 80 percent (Szokolay 1980). However, since relative humidity is closely dependent on temperature variability, relative humidity itself does not well define the degree to which the atmosphere can retard terrestrial heat loss. Vapor pressure or other measures of the quantity of water in the atmosphere are more suitable, and for present purposes, precipitable water vapor (or the equivalent linear amount of water within a vertical column of the atmosphere, global range 0.1-6 cm) estimates, as shown by Barry and Chorley (1976), have been used.

In terms of whole units, the 4-cm precipitable water vapor isohyet for the 2 months selected for Figs. 5.1 and 5.2 appears to correspond most closely to the relative humidities in the warmer regions of the world. It also coincides with the general patterns of other criteria at times used to describe warm-humid climates, including the "sultry" human classification by Terjung (1968). The 4-cm isopleth, as shown by the dotted line in Figs. 5.1 and 5.2, is therefore adopted as defining the geographic limits of the human ability to achieve thermal comfort by purely passive design features.

Thus, although in terms of simple temperature gradients a zone may be defined as "comfortably warm," comfort due to excessive humidity could be achieved only by active air cooling. Such mild, moist, temperature-humidity regimes occur in large parts of the tropical monsoon world during the high-sun season, including the Sudan, the Congo Basin, and the Sahel countries of Africa, India, and China, almost the whole of Southeast Asia, western Borneo, Papua, the Yucatan, the upper catchment areas of the Amazon, and the Australian Northern Terrritory and Queensland. Some parts are affected the whole year round. These include much of insular Southeast Asia, Guinea in Africa, Central America, and the Amazon Basin. A few areas are even more adversely affected, being also exposed to high thermal gradients. Particularly affected appear to be southern parts of the Sahara, the Nile, Red Sea, and Arabian Sea littorals, the Kalahari, the central Amazon Basin, Sonora in Mexico, and northwest Australia.

The areas without the burden of high moisture provide the most scope for passive maintenance of indoor comfort in massive constructions. Unfortunately, such buildings are unlikely to achieve reductions equivalent to 3 to 5°C over prolonged periods. In general, therefore, places warmer than 25.5°C but having the potential for passive control of positive outdoor-indoor gradients are restricted to southwestern United States, northern Africa, the Caucasus and Turkestan in July, and the plateau country of Cape Province in South Africa and much of central Australia. In Figs. 5.1 and 5.2, the shaded areas are those which appear to have little alternative to active air cooling, at least during the hottest times of the year.

The Heating Zone

It could be argued that well-designed passive systems relying on adequate insulation, mass, double glazing (where appropriate), and optimized orientation should be well able to maintain a mean monthly deficit of 10°C. To the cool side of $T_\psi = 25.5$, therefore, owing to progressively decreasing indoor warmth requirements, it should be possible economically to construct buildings without heating appliances at least down to $T_\psi = 20.5$, at which point an outdoor-indoor gradient of about 12°C would need to be maintained.

Thus, in July, active energy inputs would be required only in the polar regions, southern Argentina and the Andes, the South Island of New Zealand and Wellington Peninsula in the North, Tasmania, southern Victoria, and higher ground of the Australian Alps-New England range. In January, of course, the heating zone extends over vast areas covering all but part of the southern tip of the Iberian Peninsula, the whole of Europe, and all of Asia northward to about 30 degrees latitude. In North America also, vast areas need active heating, and excluded are only the coastal zones along the Gulf of Mexico and southern California. In South America, only Tierra del Fuego and Patagonia are below $T_\psi = 20.5$.

USER RESPONSES TO AIR-CONDITIONING

Lack of Information

The preceding climate analysis indicates that, even allowing for maximum climatic adaptation, the attainment of thermal comfort in certain parts of the world can do little other than rely on air cooling. The reality is, of course, that air cooling has spread to places well beyond the shaded portions in Figs. 5.1 and 5.2.

Surprisingly, in view of the significance of air-conditioning to everyday life and the controversy surrounding the thermal constancy hypothesis, as discussed in Chap. 1, very little work appears to have been done to investigate user perceptions and preferences with regard to air-conditioning. Some contentious issues that need to be resolved or at least illustrated in quantitative terms for obviously culturally and geographically diverse populations are as follows:

1. What conditions are subjectively definable as the most comfortable and desirable?
2. What thermal conditions should be specified as optimal for facilitating homeothermy?
3. To what extent are individuals prepared to accept discomfort and for what tradeoffs?
4. What are the measurable economic and perceived costs of air-conditioning?
5. Are there differences in preferences for air-conditioning at work and at home?
6. Does exposure to air-conditioning in one place affect preferences in the other?
7. To what extent are ideal conditions achieved in particular environments and at particular activities?
8. What practical alternatives exist to present microclimatic, especially air-conditioning, control systems?

Two recent series of studies have attempted to quantify such obvious gaps in the understanding of human responses to air-conditioning under real-life conditions. Although the results in both are likely to have been affected by cultural dimensions, the general findings probably apply to populations elsewhere. Given the general cultural, socioeconomic, and technological similarities between Australians and other Western people, especially those within the United States and England, the results of the Australian study may be directly applicable to these locations.

TABLE 5.1 Air-Conditioning and the Desire to Go Outdoors in Osaka, Japan

	People Who Use AC Only When Absolutely Necessary	People Who Use AC Considerably or All Day
Desire to go out often or occasionally	9 (18%)	62 (74%)
Do not desire to go out at all	41 (82%)	22 (26%)

Note: AC = air-conditioning. Question asked of respondents with air-conditioned apartments: "[After prolonged use of AC] do you desire to go outdoors for fresh air?"
Source: Data from Hirokawa and Horie (1985).

A Japanese Study

In a Japanese study reported by Hirokawa and Horie (1985), two groups of people, apparently drawn from a random sample of 150 individuals living in two multiple-dwelling housing developments in Osaka City, were compared on the basis of responses to questions relating to their environmental perceptions. The work was carried out during August and September of 1979. Group A consisted of 37 individuals who depended solely on natural ventilation, and group B consisted of 58 people who depended on air-conditioning. Although no indoor data are supplied, in both cases it is claimed that the environments were maintained within the range of thermal comfort. Unfortunately, no information is supplied in the report on socioeconomic status or other characteristics of the two groups.

The main results of these comparisons were as follows.

1. The residents living in air-conditioned environments expressed a stronger desire to go outdoors (see Table 5.1).
2. The air-conditioned group was less relaxed and less capable of perceiving general fluctuations within the environment.

Since no objective count was made of the number of people frequenting outdoor areas, the need for the outdoors may represent little more than a verbalized longing for an imagined ideal. In Darwin, North Australia, a frequent criticism of air-conditioning was that it actually encouraged people to spend a disproportionate amount of time indoors (Auliciems and de Dear 1986*b*).

An Australian Study

In Australia, a series of field studies was conducted to test the efficiency of current practices in air-conditioning in different geographic locations by socioeconomically similar populations. While particular aspects have been reported in Auliciems and de Dear (1986*a, b*) and de Dear and Auliciems (1985*a, b*), the most comprehensive details for this work may be found in de Dear's (1985) doctoral thesis, "Perceptual and Adaptional Basis for the Management of Indoor Climates," at the University of Queensland.

Testing was done in three cities, during both seasons of 1983 in monsoonal Darwin (latitude S12.5°), at the height of the 1983-1984 summer in subtropical Brisbane (latitude S27°), and at the height of the 1982-1983 summer in midlatitude Melbourne (latitude S37.8°). At both latter locations, two office populations,

TABLE 5.2 Summary Statistics for Indoor and Outdoor Climates in Six Australian Field Surveys

	Darwin AC		Brisbane Summer		Melbourne Summer	
	Dry	*Wet*	*AC*	*Non-AC*	*AC*	*Non-AC*
No. of respondents	174	197	211	201	186	194
No. of questionnaires	493	555	564	611	512	555
Females	99	105	101	96	63	73
Males	75	92	110	105	123	121
Median age	31	32	29	26	31	27
Mean length of residence in city of survey	9	10	18	19	24	21
T_{db}, mean	23.3	23.7	23.8	27.9	23.4	24.3
T_{wb}, mean	16.1	17.8	17.4	21.6	15.7	15.6
Air speed, mean	0.07	0.14	0.15	0.31	0.11	0.17
T_{mrt}, mean	23.7	24.6	24.1	27.8	24.0	24.6
Indoor RH, mean	47%	56%	53%	58%	44%	39%
Clo, mean	0.49	0.43	0.48	0.45	0.55	0.57
Act. (met.), mean	1.2	1.2	1.2	1.2	1.2	1.2
PMV, mean	−0.46	−0.48	−0.49	−0.66	−0.39	−0.20
T_{set}, mean	22.6	23.0	23.0	26.5	23.2	23.6
Mean monthly outdoor temp.	25.0	28.8	24.4	24.9	18.7	18.7
Mean 0900 hours outdoor RH	57%	67%	68%	71%	64%	64%

T_{mrt}—mean radiant temperature

Clo—clothing insulation in clo units

PMV—predicted mean vote

T_{set}—standard effective temperature

Source: From de Dear and Auliciems (1985*b*).

one with air-conditioning and the other without, were used. All subjects were tested with ASHRAE and Bedford and thermal preference scales, equal-interval attitude scales, open-ended attitude questions, and directed questions. Personal information gained included details on metabolic rates and clothing, and data were collected on atmospheric parameters both indoors and outdoors. Some sample means are shown in Table 5.2. Thermal data were analyzed by probit regression, and significance in differences between samples was established by the *t* test.

With regard to thermal practices and responses, some data are presented in Tables 5.3 and 5.4. In brief, the main findings on Australian thermal perceptions can be summarized as follows:

1. Despite the considerable differences climatically and geographically, air-conditioning was equally applied to all three populations. The same temperature was maintained at all locations.
2. No seasonal differences were observed in air-conditioning practices in Darwin.
3. Overall, air-conditioned offices were evaluated by respondents as excessively cool.
4. With air-conditioned samples, calculations from mean indoor and outdoor air temperatures better predicted observed neutralities than those estimated by *PMV*. In buildings without air-conditioning, neutralities were better predicted from mean monthly temperatures than from *PMV*.
5. Dissatisfaction with temperature departure from neutrality was increased in air-conditioned samples (see temperature magnitudes required for incidence of 20 percent of comfort votes greater than 1.5 in Table 5.4).

TABLE 5.3 Comparative Probit Analysis of 7-Point Scales (as in Table 1.2) Using Dry-Bulb Temperature (°C) as Predictor

	Darwin AC		Brisbane		Melbourne	
	Dry	Wet	AC	Non-AC	AC	Non-AC
ASHRAE (neutrality)	24.2	24.1	23.8	25.5	22.6	21.3
Preference (neutrality)	24.1	23.3	23.4	25.2	22.1	21.6
Bedford (neutrality)	24.2	23.9	23.9	25.6	22.7	21.8
Bedford ED20 for +1.5*	25.3	24.9	24.8	27.7	24.3	24.3
Neutrality—ED20	1.1	1.0	1.1	2.1	1.6	2.5

*ED20 is the effective-dose temperature at which 20 percent of the people cast a comfort vote greater than or equal to +1.5; i.e., the temperature causing 20 percent dissatisfaction.
Source: From de Dear and Auliciems (1985a).

TABLE 5.4 Prevailing Levels of Warmth and Comfort in Six Australian Surveys

	Darwin AC		Brisbane		Melbourne	
	Dry	Wet	AC	Non-AC	AC	Non-AC
PMV						
Mean	−0.46	−0.48	−0.49	0.66	−0.39	−0.20
SD	0.67	0.63	0.67	0.53	0.60	0.85
Fanger's preferred temperature						
Mean*	27.7	25.3	25.1	26.2	24.8	25.0
SD	1.4	1.2	1.2	1.0	1.3	1.5
Bedford						
Mean	−0.36	−0.06	0.01	0.83	0.26	0.67
SD	1.20	1.20	1.10	1.03	1.10	1.16

*Fanger's preferred temperatures calculated for values when $PMV = 0$.
Source: From de Dear and Auliciems (1985a).

The main results of attitude assessments for the three cities can be summarized as follows:

1. The open-ended questionnaire technique showed that the main perceived disadvantages of air-conditioning (Tables 5.5 and 5.6 for Darwin) were as follows:

   a. Elevated indoor-outdoor temperature gradients (i.e., thermal shock when entering or leaving cooled buildings)
   b. Lack of personal control of microclimates
   c. General concern with deleterious health effects
   d. Stuffiness
   e. Sensations of being cold
   Perceived advantages were far less numerous, and only reduction of excessive temperature showed favor (see Tables 5.7 and 5.8 for Darwin).

2. In response to specific questions, answers as in Table 5.9 indicated that there was a perceived connection between summer air-conditioning and adverse health effects. Suggested ailments are listed in Table 5.10, with colds and sniffles identified as the main problem. It should be noted, however, that these responses were not objectively validated, nor were states of health

TABLE 5.5 Perceived Disadvantages of Air-Conditioning in Darwin Offices*

Response Category	Response Tally					
	1st	2d	3d	Total (1 + 2 + 3)	Weighted Score	Weighted Rank
No response	40	128	244	412	620	1st
Too cold	62	37	6	105	266	4th
Thermal gradients indoors/outdoors	41	37	23	101	220	5th
General health problems	65	64	38	167	361	2d
No control of climate	47	31	7	85	210	6th
Lack of fresh air, odors, stuffiness, smoke	63	43	17	123	292	3d
Breakdowns of air-conditioning plant	29	17	9	55	130	7th
Deprivation of acclimatization	13	5	11	29	60	9th
Other/miscellaneous	11	9	16	36	67	8th

*"What are the disadvantages, in your opinion, of air-conditioning in your workplace?"
Source: From Auliciems and de Dear (1986b).

TABLE 5.6 Perceived Disadvantages of Home Air-Conditioning in Darwin*

Response Category	Response Tally					
	1st	2d	3d	Total (1 + 2 + 3)	Weighted Score	Weighted Rank
No response	74	198	308	580	926	1st
Running costs	157	26	14	197	537	2d
Health issues	26	23	10	56	134	4th
No fresh air, odors, stuffy	22	35	10	67	146	3d
Have to seal up house	20	19	6	45	104	7th
Thermal gradients indoors/outdoors	24	22	7	53	123	6th
Noise of machine	2	6	0	8	18	9th
Deprives acclimatization	29	19	7	55	132	5th
Too cold	4	3	0	7	18	10th
Others	13	20	9	42	88	8th

*"What are the disadvantages, in your opinion, of air-conditioning in the house?"
Source: From Auliciems and de Dear (1986b).

TABLE 5.7 Perceived Advantages of Air-Conditioning in Darwin Offices*

Response Category	Response Tally					
	1st	2d	3d	Total (1 + 2 + 3)	Weighted Score	Weighted Rank
No response	20	208	345	573	821	1st
Productivity, concentration, ease of paperwork	87	89	8	184	447	3d
Coolness, comfort	235	42	3	280	792	2d
Clean environment	5	12	6	23	45	6th
Quietness with shut windows	3	2	5	10	18	7th
Constant temperature	11	6	1	18	46	5th
Others	10	12	3	25	54	4th

*"What are the advantages, in your opinion, of air-conditioning in your workplace?"
Source: From Auliciems and de Dear (1986a).

clinically compared between people from air-conditioned and uncooled buildings.

3. Even in the hottest location, Darwin, there was no overwhelming demand for air-conditioning. Certainly, as shown in Fig. 5.4, during much of the year, air-conditioning was not preferred either in the workplaces (which were conditioned at all times) or homes. Equal-interval Thurstone attitude scales did demonstrate, however, that on average air-conditioning in offices at the peak of summer had certain advantages (Fig. 5.5). For the home, there was overwhelming rejection of air-conditioning, even for the stressful monsoonal buildup period (see Fig. 5.6).

TABLE 5.8 Perceived Advantages of Home Air-Conditioning in Darwin*

	Response Tally					
Response Category	1st	2d	3d	Total (1 + 2 + 3)	Weighted Score	Weighted Rank
No response	97	276	352	725	1195	1st
Better night's sleep	121	33	3	157	432	2d
Clean, dust-free house	10	8	4	22	50	4th
Cool, comfort	120	25	1	146	411	3d
Kills outdoor noise	4	11	2	17	36	7th
More microclimate control	12	6	2	20	50	5th
Better moods, temper	1	5	1	7	14	8th
Others	6	7	6	19	38	6th

*"What are the advantages, in your opinion, of air-conditioning in the house?"

Source: From Auliciems and de Dear (1986*b*).

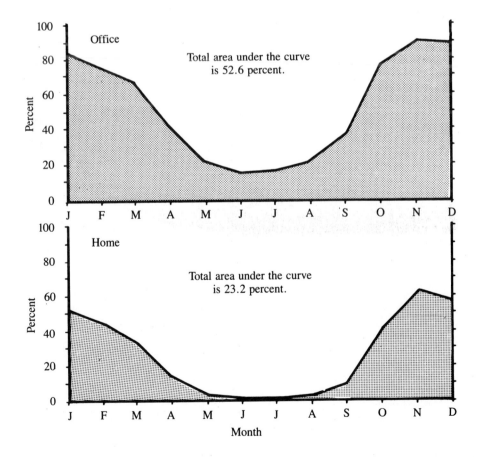

TABLE 5.9 Ailments Attributed to Summer Air-Conditioning

Ailment	Melbourne AC Survey	Brisbane AC Survey
Colds or "sniffles"	57 (15.1%)	65 (13.9%)
Nasal or sinus	12 (3.2%)	41 (8.8%)
Respiratory infections	37 (9.8%)	37 (7.9%)
Headaches	9 (2.4%)	8 (1.7%)
Bronchial asthma	6 (1.6%)	12 (2.6%)
Dry skin or scalp	16 (4.2%)	4 (0.9%)
Rheumatism or aches	2 (0.5%)	3 (0.6%)
Thermal "shock" entering or leaving building	3 (0.8%)	2 (0.4%)
Hayfever	6 (1.6%)	10 (2.1%)
Tiredness	11 (2.9%)	4 (0.9%)
Sore eyes	0 (0%)	5 (1.1%)
Miscellaneous	54 (14.3%)	55 (11.8%)
TOTAL	378 (100%)	468 (100%)

Source: From de Dear (1985).

TABLE 5.10 Typical "Sick Building" Symptoms

Mucous membrane irritation	Sore throat
Eye irritation	Shortness of breath
Headaches	Wheeze
Odor irritation	Lethargy/fatigue
Skin irritation	Abnormal taste
Cough	Dizziness
Sinus congestion	

Source: Compiled by de Dear (1985).

FIGURE 5.4.
Preference for air-conditioning in tropical Darwin. (From Auliciems and de Dear 1986*b*.)

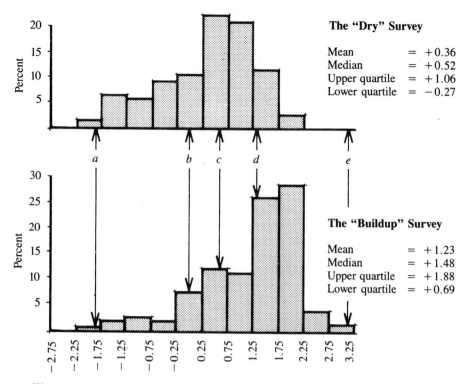

FIGURE 5.5.
Perceived disadvantages of air-conditioning in Darwin offices. (From Auliciems and de Dear 1986b.)

a: We would all be a lot better off without air-conditioning in the office.
b: I don't strongly like or dislike air-conditioning in the office.
c: There are a few disagreeable aspects of office air-conditioning, but I am still in favor of it.
d: Office life before the invention of air-conditioning was primitive.
e: Air-conditioning is absolutely necessary in offices.

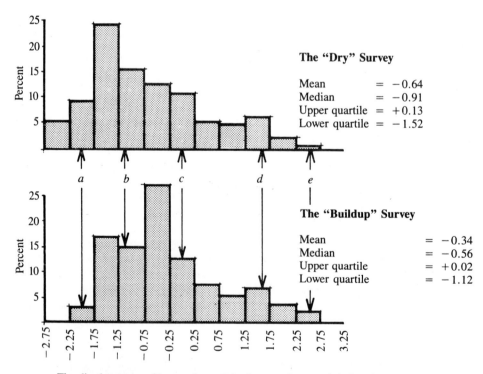

FIGURE 5.6.
Perceived disadvantages of air-conditioning in Darwin homes. (From Auliciems and de Dear 1986b.)

a: The disadvantages of home air-conditioning greatly outweigh the advantages.
b: There are plenty of things about home air-conditioning that annoy me.
c: I am neither pro nor con air-conditioning in the home.
d: An air-conditioner in the home keeps you feeling fresh.
e: Home air-conditioning is no longer a luxury but an essential part of modern living.

HUMAN PROBLEMS IN AIR-CONDITIONING

From the preceding studies, several broad conclusions can be reached:

1. Air-conditioning is not the self-evident and universally desired "optimum" often portrayed by people representing vested interests.
2. There does not exist an overwhelming demand for constantly cool indoor temperatures. People do, however, become habituated to persistent thermal levels without a lowering in overall satisfaction.
3. People are attuned to outdoor events, and thermal satisfaction is maximized when indoor conditions vary according to seasonal and weather conditions.
4. In general, people are not satisfied with the quality of conditioned air. Whether real or imagined, air-conditioning is believed to adversely affect health.
5. Lack of individual controls, including sealed windows, are seen as highly undesirable features.
6. Home air-conditioning is overwhelmingly rejected in favor of other technologies or changes in lifestyles.
7. The typical manager of air-conditioned buildings in general lacks sufficient understanding of human responses to thermal environments. Many practices are outdated and too inflexible in response to complaint. Too much uncritical reliance is placed on the air-conditioning equipment manufacturer's recommendations.
8. Overall there is an unnecessary wastage of energy in an oversupply of cooled air.

These observations are consistent with some general and well-known concerns. First, the recirculation of air in air-conditioning systems is not particularly efficient at removing certain noxious substances. Indoor pollutants, including tobacco smoke, may accumulate or become distributed away from the source. Moreover, with the generation of large quantities of condensation, there is increased potential for the growth of harmful microorganisms within the machinery and ducting. Because of these phenomena, recently there has been an increasing reference to the "sick building" syndrome (Stolwijk 1984), with typical symptoms as shown in Table 5.10 and discussed in Chap. 2.

The second problem, while somewhat more theoretical, is nevertheless an important one to the long-term well-being of an increasingly indoor and sedentary population. Both practice and adaptation theory (as described in Chap. 1) demonstrate that people will "get used to" given environmental conditions. Air-conditioning systems governed by thermostats produce constant and, at their best, dependable flows of cool air. This will tend to promote an increasing reliance on their efficiency and particular set temperatures. This, in turn, is likely to encourage standard clothing ensembles that address the prevailing levels of warmth, with minimum capacity for adjustment.

Such reliance on air-conditioning technology can do no more than necessitate reinforcement of "zoo keeper" (Cooper 1982) efficiency in controls with an ever-increasing precision and invariability. In the long term, and in theory at least, this is further likely to reduce the occupants' flexibility in physiological thermoregulatory mechanisms and increase dependence on this technology. The extension of air-conditioning to vast volumes of air in sports arenas and shopping complexes probably represents a spiral of events with inbuilt positive feedbacks: the greater the cooling, the larger the impetus for more. The sum total of human contact with the unceasing variability of natural perturbations of outdoor air, in the highly developed technological world at least, is likely to progressively decrease.

Whatever the inevitability or arguments in favor of such developments, looking ahead to the future, there is potential for a progressive degeneration of thermoregulatory capacity in people living in such environments. Should sudden disruptions occur in the maintenance of such environments, the habituated populations are likely to be exposed to unusual degrees of thermal strain.

Traditionally, coping with thermal inclemency has involved modifications

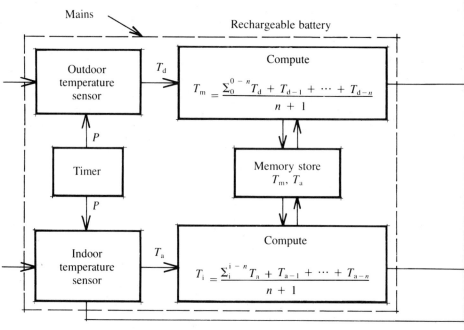

Inputs: Pulse interval P (suggest hourly)
Averaging period N (suggest $300P$)
Temperature:
Indoor T_a
Outdoor T_d

FIGURE 5.7.
Basic design for thermobile control system. (From Auliciems 1984.)

in lifestyles to suit particular climatic rhythms and a reliance on more passive energy technologies. Indoor thermal regulation has depended more on natural ventilation, shading, sizes and positions of openings, orientation, materials, and thermal mass. There exists more than just prima facie evidence to suggest that, at least in some instances, such passive methods also may be preferable to individuals in today's energy-conscious world. At least the rationale and economy of air cooling in latitudes above N35° needs to be put under close scrutiny. What is questioned here, therefore, is not air-conditioning per se, but its uncritical usage in areas and at times when other technologies are available.

VARIABLE TEMPERATURE-CONTROL STRATEGIES

From the preceding theoretical discussions of human thermal responses, general considerations of air-conditioning management practices, and user survey results, there would seem to exist an overwhelming case for reconsidering microclimatic management strategies. The first problem that needs to be tackled is how economically to achieve indoors atmospheric variability that more closely corresponds to the outdoors and without loss in thermal satisfaction.

One approach could be as follows: for populations already residing in buildings with close climate control, a program of deregulation could be instituted by allowing a gradual but progressively increasing indoor temperature drift toward those outdoors. This is likely to encourage greater adjustment by individual clothing changes and the wearing of ensembles that better reflect weather conditions outdoors. This, in turn, is likely to further reduce the thermal "shock" of moving from one environment to the other and presumably promote outdoor activities. The prospects for the processes of seasonal acclimatization also would become enhanced.

The target indoor level of warmth would be determined by Eqs. (1.4) to

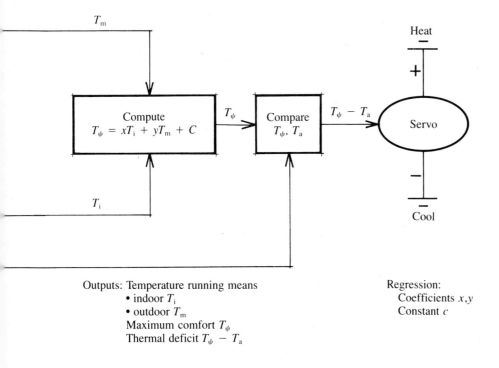

Outputs: Temperature running means
- indoor T_i
- outdoor T_m

Maximum comfort T_ψ

Thermal deficit $T_\psi - T_a$

Regression:
Coefficients x, y
Constant c

(1.10), depending on available data and the type of building. Generally, in old buildings, providing that adequate temperature information exists, Eqs. (1.5) and (1.7) would be preferable; in new buildings, Eqs. (1.4) and (1.6) would be appropriate. In places such as schools, where populations are likely to spend much time outdoors, shorter time periods should be considered. Optimally, of course, continuing surveys and ongoing monitoring of thermal preferences should be carried out to enable further adjustment as determined by local climates and cultures. Initially at least, the limits of indoor temperatures should be those discussed earlier and shown in Fig. 1.11.

The problem remains a relatively simple one: How best to continuously adjust air-conditioning or, for that matter, heating machinery?

THERMOBILE CONTROL

During the past decade, microprocessor developments have enabled the already versatile thermostat to undertake remarkably sophisticated control roles. In an effort to conserve energy and decrease the adjustment time of conditions indoors, the thermostat has been variously modified to anticipate changes in heat flows to and from the outdoors, to set back to reduced levels of heating and cooling at specified times, and in general to achieve an increased precision in operation. By definition, however, and irrespective of the complexity of design or designation of purpose, any thermostat can aim only to achieve a *stasis* within the environment. Its gyrations per se are no more than a means of achieving temperature constancy according to some predetermined comfort criteria.

However, the thermostat function can be readily altered to a mobile one, either by programming or by designated chip, to allow additional inputs and integration by the algorithms of any of the adaptation equations discussed in Chap. 1. The resulting control systems should be referred to as "thermomobiles" or, more simply, "thermobiles." The basic design is shown in Fig. 5.7.

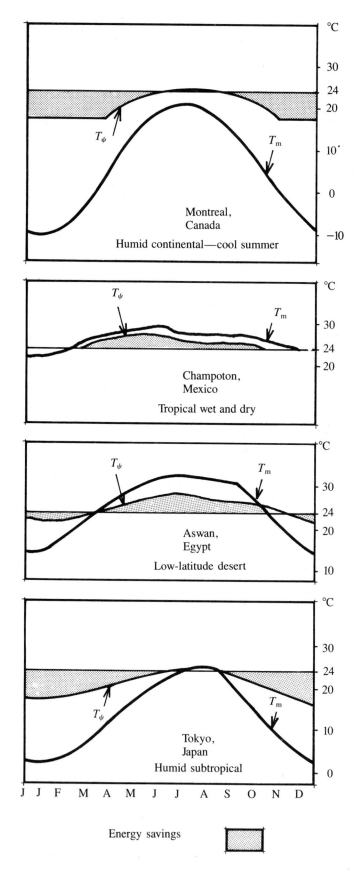

FIGURE 5.8.
Energy savings possible with various neutralities in comparison to constant 24°C. The unshaded portions between curves T_m and T_ψ indicate the indoor-outdoor temperature gradient that needs to be compensated for by heating or cooling. The shaded areas between T_ψ and 24°C show the theoretically possible reductions in active energy usage by maintaining variable indoor temperatures. (From Auliciems 1983.)

Energy savings

T_ψ = indoor neutral temperature as predicted by $T_\psi = 0.31T_m + 17.6$
T_m = monthly mean outdoor temperature

Such thermobiles, with already existing setback features, could be incorporated into all new housing designs where active energy usage for people is required and, indeed, could readily replace existing thermostats. Alternatively, the existing devices may be amenable to conversion to achieve the desired mobility in control. In any case, the cost of this modification should be minimal.

In existing buildings with adequate temperature records, thermobiles should be programmed according to the algorithms in Eq. (1.7) or some modified version to preference or special local circumstances. In the design of new buildings, initially at least, indoor temperatures could be maintained by programming according to Eq. (1.6). With further verification of shorter-term weather effects on indoor comfort, other equations could be established.

The resulting benefits would be twofold. First, overall comfort levels would improved. Second, and more important, as shown in Fig. 5.8 [where comparison is made between energy requirements to maintain a set 24°C and those permitting variability according to Eq. (1.6)], the reductions in thermal gradients would lead to savings in energy. On a worldwide basis, the potential for savings is enormous.

SUMMARY

1. From a climatic viewpoint, air cooling at the present time is largely overused both with respect to geographic location and to prevailing warmth.
2. General human responses to air-conditioning are inadequately understood, and little is known of cultural differences in preference.
3. Dissatisfaction with the performance of air-conditioning systems exists within both homes and office buildings.
4. Air-conditioning management strategies are often outmoded, overly reliant on simple engineering solutions, and not commensurate with either environmental or user requirements.
5. New approaches to indoor climate management should abandon thermostatic control and be based on a more harmonious synchronization of indoor and outdoor levels of warmth and the achievement of controlled variability in indoor atmospheric conditions to enable increased human adjustment by adaptation.

REFERENCES

Auliciems, A. 1983. Psycho-physiological criteria for global thermal zones of building design. *Int. J. Biometeorol.* 26 (Suppl.): 69-86.

Auliciems, A. 1984. Thermobile controls for human comfort. *Heat. Vent. Eng.* 58: 31-33.

Auliciems, A., and de Dear, R. J. 1986a. Air-conditioning in Australia: I. Human thermal factors. *Arch. Sci. Rev.* 29: 67-75.

Auliciems, A., and de Dear, R. J. 1986b. Air-conditioning in a tropical climate: Impacts on European residents in Darwin, Australia. *Int. J. Biometeorol.* 30: 259-282.

Ballantyne, E. R., Hill, R. K., Spencer, J. W., and Bartlett, N. R. 1979. A Survey of Thermal Sensation in Port Moresby, Papua, New Guinea. CSIRO Division of Building Research, Technical Paper No. 32, Melbourne.

Barry, R. G., and Chorley, R. J. 1976. *Atmosphere, Weather and Climate.* 3d Ed. Methuen, London.

Budd, G. M., Fox, R. H., Hendrie, A. L., and Hicks, K. E. 1974. A field survey of thermal stress in New Guinea villagers. *Philos. Trans. R. Soc. Lond.* [*Biol.*] 268: 393-400.

Cooper, I. 1982. Comfort theory and practice: Barriers to the conservation of energy by building occupants. *Appl. Energy* 5: 243-288.

de Dear, R. J. 1985. Perceptual and Adaptational Basis for the Management of Indoor Climate. Ph.D. thesis, University of Queensland.

de Dear, R. J., and Auliciems, A. 1985a. Validation of the predicted mean vote model of thermal comfort in six Australian field studies. *ASHRAE Trans.* 91: 452-468.

de Dear, R. J., and Auliciems, A. 1985b. Thermal Neutrality and Acceptability in Six Australian Field Studies. In Fanger, P. O. (Ed.), *Clima 2000*, Vol. 4, pp. 103-108. VVS Kongres-VVS Messe, Copenhagen.

Hirokawa, Y., and Horie, G. 1985. Effect of Air-Conditioned Indoor Environment in Summer on the Living Behaviors in the Immediate Outdoor Surroundings. In Fanger, P. O. (Ed.), *Clima 2000*, Vol. 4, pp. 21-26. VVS Kongres-VVS Messe, Copenhagen.

Mookerjee, G. C., and Murgai, M. P. 1952. A preliminary report on the determination of comfort zone of Indian subjects during North Indian summer. *J. Sci. Ind. Res. India* 11a: 14-16.

Rao, M. N. 1952. Comfort range in tropical Calcutta: A preliminary experiment. *Ind. J. Med. Res.* 40: 45-52.

Stolwijk, J. A. J. 1984. The Sick Building Syndrome. In Berglund, B., Lindvall, T., and Sundell, J. (Eds.), *Indoor Air*, Vol. 1, pp. 23-29. Swedish Council for Building Research, Stockholm.

Szokolay, S. V. 1980. *Environmental Science Handbook for Architects and Builders.* Construction Press, London.

Terjung, W. H. 1968. World pattern of the distribution of the monthly comfort index. *Int. J. Biometeorol.* 12: 119-151.

Woolard, D. S. 1979. Thermal Habitability of Shelters in the Solomon Islands. Ph.D. thesis, University of Queensland.

Lighting Design

6

Nancy Ruck

THE IMPORTANCE OF LIGHTING

It is meaningful to consider the place of lighting in the overall evaluation and design of buildings, for the luminous environment and the human response to it are of importance both to the lighting engineer and to the architect, who is concerned with both the aesthetic and functional aspects of lighting. Among the environmental conditions considered important for occupant satisfaction in buildings are heat, light, acoustics, ventilation, air-conditioning, and an outside view. Lighting is considered to be one of the most important features of the work environment that affect worker satisfaction (Wineman 1982). It was found in a recent survey of worker satisfaction in office buildings by Ne'eman, Sweitzer, and Vine (1984) that over 90 percent of the office workers felt the amount and quality of light for reading and writing were very important, as were the amount and quality of light for computing, filing, and other tasks. The relationship between importance and satisfaction of workspace features from this survey is shown in Fig. 6.1. Each data point represents one of the 24 workspace variables listed. Among the most important are the right amount of light for reading and writing, the ability to control summer temperatures and privacy of conversation.

The basic aim of good lighting design is to create a luminous environment that provides both good task visibility and visual satisfaction without eyestrain. The relative importance of these two criteria depends to a large extent on the function of the interior space, for what is important for a particular interior

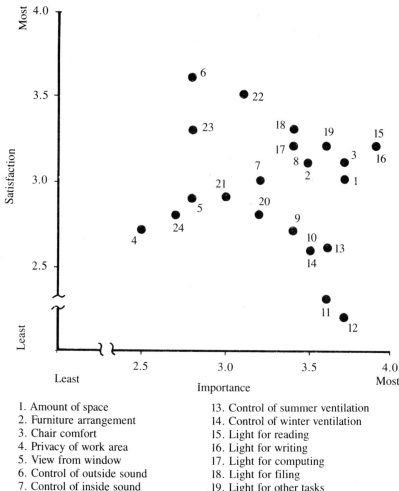

FIGURE 6.1.
The relationship between importance and satisfaction of workspace features. (After Ne'eman, Sweitzer, and Vine 1984.)

1. Amount of space
2. Furniture arrangement
3. Chair comfort
4. Privacy of work area
5. View from window
6. Control of outside sound
7. Control of inside sound
8. Control of lighting glare
9. Private phone conversations
10. Private office conversations
11. Control of summer temperatures
12. Control of winter temperatures
13. Control of summer ventilation
14. Control of winter ventilation
15. Light for reading
16. Light for writing
17. Light for computing
18. Light for filing
19. Light for other tasks
20. Control of work surface lighting
21. Control of ceiling lighting
22. Control of venetian blinds
23. Control of drapes
24. Control of other lights

such as an office would not be appropriate for a reception area or a car sales space. The method of lighting and the type of light source used also have an impact on electrical energy consumption and hence energy savings.

Visibility is concerned with the ease, speed, and accuracy with which a task such as reading text in a book or on a display monitor can be performed. The degree or measure of task visibility is generally determined by the visibility of the smallest or most difficult detail that must be recognized, and this recognition depends on the illuminance on the task, the task's apparent size, its contrast with the immediate surroundings, the available time for viewing, and many other factors, including the capabilities of the visual system.

However, good visibility is not always the only requirement a person needs to perform a task or activity easily and in comfort. Distractions from veiling reflections, discomfort glare, and other subjective considerations such as an inappropriate balance of luminances for surrounding surfaces are all qualitative issues that can affect task performance.

There exist criteria which set limits of tolerance with regard to illuminance and luminance balance, and these can be found in code recommendations. However, over the past few years, more and more complaints have arisen as a result of the use of screen-based equipment in the workplace. One concern is that of the luminance of environmental lighting, particularly the location of luminaires with respect to the display terminals and the luminance distribution on vertical and horizontal surfaces. In addition, the effects of flicker between environmental fluorescent lighting and display terminals can result in visual fatigue for operators.

The integration of daylighting and electric lighting also can produce problems of luminance balance and response to light controls. A significant amount of lighting energy can be saved by incorporating in the design of new and retrofitted facilities lighting hardware controls and operating strategies such as automatic exterior shading devices and electric lighting dimmer controls. However, evaluations of such sophisticated systems have primarily emphasized technical and economical criteria without considering their impact on building occupants and their productivity. It should be noted that the monetary value of productivity currently exceeds energy costs, so that any solution that maximizes energy savings but reduces productivity, e.g., by the presence of glare, will be counterproductive.

This chapter discusses those aspects of the interior visual environment which have an impact on human well-being and productivity: factors relating to task performance, discomfort glare, screen-based equipment, the use of daylight and its integration with electric light, and associated criteria.

VISUAL PERFORMANCE AND PRODUCTIVITY

In lighting design, it is very difficult to isolate and measure the effects of a lighting system (e.g., illuminance) on productivity. Simple correlations are inadequate because of the lack of control over such variables as work load, age, motivation, and temperature. One approach that has been used to get around this difficulty is to measure visibility and visual performance under laboratory conditions, where confounding variables can be better controlled (Clear and Berman 1980).

In these experiments, visibility is defined with respect to the "threshold" detection levels of a reference task and is generally measured by visibility level (VL), which is the ratio of equivalent contrast to threshold contrast (i.e., the luminance contrast of a target when it can only just be seen). In the United States, to determine the extent to which a lighting installation satisfies the Illuminating Engineering Society's (IES) visual performance criteria, one must calculate the equivalent sphere illuminance, which is the level of sphere illuminance that will produce task visibility equivalent to that produced by the lighting environment. The latter concept requires calculation by computer and is difficult to measure.

Using the measure of VL, the types of tasks examined have ranged from identifying the correct orientation of Landolt rings, the size and contrast of the gaps in the rings being taken to describe the visual difficulty of the task, to proofreading checks or even scores on the Davis Reading Test (CIE 1980) with the task directly fixated. Visual performance criteria for these experiments have generally been considered to be some combination of speed and accuracy, attributes that determine, in part, productivity.

However, although there are many tasks in which fine visual discrimination is a critical component, in the majority of workplaces it is not. The acceptance and application of criteria derived from threshold performance experiments have led to the present tendency of installing very high levels of lighting, i.e., 1000 lx and above. Such high levels are undesirable because of their side effects, e.g., on arousal and discomfort.

If visual performance is the dominant limiting process in a real environment, and if lighting does not influence productivity in any other manner, then changes in relative visual performance calculated from changes in visibility will be approximately proportional to changes in real productivity. The cost-

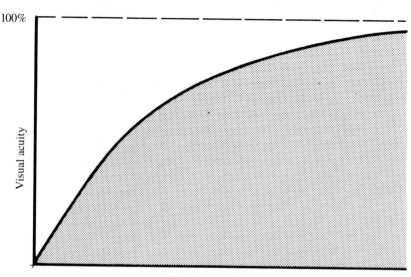

FIGURE 6.2.
Illuminance versus visual acuity.

effectiveness of a lighting system can then be examined by analyzing the visibility under the lighting. However, this analytical approach, based on quantifying how difficult tasks are to see, has yet to be fully validated, since the qualitative aspects of glare in the work environment have not been fully considered.

The relationship between illuminance and task performance follows a law of diminishing returns, as shown in Fig. 6.2, although the illuminance at which task performance saturates depends on the visual difficulty of the task. This example indicates that visibility should be the criterion for task lighting rather than general lighting.

Recommended Illuminances

Code recommendations for illuminance have traditionally been made in the form of minimum levels that vary from one country to another. This may be interpreted as a rational standard for a time period characterized by rising productivity and consistently falling electricity costs, since the cost of overlighting tends to be insignificant. However, with rising electricity costs, there could be a substantial cost in overlighting, and in this situation, target rather than minimum levels such as the CIE-recommended illuminances might be a more logical choice. However, these recommended illuminances are valid only if other factors that affect visibility and visual comfort have been accounted for in a satisfactory manner.

Much research has been carried out to determine a preferred range of horizontal illuminances in working interiors. The recommended illuminances of the Commission Internationale de l'Eclairage (CIE) are given in the form of ranges of three values of horizontal illuminance for each type of interior or activity, as shown in Table 6.1. The middle value of each range represents the illuminance recommended for normal situations; the highest, where there are low reflectances or contrast; and the lowest, where speed and accuracy are not important or where the task is executed only occasionally.

However, it is suggested that horizontal-plane illuminances are not always appropriate. In modern electronic data-processing for three dimensional objects and in some cases of industrial lighting, the horizontal plane has no significance. A skilled toolmaker, for example, would find it difficult to see tasks in positions other than on the horizontal plane even with an increase in illuminance. Therefore, lighting levels in some situations would be more appropriate and more cost-effective if calculated for vertical or inclined planes or if other measures such as mean spherical illuminance or mean cylindrical illuminance are

TABLE 6.1 CIE Recommended Illuminances

Ranges of Illuminance	Type of Task or Activity
20 — 30 — 50	Outdoor entrance areas
50 — 75 — 100	Circulation areas, simple orientation, or short temporary visits
100 — 150 — 200	Rooms not used continuously for working purposes, e.g., industrial surveillance, storage areas, cloakrooms, entrance halls
200 — 300 — 500	Tasks with simple visual requirements, e.g., rough machining lecture theaters
300 — 500 — 750	Tasks with medium visual requirements, e.g., medium machining, offices, control rooms
500 — 750 — 1000	Tasks with demanding visual requirements, e.g., sewing, inspection and testing, drawing offices
750 — 1000 — 1500	Tasks with difficult visual requirements, e.g., fine machining and assembly, color discrimination
1000 — 1500 — 2000	Tasks with special visual requirements, e.g., hand engraving, inspection of very fine work
2000	Performance of very exacting visual tasks, e.g., minute electronic assembly, surgical procedures

used (Cuttle et al. 1967). It is recommended that a simple process of visual task analysis be applied to the design by asking the following questions:

1. Which are the working planes?
2. How much light is needed?
3. From which direction does the light need to come?

Such an approach is an example of the lighting principles practiced in earlier years when the most common lamp in use was the tungsten lamp and when the cost of lighting followed the cost of living index. The oil crisis of the 1970s has provided the incentive for more effective lighting design.

Illuminances from daylight will be sufficient if they equal or exceed the illuminance requirements of the same visual task or activity under electric lighting. However, as mentioned in Chap. 3, for visual comfort, the average illuminances over general interior work areas normally should be not less than a third of the average illuminances of a specific task area. If this cannot be achieved by daylight alone, then additional electric lighting will be necessary.

Contrast and the Directionality of Lighting

The illuminances recommended for the various activities listed in Table 6.1 may be considered valid only on the condition that other factors that affect visibility and visual satisfaction have been accounted for in a satisfactory manner. One of these factors is contrast. A task or object can only be seen because of its contrast, or relative luminance difference, with its background, and the amount of contrast also will govern the amount of visibility. For a given illuminance and for matt surfaces, task contrast, and hence visual performance, relies completely on the reflectances of the task and background (Fig. 6.3). For tasks that are not completely matt, as is usually the case in practice, the luminance of the task detail and background in the viewing direction depend on the luminances of the surfaces and on the illuminance and directionality of the task lighting.

If lighting is designed with insufficient attention to appropriate directionality, high luminances from the light source can produce reflections in the direction of viewing that result in contrast reduction or glare caused by veiling reflections, as shown in Fig. 6.4.

When a specular surface is viewed from such an angle that a high-luminance source such as a luminaire is reflected from it into the eyes of the observer, veiling reflections occur. These can cause a reduction in task efficiency.

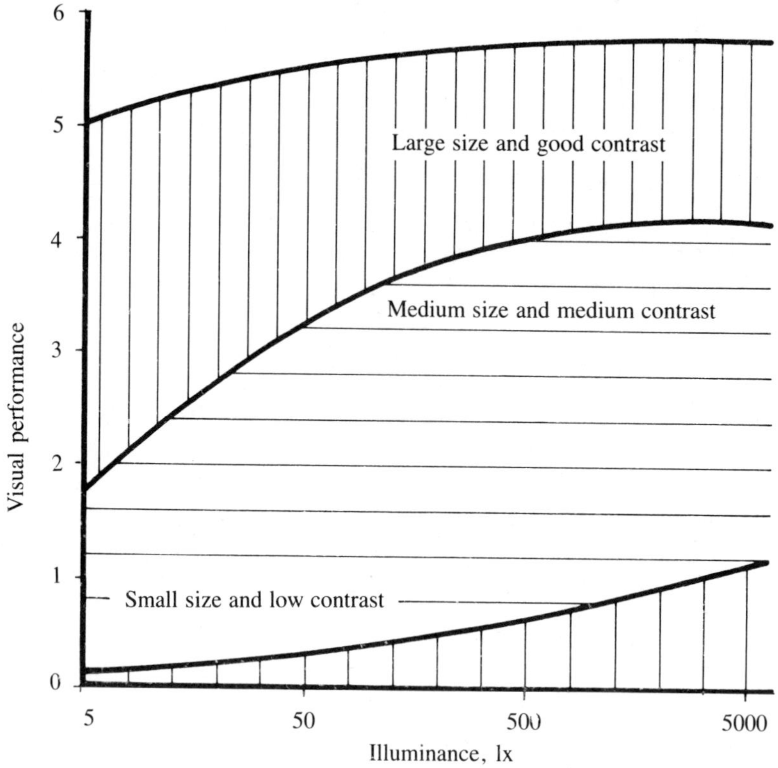

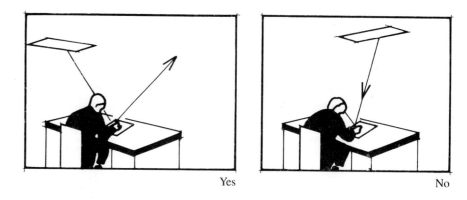

FIGURE 6.3.
Illuminance versus visual performance.

FIGURE 6.4.
Veiling reflections.

Veiling reflections superimposed on a task decrease contrast, and a reduction in contrast from reflected glare causes a reduction in performance. There are many ways to deal with contrast reduction and glare by reflection, and these are mentioned in the section on glare control (page 95).

To specify the contrast between a task and its background, the term "contrast-rendering factor" (*CRF*) was introduced (CIE 1986). The *CRF* is the ratio for the actual task contrast (relative visibility under actual lighting conditions) to the theoretical task contrast (relative visibility under reference lighting conditions). As a reference environment, the CIE uses a sphere of uniform illuminance. The values of *CRF* to be expected are as follows:

$CRF > 1$:   Lighting installations that direct light onto a glossy task from the sides

$CRF < 1$:   A totally indirect lighting installation (luminances approach those of the reference environment)

$CRF < 1$:   Conventional arrangement of luminaires in which luminances of luminaires in offending zone exceed those of reference environment of same average luminance

Glare and Its Control

The effect of a lighting system on productivity should take into account the negative impact from the presence of glare. As mentioned in Chap. 3, distinction is made between glare that reduces the visibility of objects but does not necessarily cause discomfort ("disability glare") and glare that causes discomfort but does not necessarily impair the visibility of objects ("discomfort glare"), and in interior lighting practice, discomfort glare from lamps and luminaires and windows within the visual field and reflected glare from glossy surfaces are more likely to be a problem than disability glare.

Windows constitute light sources that may subtend quite large solid angles at the eye, depending on how close the observer is to the window or windows. The luminance of the sky visible through a window may rise to 10,000 cd/m^2 even on overcast days and may be several times greater if bright sunlit clouds are seen, even though no direct sunlight may be entering the room. Such a source is likely to be seen directly or to one side of the line of sight and therefore is more likely to cause disability or discomfort glare than electric light sources (which are usually mounted overhead). The avoidance of discomfort glare from the sky or sun seen through windows or skylights is therefore an important consideration in daylighting design.

Glare Control with Windows

In the case of windows, there are several practical methods of glare control:

1. Arrangement of windows and surrounding interior surfaces to reduce the contrast between the latter and the observed sky.
2. Reduction of sky luminance by screening the windows with blinds or louvers.

Screening is necessary for windows with other than a north orientation (or south in the Southern Hemisphere) to reduce the entry of direct solar radiation at times when it can cause overheating. Screens should be adjustable so that at times of low daylight illuminance, the maximum amount of light can be admitted.

Figure 6.5 presents some solutions for glare control for sidelit windows. There are several ways to deal with the problem of controlling window luminance and at the same time retain a view out. In some instances, these solutions can be treated as temporary measures for use only on the brightest days, e.g., curtains or blinds under the control of the users. Structural solutions, on the other hand, impose permanent conditions. Overhangs, canopies, and awnings can reduce sky glare by cutting down sky view and reducing contrast, but daylight penetration into the interior is also reduced unless ceilings have highly reflective surfaces.

Glare Control with Electric Lighting

With luminaires, there are basically two methods for controlling source luminance. Usually, it is assumed that the viewing directions are horizontal and parallel to the walls and that the field of view is limited upward at an angle of elevation of 45 degrees, as shown in Fig. 6.6. The minimum angle of elevation to be considered is that of the most distant source in the field of view. The two methods for controlling source luminance are:

1. Control by shielding (Fig. 6.6).
2. Control with translucent materials, e.g., diffusing or prismatic material (Fig. 6.7).

These methods may be used singly or in combination, e.g., a luminaire with translucent louvers.

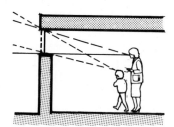

Glare
This unmodified window gives good light to the rear of the room, but it will be glaring on bright days.

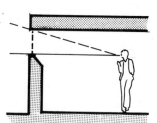

Splayed sill
The splayed sill can help by buffering contrast between sky luminance and the comparatively low luminance of the window wall.

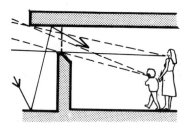

Ceiling—high reflectance
Overhangs reduce daylight penetration, but if the ceiling is of a high reflectance and is aided by reflections, say, from light-colored external paving, sky glare is reduced.

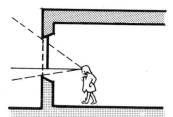

Screening with translucent blinds
Screening with translucent blinds can reduce the zone of high sky luminance (2) while still allowing view out (1).

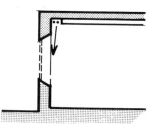

Electric light
Artificial lighting can be used to illuminate the window surrounds and to reduce glare potential by reducing the contrast between the sky luminance and the luminance of the window wall.

Window detailing
Splayed window detailing, i.e., jambs, head, sills, mullions, and transoms, reduce hard-edge contrast.

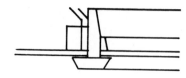

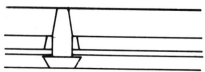

FIGURE 6.5.
Methods of glare control for sidelit windows. (After CIE 1989.)

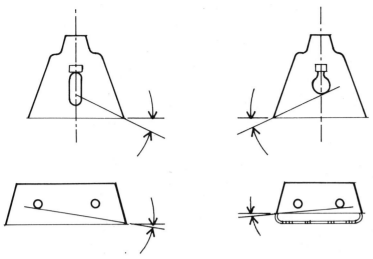

Shielding angles for various types of luminaires in which the lamps or parts thereof are visible when viewed at critical angles.

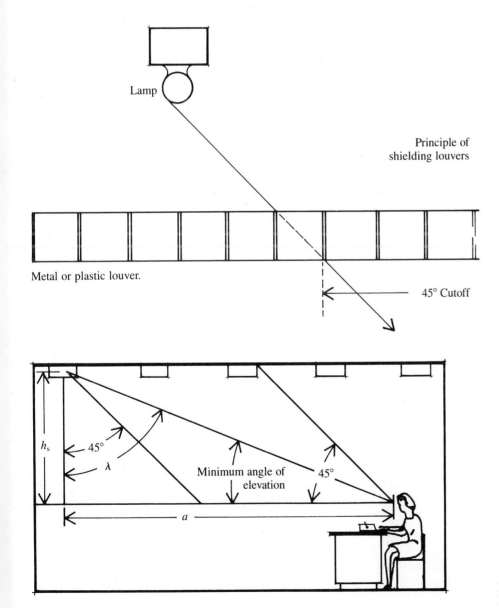

Lamp

Principle of shielding louvers

Metal or plastic louver.

45° Cutoff

h_s

45°

λ

Minimum angle of elevation

45°

a

Source luminance control for angles γ from 45° upward.

FIGURE 6.6.
Luminance control by shielding. (After CIE 1986.)

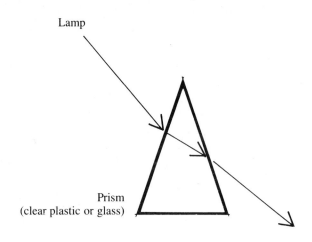

Lamp

Prism
(clear plastic or glass)

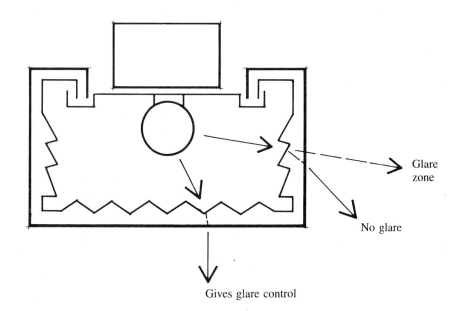

Glare
zone

No glare

Gives glare control

FIGURE 6.7.
Luminance control by refraction.

There are many ways to deal with contrast reduction and glare by reflection. The most effective methods are to arrange the location of the worker and the light source such that reflections are not toward the eyes but are directed away or to the side (see Fig. 6.4). Alternative solutions involve the use of diffuse luminaires that have large surface areas, luminaires with reduced luminance in the critical direction, or indirect lighting installations.

DESIGNING FOR SCREEN-BASED EQUIPMENT

Although microimage technology has considerable potential for positively changing office working conditions, over the past few years, operator complaints involving the use of screen-based equipment in the workplace have been increasing. Determining the causes of these complaints is a difficult task because such technology is a relatively new addition to the workplace and because the complaints may be a result of many interacting and compounding factors. The workspace's ambient illuminance and the operators' visual capabilities are only two of the many factors that influence the ease and speed of performance of such systems, as shown in Fig. 6.8. Complaints of eyestrain and fatigue when performing close visual tasks with SBEs have become more common with

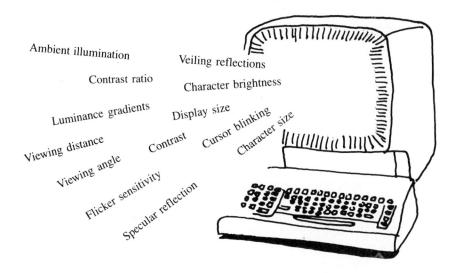

Ambient illumination

Veiling reflections

Contrast ratio

Character brightness

Luminance gradients

Display size

Viewing distance

Contrast

Cursor blinking

Character size

Viewing angle

Flicker sensitivity

Specular reflection

The workroom's ambient illumination and the operator's ophthalmologic status are but two of the many factors influencing the easiness of reading CRT pictures.

Reflected glare

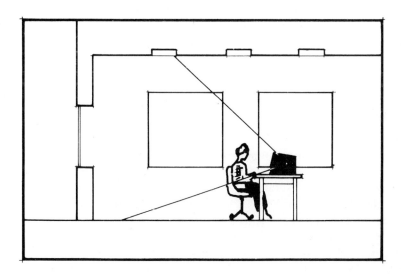

Area of glare sources

FIGURE 6.8.
Factors influencing visual performance with screen-based equipment.

the increased use of this type of equipment in office work routines. Research on visual fatigue has as yet not resulted in any realistic and useful measures. The major obstacle in sorting out the causes of visual fatigue is that it cannot be separated by measurements from general fatigue.

Strain due to accommodation and convergence, however, can be predicted. With age, another factor becomes important, presbyopia (a decrease in the ability to accommodate), whereby additional optical power is needed. The correction needed increases with age and with the nearness of the visual task to the eyes.

Maintaining a proper muscular balance for eye convergence to enable an object on the screen to be perceived as a single image may be further complicated if there are refractive errors such as hyperopia (long sight), myopia (short sight), or astigmatism. Performing a visual task that requires binocular fusion has been found to cause a temporary change in the intraocular and extraocular muscles that constitutes, in effect, a temporary myopia. However, the most common problems are due to either the excessive difference between the dark screen and the brightness of the visual field or reflections of high luminances in the immediate environment on the surface of the screen. Guidance on how to avoid these problems can be found in most National Codes (Standards Association of Australia 1987).

Only the physical aspects of the workplace, including the accommodation range of the operator, the inherent properties of the screen-based equipment, and the visual environment that surrounds the equipment, will be discussed here. Assuming that such physical attributes as crowded work areas, poor air quality, lack of functional furniture (especially seating), and noise pollution do not significantly contribute to visual problems, the first component of the visual environment is the SBE itself, and here several factors are controllable by the manufacturer, such as the form of the letters and how well they reproduce typed print, letter size, spacing between letters and lines, instabilities such as flicker, color of print and contrast in amount and choice of dark or light characters, background luminance, the visual properties of the keyboard, and screen material composition, including reflection controls and the geometric relationship between the keyboard and screen.

The second component is the operator's visual system. This is not controllable by either the equipment manufacturer or designer. However, the accommodation range or range of distances must be considered in the visual environment of the equipment.

The third component of the visual environment is the surrounding lighting. The major aspects of this component are illuminance levels and luminance distribution (especially on horizontal and vertical surfaces), the location of luminaires with respect to the equipment, color, flicker, lighting controls, and window management.

Illuminance

Although the most common way to achieve good visibility for normal office tasks is to have a reasonably high level of illuminance, for SBE readability, a low level of illuminance promotes greater visibility. These different needs are clearly shown in the *U.S. Lighting Handbook* (IES 1972), in which the recommended level of illuminance is 700 to 1000 lx for general offices and 300 to 500 lx for microimage reading areas. Several studies support such recommendations (Seppala and Turunen 1974, Lee and Buck 1975). However, it is not enough to reduce the level of illuminance. The luminances in the immediate surroundings and the background to the SBE are important, since the raised gaze of operators makes them susceptible to glare from windows and luminaires and high reflectances from adjacent wall surfaces.

Luminances and Luminance Ratios

It is necessary here to separate preference from performance. DesRosiers (1976) has studied the effect of image-surround brightness ratios and has found that when the ratio is high, attention is greater and visual comfort is lower. There is therefore a need to separate one from the other.

A common mistake of operators is excessive screen brightness (Dierich and Hilbert 1975). This may seem necessary to overcome veiling reflections on the screen caused by the location of the equipment in relation to windows or luminaires (Hultgren and Knave 1974). However, excessive screen brightness results in a task-surround brightness ratio of say 100:1 compared to an optimal range of 10:1 to 1:1 and has led to the recommendation that negative images be used in conjunction with rear-projection readers equipped with tinted screens (Criteria 1974).

Standards for visual display units must deal with luminance distribution in a more realistic way than that specified for normal office tasks. The recommended luminance ratios between a task, its immediate surroundings, and the background is 10:3:1. This is inappropriate for equipment operators because the screen is darker than the surroundings (ratio of 1:5 rather than 5:1). The rules might be valid when reading a moderately bright book page, but they are not valid for viewing small or large areas nor for low or high luminances. Recommendations on brightness ratios must be supplemented with methods for the assessment of visual discomfort in situations where these recommendations have been violated. An operator's eyes must switch between a dark screen and bright copy, which is a situation that also must be evaluated in terms of visual discomfort.

Viewing Distance

The visual angle subtended by the characters on the screen is seldom the critical factor in deciding the viewing distance (Ostberg 1976). Because of the physical dimensions of the screen-based equipment and the office furniture and the necessity to take notes, a realistic viewing range would be 50 to 100 cm. For example, in the performance of close visual work, there is, in addition to the need to maintain convergence (focus), a need to concentrate, and thus a concomitant mental effort is required. Some abnormalities in the visual mechanism that have been shown to give rise to visual fatigue are refractive errors, convergence insufficiency, and oculomotor imbalance. When looking at objects closer than 6 m, the eyes are converging, and the nearest convergence point is around 10 cm in front of the eyes. Failure to maintain or achieve this degree of convergence without undue effort is known as "convergence insufficiency" which can be a major cause of headaches and eyestrain in persons between 15 and 40 years of age (Mahto 1972).

The selected viewing distance in many instances is a result of attempts to establish a luminance balance between the operator, the screen, and the surroundings. A proper luminance balance has a great influence in the reduction of eyestrain. In some cases, it is the index of blur (rather than the character size and contrast) that determines readability (Fry 1963). It has been suggested that visual acuity may increase when the eyes are out of focus (Green and Campbell 1965). The distance between the operator and the screen cannot be adjusted easily, and research to determine the optimum viewing distance in relation to type size, screen size, and luminance ratios would be extremely valuable.

Flicker

There is also possible eye strain from visible flicker in the screen-based equipment environment. Two flickering visual sources are the environmental lighting (usually 50- or 60-Hz fluorescent lamps) and the SBE characters, which are refreshed at near the line frequency. Although the effects of flicker are not understood, it is a possible source of quality reduction in the visual environment by the distraction produced and systematic variations in the electrical activity of the brain. From experiments with flickering light, it is known that about 40 percent of all operators will find a flicker frequency below 60 Hz uncomfortable (Berman et al. 1983).

The operator of screen-based equipment is confronted with SBE phosphor with its inherent refresh rate and the general fluorescent lighting reflected from the screen. Owing to differences in frequency, the two sources of flicker can produce low-frequency beats in the net light intensity presented to the SBE operator.

Experimental studies relating environmental lighting and flicker to visual fatigue show that a decrease in contrast sensitivity can be attributed to the combined presence of flickering illumination and the properties of the SBE (Berman et al. 1983). The results indicate that operators suffer under 69-Hz fluorescent illumination as compared with nonflickering light sources. Accuracy decreases and the average time required for the task increases. However, more experiments are needed to provide statistically significant results and to determine if subjective discomfort is increased by the combination of SBE and standard low-frequency lighting.

However, solutions are possible. Environmental lighting can be provided at high frequency in the range of 30,000 Hz by using high-frequency ballasts, and the refresh rate can be increased to a higher value, perhaps in the range of 300 Hz, which is beyond the critical fusion frequency. This is especially important in instances of negative contrast, where dark letters have a large, bright background and this flickering bright area stimulates the peripheral vision, which is more sensitive to flicker.

Care should be taken to ensure that improvement in one area does not degrade another. For example, in adjusting the refresh rate, one could use phosphors with longer decay times, but this could cause increased blur at the edges of the characters. Another example of this kind would involve decreasing luminaire glare by having vertical screens. However, the accommodation distance would then change as operators read down the screens, and this has the possibility of reducing visual performance.

Recommendations for the Screen-Based Equipment Lighting Environment

1. General room illuminance should be low, about 300 lx, and each workplace should be provided with supplementary local lighting so that 500 lx is provided at work stations.
2. Unless windows are shaded with drapes or blinds, the great difference in luminance between the SBE screen and the daylit surfaces in the room would indicate that a good solution would be to place the terminals far away from windows or in an electrically lit environment.
3. Reflections from the bright sky through windows can be considerably reduced by placing comparatively low screens around the workplaces.
4. Screening also can assist in eliminating reflections from ceiling lighting installations.
5. High luminances from ceiling mounted luminaires can be reduced by the use of luminaires with sharp cut-off batwing distributions with semi-specular louver reflectors or the use of uplighting to produce large, relatively low luminance sources of indirect light.

INTEGRATION OF DAYLIGHT WITH ELECTRIC LIGHT

The integration of daylight with electric light in buildings during daytime hours is a key element in energy conservation. However, it is necessary to understand the integration process in order to achieve energy-conserving results in the context of a comfortable visual environment. An optimal solution depends on the type of building, its activity patterns, and the local visual environment; such a solution can be achieved only through simultaneous design of the daylight and electric light systems.

A number of factors need to be considered with regard to the design of an integrated system. First, electric lighting can be used to bring daylight illuminance up to adequate levels for efficient and comfortable vision. The second requirement is to reduce the extreme variations in illuminance and surface luminance that occur with distance from the window wall in sidelit buildings. The high luminance of the sky causes a state of adaptation in the observer's eye that produces an impression of gloom at the back of a room. One objective of using electric lighting in the daytime is to reduce this contrast, and in theory, the illuminance supplement provided should vary with changes in sky illuminance. However, the eye can tolerate illuminance and luminance differences within a limited range.

It should be noted that an interior can appear underlit or gloomy even though integrated horizontal illuminances have been provided. What is important is not horizontal illuminance but the surface luminances of the space, which are a result of the illumination of walls, ceiling, and work plane. The windows will establish a high adaptation luminance against which other surface luminances will be judged.

Originally, the concept of integration was based on the assumption that daylight was the major light source (Building Research Station, 1960). It also was assumed that room depths were small and required illuminance levels were low. Since then, recommended illuminance levels have increased and deeper and larger buildings have been built, requiring extension of the integration concept. It has been suggested (Ne'eman, 1984) that integration be extended to various types and depths of buildings with either daylight or electric light being the dominant light source in a single space, or in adjacent spaces, i.e., interspace integration (Figs. 6.9 to 6.11).

FIGURE 6.9.
Single space integration with daylight dominant. (After Ne'eman 1984.)

FIGURE 6.10.
Single space integration with electric light dominant. (After Ne'eman 1984.)

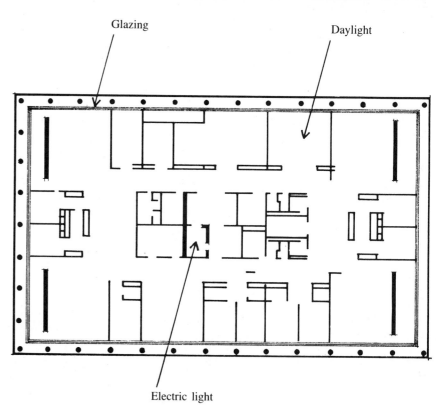

Glazing

Daylight

Electric light

FIGURE 6.11.
Interspace integration. A typical design of a deep-plan hospital with daylight dominant in peripheral areas. (After Ne'eman 1984.)

Qualitative Aspects of an Integrated System

Several factors must be taken into consideration in choosing light sources for supplementary electric lighting. First, the color qualities of the lamps must be compatible with available daylight. In a daylighted room with supplemental electric light, differences between their color appearances tend to be accentuated because they are seen simultaneously. If the differences are too great, complaints may arise, especially in situations where color judgments are critical. Second, the color rendering of the surfaces and objects in the space must be similar to that of daylight to ensure that color judgments are not confused. There may be some conflict here. At night, most people prefer a warm light of the traditional incandescent type, whereas during the daytime, a bluish light similar to daylight is more stimulating in a work situation. In buildings that have primarily daytime use patterns, this is not a problem. The color appearance and color-rendering properties of the supplemental electric light sources should be chosen in relation to the activities undertaken and the use of the room after daylight hours.

Daylight is continuously variable in spectral characteristics as well as in quantity and direction (Henderson and Hodgkins 1964). The light from a fluorescent lamp or discharge lamp does not have the continuous spectral characteristics of daylight, so that the colors of objects and surfaces may be distorted to some extent from their appearance under daylight.

In addition, the color appearance of light sources, particularly a discharge lamp or fluorescent lamp, may not necessarily indicate its ability to render colors. Fluorescent lamps with a color appearance and color-rendering characteristics similar to natural light are now available. Color-corrected discharge lamps also are available, although their color-rendering ability is more dissimilar to daylight. The daylight type of fluorescent lamp with a correlated color temperature of 6500 K has been found to be an adequate match with sky light visible through windows, although it is not the best choice in terms of color appearance for nighttime lighting.

The nonvisual and visual effects of the daylight type of fluorescent lamp have been studied recently. Results show that persons working in the daylight type of lighting show less discomfort than persons working under the standard white fluorescent lamp. In certain special areas where color-rendering properties may be linked to certain specific color judgments, e.g., hospitals, other types of lamps may be advisable.

Window Design in an Integrated System

The design and shape of windows in an integrated system will depend on the daylight penetration required and the need for an adequate view of the outside world. These two considerations can be in opposition, since the penetration of daylight can be better achieved with the use of tall or high windows and occupants generally prefer horizontal windows for a view of the landscape rather than an extensive view of the sky. In addition, considerable discomfort can result from views of the bright sky in the vicinity of the zenith. Therefore, an important aspect of daylighting requirements is the provision of adjustable shading to reduce the luminance of the sky when the sky is bright.

Design Recommendations

The following recommendations have been suggested by Ne'eman (1984).

Sidelit Interiors

1. Choose shape and size of windows according to the function of the interior, the needs of occupants, and the necessity for energy conservation.
2. Design the window system including necessary shading devices to provide a comfortable visual environment.
3. Design the electric lighting to meet the quality and illuminance recommended in the code, giving special consideration to illuminance on the window wall, color of light source, and modeling.
4. Arrange the grouping of the luminaires parallel to the window wall. In some cases, it may be an advantage to control each luminaire separately.
5. Considering the occupation and activity pattern in the interior, choose a control system to ensure that the lighting in each area can be maintained (see next section, "Lighting Control Systems").

Rooflighted Interiors

Rooflighting is generally designed to give at least a daylight factor of 5 percent and a uniformity of 0.7. This will provide an illuminance of 500 lx for 70 percent of daytime hours in a temperate climate such as experienced in England. Higher levels can be expected in hot humid and hot arid climates. Since this level is adequate for most visual tasks in industry, the daylight must be supplemented with electric light only during those times (30 percent) when the illuminance falls below the recommended design level.

1. Design rooflights to minimize direct entry of sunlight (through proper orientation and provision of shading devices).
2. Consider the pattern of activity in the area, and design the electric lighting appropriately. If there is a wide variation in activity patterns, a number of different areas may need to be controlled separately; otherwise, the general lighting can be connected to a single control system.

In a sidelit room, for example, areas remote from the window wall receive little illuminance on horizontal working surfaces, but daylight on the vertical surface assists in modeling and counteracts any dull effect of light coming exclusively from above provided the reflectances of the vertical surfaces are reasonably high.

ELECTRIC LIGHTING CONTROL SYSTEMS

Many types of controls are now available, ranging from the simple cord-pull switch that operates a single luminaire to photoelectrical dimming controls that ensure that as the daylight fades the total illuminance at a given point on the working plane will not fall below the recommended design value (Crisp 1978; Matsuura 1979; Rubinstein 1984). However, whatever type of control is used, design methods and guidance must incorporate what we know about the switching behavior of building occupants in response to available daylight, particularly as manifested in their use of electric lighting (Crisp 1978). This approach moves the consideration of daylighting into the realm of the electric lighting designer and allows for a more rational integration of the two forms of lighting. It has been suggested that the design of appropriate lighting controls, including those for users (occupants), should be related to occupancy patterns of space use and daylight availability in the space and that provision of such systems should be seen as "active" daylighting.

From the studies quoted earlier, it is evident that daylight is used not merely because it is available, but rather because of its availability at the beginning of the working day or occupation period. It can be surmised, therefore, that any system that relies on subjective assessment is unlikely to succeed economically until switching arrangements can be made more localized and thus local switching can be related to the worst-lit daylit area. It is possible that a general strategy of centralized switching related to known occupancy patterns with convenient local switches available to occupants can be enhanced by the addition of automatic photoelectric controls or occupancy detector controls to provide a range of potential uses. Such applications could be related essentially to the type of occupation of the spaces under consideration (Fig. 6.12).

Controls and Control Options

The operation of controls must be consistent with the visual performance requirements and the perceived needs of the building occupants, and the control system design must be appropriate for the overall lighting strategy for the space. The relationship between task illuminance and general lighting also must be considered, and the lighting control system should be cost-effective, so that the system meets minimum cost-recovery criteria.

A lighting control system has three components:

1. A light detector, e.g., a cadmium sulfide cell that translates light energy into change of electrical resistance.

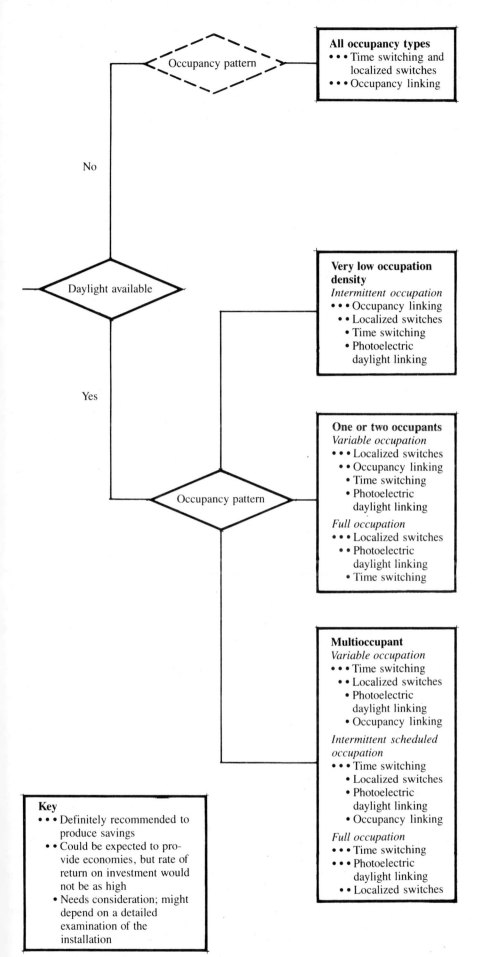

Occupancy pattern

All occupancy types
- ••• Time switching and localized switches
- ••• Occupancy linking

No

Daylight available

Yes

Very low occupation density
Intermittent occupation
- ••• Occupancy linking
- •• Localized switches
- • Time switching
- • Photoelectric daylight linking

Occupancy pattern

One or two occupants
Variable occupation
- ••• Localized switches
- •• Occupancy linking
- • Time switching
- • Photoelectric daylight linking

Full occupation
- ••• Localized switches
- •• Photoelectric daylight linking
- • Time switching

Multioccupant
Variable occupation
- ••• Time switching
- •• Localized switches
- • Photoelectric daylight linking
- • Occupancy linking

Intermittent scheduled occupation
- ••• Time switching
- •• Localized switches
- • Photoelectric daylight linking
- • Occupancy linking

Full occupation
- ••• Time switching
- ••• Photoelectric daylight linking
- •• Localized switches

Key
- ••• Definitely recommended to produce savings
- •• Could be expected to provide economies, but rate of return on investment would not be as high
- • Needs consideration; might depend on a detailed examination of the installation

FIGURE 6.12.
Control options as related to occupancy patterns. (After Ne'eman 1984.)

2. A control logic that compares the instantaneous measured value to some present design criteria.
3. A control device such as an electronics package or relay that acts to control the light output based on signals from the control unit.

In any system, one or more of these units may be combined, but the functional requirements will be present in all systems. However, where a group of luminaires is controlled by a single sensor, it is essential that different regions in the space or zone be illuminated in a similar way, with little variation longitudinally across that space. If the space is divided into individual offices, each with its own drapes or blinds, a single sensor located in one of the offices may produce very misleading control signals.

The appropriate location of a sensor controlling a single zone is the subject of continuing research. A sensor located in an interior space can be mounted on the wall or ceiling or placed at a task location. Research results to date suggest that a ceiling-mounted sensor with a limited field of view and one that avoids a direct view of a window is a preferred design.

The sensor needs to see approximately equal quantities of natural and electric light and be free from reflection. In practice, a time response on the order of 30 seconds is sufficiently long to damp out any feedback problem between adjacent sensors and sufficiently short to respond to most of the time variations in daylight levels.

Subjective Problems

However, most existing ceiling-mounted sensors, which respond indirectly to changes in horizontal illuminance in a space, are not compatible with occupants' perceptions of lighting quantity and quality, which are based on wall luminance and the perceived brightness of other objects in the space. This can result in a situation in which the desired illuminance is maintained on a horizontal work plane while at the same time relatively large changes in illuminance distribution occur around the space. Furthermore, as stated previously, horizontal illuminance is known to be a poor indication of visual performance.

The spatial distribution of that illuminance and the resulting luminances and task contrast as perceived by occupants are important issues. A system that drives electric lighting systems so that equivalent visibility is maintained at all times would produce improved performance and might justify the increase in cost and complexity. The time delay as the level of natural light changes should be at least 10 seconds, and preferably longer, so that the changes in electric lighting are virtually imperceptible. Although the eye and the brain can be considered to be a sensor-logic system (for manually operated lighting, so that the desired visual performance requirements can be met precisely in terms of the time required), humans are fallible, and most will not spend their time in a working environment worrying about whether lights should be turned on or off, as suggested by recent field surveys. Thus savings from manually operated systems would be less than those from automatically operated systems. However, manually operated systems are likely candidates for one- or two-person offices, where the lighting to be controlled is for a personal task area.

The level to which the lights can be dimmed is dependent on a number of factors. An installation for an office or other critical work area will be satisfactory to a minimum of 40 percent power. Installations that have been dimmed to zero or near zero have been found unacceptable aesthetically, despite the fact that light-meter readings have demonstrated that the amount of available light has been far more than required. The fact that people can look up and see that the lights are dim is sufficient to incur criticism.

On the other hand, there are many automatically dimmed indoor tennis courts in Europe that turn lights completely off when the incoming daylight reaches twice the required minimum acceptable light level. It appears from field surveys that, provided the system is part of the original structure, there are no complaints. There are also many warehouse and storage types of operations which are noncritical in which complete shutdown has been successful.

EVALUATION OF LIGHT IN INTERIORS

Subjective Appraisal of the Interior Environment

Emphasis in lighting literature has been directed toward considerations of objective measurement. Reference is generally not made to individual impressions, but to standards of physical measurement. Recommendations of lighting levels in lux and power consumption in watts are examples of this approach. Other criteria are now being recognized. Subjective studies of the requirements for preferred and efficient lighting (Rowlands et al. 1983) and surveys of office worker response to lighting and daylighting issues in workspace environments (Ne'eman, Sweitzer, and Vine 1984) are encouraging a more integrated approach to field appraisals.

It is evident that lighting in a space contributes to the occupants' impressions of that space, and these impressions can be either positive or negative depending on the degree of discomfort or satisfaction experienced. This section considers some results from available techniques for evaluating subjective responses to lighting and analyzes available information from field studies.

Methods of Subjective Appraisal

One method to obtain a much clearer understanding of the ways in which different lighting conditions influence people's attitudes has been demonstrated by Kimmel and Blasdel (1973) using semantic differential scaling (multidimensional scaling) and using rating scales. The sets of scales are derived from lists of descriptive words that apply to the built environment, for example, glare and color of light. An attempt is made to understand the psychological "sense" that people make of the physical environment by using a statistical technique called "factor analysis" to generate the clusters of adjectives used in similar ways by those who responded to the environment using the scales (Canter 1974; Harman 1970). By using this technique, an enormous quantity of data can be collected on rating scales and a minimum number of underlying independent dimensions to explain the patterns of differences identified by multi-dimension scaling (Hawkes 1970). Advice on the selection of rating scales for assessing the lighted environment is available from Kuller (1972).

Although many factors may be generated as variables (or pairs of adjectives in the case of semantic differential studies), it is usual to reduce the variables to a relatively small number that are similar in conceptual framework. For example, in a study by Flynn et al. (1973), 96 observers experienced a conference room lighted in six different ways. The observers' responses were collected on rating scales and then factor analysis was applied to the data (Harman 1970, Canter 1974).

In addition to knowledge of these procedures, some imagination and initiative are required in the selection of the range of interiors to be investigated, the type of lighting studied, and the scales to be used. It also should be noted that lighting is but one aspect of the physical environment. Since the architect is concerned with all the functional aspects of the building, there is a need to study the relative importance people give to different facets of the environment.

Recent research has indicated that worker satisfaction with office environments is associated with job performance (Harris et al. 1978). Among the physical environmental conditions considered important are lighting, outside view, and the design of the workspace (Wineman 1982). In a recent survey by Ne'eman et al. (1984), office workers were asked to rate the importance of selected features as described in Fig. 6.1. Over 90 percent of the office workers felt that the amount and quality of light for reading and writing were very important. Moreover, the amount and quality of light for computing and filing and the ability to control glare from light sources by drapes or blinds also were considered of some importance by most of the respondents.

One objective of the above research project was to see how questions of importance and satisfaction were related to each other, and by examining questions and conducting factor analysis, distinct groups of variables (factors) were found. The first factor (thermal controls) contained variables concerned with access to controls; the third factor (light for specialized tasks) was concerned with the amount and quality of light for specialized tasks such as computing and filing; and the fourth factor (sound controls) contained variables concerned with the control of sounds from outside and within the building. Other factors included were light for general tasks (relating to the amount and quality of light) and shading devices (relating to the control of venetian blinds).

The purpose of the factor analysis method is to isolate office workers' impressions of their working environment. One of the most important findings is the negative relationship between how people evaluate the importance and satisfaction of their working environment. It appears that workers who are least satisfied with features of their workspace consider these features to be very important, and vice versa. It also has been found that floor location and window orientation are statistically significant correlates of many attitudes toward workspace features.

The above project was one of the first to analyze office worker responses to lighting controls. Improvements in other environmental conditions can be addressed in the types of questions asked, particularly as regards the roles of office workers and the status of job satisfaction. Until further work is carried out in this area, no positive conclusions can be drawn about the relationship between lighting and the attitudes and behavior of office workers. However, from the results achieved, it is evident that improving occupant satisfaction and productivity should receive as much, if not more, attention than energy conservation.

User Responses to Windows and Daylighting

Because windows can lead to excessive heat gain in a building in summer and equally undesirable heat loss in winter, their use and size have been questioned. The contention in recent years that the need for natural light and ventilation is no longer relevant with the increasing use of air-conditioning and electric lighting in buildings has resulted in the building envelope becoming a barrier to the external world. In such buildings, the use of windows appears unnecessary. However, it is debatable whether total internal environmental control has resulted in any significant increases in human comfort. It is also doubtful whether ignoring the external environment rather than utilizing its assets to limit nonrenewable energy use has resulted in a substantial increase in energy savings.

The variations in the natural environment, which have been eliminated by this approach to building design, are now being considered as providing an improvement in the quality of the interior environment (Chaps. 1 and 5). It is therefore appropriate to regard the building envelope as a filter rather than a barrier to the external world and design the building envelope accordingly, taking into account the psychological needs of occupants.

Windowless Environments

Consideration of the range of reactions to windowless spaces has indicated that the smaller and more restricted a windowless space, the greater the reduction of freedom of movement, and the more repetitive and monotonous the task, the more unpleasant and oppressive the space will be. Such static and confined environments appear to generate tension, and this tension is accentuated by the absence of windows.

However, research results vary according to the function of the space. In schools it has been found that the absence of windows neither impairs nor improves pupil performance, although there is an absence of enthusiasm and some absenteeism from the windowless classrooms (Larson 1975). In factories, complaints of headaches and general depression have been reported, and problems with absenteeism and vandalism have increased. Plant (1970), in a review of existing literature, has noted from Czechoslovakian and Russian research that not only is absenteeism higher in windowless factories, but the occupants are more subject to sickness. These comments suggest dissatisfaction with windowless environments, although such dissatisfaction has not been confirmed by any experimental procedures. Windowless offices are less common than factories, and although relatively few studies have been assessed, there is a widespread opinion that people prefer windows in the office environment.

Another relationship that has been explored is that between lighting and the recovery rates of hospital patients. This research indicates that patients treated in windowless units have slower rates of recovery and greater post-operative complications than those patients treated in units with windows. It is considered that depression, fueled by a lack of contact with the outside world, may be the psychological parameter that causes these adverse effects (Wilson 1972).

Decreased productivity and inefficiency are often the result of depression and job dissatisfaction; each of these parameters reinforces the other. Wurtman (1975) suggests that parallel to the regulatory laws for food and drugs that currently exist there should be regulation of lighting environments to ensure that physical and psychological health is not threatened.

View

Evaluation of research on spaces with windows has resulted in the specification of window qualities that must be preserved. One of the most important is view. Although the need for contact with the outside world has been established, the desirable window characteristics to fulfill this need are less well understood. The basic requirements of area and shape have been researched in many countries. Among the studies that have dealt with this question are those of Markus (1967), Ne'eman and Hopkinson (1970), and Keighly (1973). All are in general agreement that office workers prefer to work by daylight and that minimum acceptable window size for 85 to 95 percent of office workers is approximately 30 percent of window-to-wall area. This recommendation is for temperate climates and is also compatible with window area required for optimal energy savings. It has been found that effective apertures (ratio of window-to-wall area multiplied by transmittance) of 0.1 to 0.2 (depending on climate and orientation) produce optimal energy savings. This infers, for example, a 30 percent window-to-wall area and a visible transmittance of 40 to 60 percent.

The preferred shape was found to be that of a horizontal aperture as opposed to a vertical one for restricted window area, the elevation being determined by the skyline of the view. The use of several small windows of different shapes was found to be particularly unsatisfactory. However, it should be noted that findings based on scale models, as in the preceding project, represent only one specific situation. Satisfaction may vary with viewer distance from the window, office layout, and variations in view.

Other functions also can influence these results. The thermal implications must be considered. For example, increasing glazing area to a high percentage of window-to-wall area (100 percent) can produce extensive over-heating problems near the windows unless suitable controls are provided. Although the use of solar-control glass preserves the view through the windows, such glass also reflects or absorbs a large portion of the visible spectrum and therefore reduces the ability of windows to admit daylight. The use of new advanced glazing materials such as low-E glazing will help solve this problem of light transmittance.

In some cases, because the glass is tinted, it also can change the spectral distribution of the light admitted. Research results, however, have suggested that, generally, adaptation can accommodate the color change, provided there is no clear glazing in the immediate proximity.

The recommendation, therefore, for temperate climates is for a window size between 20 and 40 percent of the window-to-wall area. Above the 40 percent level, special measures should be taken, such as the use of low-E glazing, to control the incidence of thermal discomfort. Climate will have a substantial influence on this recommendation. In tropical climates, the dangers of overheating will place greater restrictions on the admission of daylight and hence on window design.

Light-Source Preferences

Studies on light-source preferences by Markus (1967), Wells (1965), and Manning, (1965) have found high percentages of occupants preferring daylight to electric light. Although Wells's study demonstrated that clerical personnel seated beyond 6 m from windows overestimated the proportion of daylight in the interior space, the personnel still preferred daylight as the illuminant. Other studies have investigated the impact of light on behavior. In a review of the effects of lighting on office environments, for example, Wineman (1982) cited several studies that found that the use of full-spectrum lighting can improve performance and reduce fatigue. Research therefore suggests that workers desire day-lighting independent of its contribution to task visibility.

Sunlight creates a number of psychological reactions apparently related to its warmth and brightness. It has been found that the desire for sunlight depends on the kinds of activities performed by a building's occupants. In a study of a 12-story building it was found that 86 percent of the occupants preferred sunlight in the office all year, but those who sat nearest the windows were less enthusiastic (Markus 1967). In other studies there was also a strong desire for sunlight, with underlying overtones with regard to visual and thermal discomfort. The complaints depended on the facilities available for sun control. Few problems were experienced when appropriate sun controls were provided and the air-conditioning system was effective.

Then again, the desire for sunlight may be lessened by the type of building and the kinds of activities performed by the occupants. A survey by Longmore and Ne'eman (1973) found that preferences for sunshine varied from 90 percent for dwelling residents and hospital patients, to 73 percent for office personnel, to only 42 percent for school occupants.

The desire for sunlight is also influenced by climatic and regional considerations. While there is a strong desire for daylight and sunlight in temperate climates, these conclusions are unlikely to be applicable in tropical climates. This balancing process indicates the difficulty in specifying window size on the basis of the provision of daylight and sunlight.

It is evident that windows do perform desirable functions for people in buildings. However, substantive conclusions are limited by the paucity of research. Although windows provide psychological benefits such as view, stimulation, sunshine, and daylight, the relative importance of these benefits requires further research. Although work has been carried out on the appropriate size, shape, and location of windows in temperate climates, these findings may not necessarily apply to other cultures and more extreme climates.

SUMMARY

Both the functional and qualitative requirements of lighting need to be considered when designing buildings. An analysis of the task is necessary to determine appropriate illuminances on horizontal or vertical work planes, and consideration must be given to the avoidance of disability or discomfort glare.

In lighting design, it is difficult to isolate and measure the effects of specific aspects of a lighting system on productivity. One approach that has some merit uses the measure of visibility level to determine ease of seeing, and this is based on contrast reduction. However, this measure is subject to individual differences.

The directionality of light can have an adverse effect on visual performance. Wrongly directed light will cause a reduction in contrast in the task area and hence a reduction in visibility from reflected glare. Discomfort rather than reduced visibility from glare also can affect performance by causing fatigue, particularly from very bright light sources over time. This may change the adaptation level of the eye or cause a distraction from the task.

Although electric light sources can meet all of a building's needs and daylight cannot, it has been demonstrated by field studies on occupant responses to their environment that daylight is the preferred light source. In the integration of daylight and electric light, supplementary electric lighting not only provides additional working light, but also reduces window glare by reducing the contrast between the sky seen through the window and the interior surfaces of the room. In integration, design consideration should be given to the color appearance and color-rendering ability of the electric light source and the role of the ceiling in the integration process.

Many types of electric controls are now available to switch off electric lights when daylight levels are sufficient. Surveys of switching behavior have shown that to a large extent the type of control used is related to occupancy patterns. In large offices or work spaces, the only effective answer is to use an automatic system.

It is possible to quantify visual comfort by such objective measures as speed of performance, but ultimately, such measures have to be calibrated against a subjective assessment of the lighting conditions. A suitable method involves multidimensional scaling together with factor analysis to statistically analyze the relationship and reduce the number of variables. However, it is essential in using any subjective appraisal method to evaluate lighting to take into account the other variables that are relevant to the work situation, and to include these parameters in questionnaires.

ACKNOWLEDGMENT

Part of the work on the integration of daylight with electric light was supported by the Assistant Secretary for Conservation and Renewable Energy, Office of Building Energy Research and Development, Building Systems Division of the U.S. Department of Energy, U.S.A.

REFERENCES

Berman, S. M., Greenhouse, D. S., Bailey, I. L., and Bradley, A. 1983. Experimental Studies Relating Environmental Lighting and Flicker to Visual Fatigue in VDT Operators. LBL Report No. LBL-16299. Lawrence Berkeley Laboratory, Berkeley, Calif.

Boyce, P. R. 1981. *Human Factors in Lighting.* Applied Science, London.

Building Research Station. 1960. The Permanent Supplementary Artificial Lighting of Interiors (PSALI). Building Research Station Digest No 135, H.M.S.O., London.

Canter, D. A. 1974. *Psychology for Architects.* Applied Science, London.

Clear, R., and Berman, S. 1980. Cost-Effective Visibility-Based Design Procedures for General Office Lighting. LBL Report No. LBL-11863. Lawrence Berkeley Laboratory, Berkeley, Calif.

Commission Internationale de L'Eclairage (CIE). 1980. An Analytical Model for Describing the Influence of Lighting Parameters on Visual Performance. CIE Publication 19/2.

Commission Internationale de L'Eclairage (CIE). 1986. Guide on Interior Electric Lighting Publication. CIE No 29/2 (TC 4.1).

Commission Internationale de L'Eclairage 1989. Guide on Daylighting of Building Interiors. Final Draft.

Crisp, V. H. C. 1978. The light switch in buildings. *Light. Res. Technol.* 10: 2, 69-82.

Criteria for the Procurement and Use of Microforms and Related Equipment by the Libraries of the California State University and Colleges, Los Angeles. 1974.

Cuttle, C. C., Valentine, W. B., Lynes, J. A., and Burt, W. 1967. Beyond the working plane. *Proc. CIE,* 471-482.

DesRosiers, E. V. 1976. The Effect of Image/Surround Brightness Contrast Ratios on Student Preference, Attention, Visual Comfort, and Visual Fatigue. Ph.D. thesis, Boston University School of Education.

Dierich E., and Hibert, D. 1975. Arbeitswissenschaftliche Aspekte bei der Gestaltung eines, Mikrofilm- Lesearbeitsplatzes. *Informatik* 22: 40-43.

Florence, N. 1979. The energy effectiveness of task-oriented lighting systems. *Light. Des. Appl.* Vol 9: pp. 28-39.

Flynn, J. E., and Spencer, T. J. 1977. The effects of light source color on user impression and satisfaction. *J. Illum. Eng. Soc.* 6: 167-175.

Flynn, J. E., Spencer, T. J., Martyniuck, O., and Hendrick, C. 1973. Interim study of procedures for investigating the effect of light on people and behavior. *J. Illum. Eng. Soc.* 3:87.

Fry, G. A. 1963. The Relation of Blur and Grain to the Upper Limit of Useful Magnification. In Bennette, E., Degan, J., and Spiegel, J. (Eds.), *Human Factors in Technology.* McGraw-Hill, New York.

Green, D. G., and Campbell, F. W. 1965. Effect of focus on the visual response to a sinusoidally modulated spatial stimulus. *J. Opt. Soc. Am.* 55: 1154-1157.

Harman, H. M. 1970. *Modern Factor Analysis.* University of Chicago Press, Chicago.

Harris, L., et al. 1978. *The Steelcase National Study of Office Environments: Do They Work:* Steelcase, Inc., Grand Rapids, Mich.

Hawkes, R. J. 1970. Multi-dimensional scaling: A method for environmental studies, *Building,* June 19, p. 69.

Henderson, S. T., and Hodgkiss, D. 1964. The spectral energy distribution of daylight. *Br. J. Appl. Phys.* 155. 15: 947-952.

Hultgren, G. V., and Knave, B. 1974. Discomfort glare and disturbances from light reflection in an office environment with CRT display terminals. *Appl. Ergon.* Vol. 5: 2-8.

Illuminating Engineering Society of North America. 1972. *IES Lighting Handbook,* 5th Ed. IES, New York.

Keighly, E. C. 1973. Visual requirements and reduced fenestration in office buildings — A study in window shape, *Building Science* 8: 311.

Kimmel, P. S., and Blasdel, H. E. 1973. Multidimensional scaling of the luminous environment. *J. Illum. Eng. Soc.* 2: 113-120.

Kuller, R. 1972. A Semantic Model for Describing the Perceived Environment. National Swedish Building Research, D12, Stockholm.

Larson, C. T. 1975. *The Effect of Windowless Classrooms on Elementary School Children.* Architectural Research Laboratory, Department of Architecture, University of Michigan.

Lee, D. R., and Buck, J. R. 1975. The effect of screen angle and luminance on microform reaching. *Hum. Factors* 17: 461-469.

Mahto, R. S. 1972. Eyestrain from convergence insufficiency. *Br. Med. J.* 2: 564-565.

Manning, P. 1965. *Office Design: A Study of Environment,* Pilkington Research Unit, Liverpool University, Liverpool.

Markus, T. A. 1967. *The Significance of Sunshine and View for Office Workers,* in Hopkinson, R. G. (Ed.), *Sunlight and Buildings.* Boewcentrum International, Rotterdam.

Matsuura, K. 1979. Turning-off line in perimeter areas for saving lighting energy in side-lit offices. *Energy Build.* 2: 19-26.

Ne'eman, E. 1984. A comprehensive approach to the integration of daylight and electric light in buildings. *Energy Build.* 6: 97-108.

Ne'eman, E. and Hopkinson R. G. 1970. Critical minimum acceptable window size, a study of window design and provision of view. *Ltg. Res. and Technol.* 2: 17.

Ne'eman, E., Sweitzer, G., and Vine. E. 1984. Office worker response to lighting and daylighting issues in workspace environments: A pilot survey. *Energy Build.* 6: 159-173.

Ostberg, O. 1976. Office Computerization in Sweden: Worker Participation, Workplace Design Considerations, and Visual Strain Abatement. In *Proceedings of NATO Advanced Study Institute on Man-Computer Interaction, Mati, Greece.*

Plant, G. G. H. 1970. The light of day. *Light and Lighting* 63: 292.

Rowlands, E., Low, D. L., McIntosh, R. M., and Mansfield, K. P. 1983. The Effect of Light Patterns on Subjective Preference. In *CIE Proceedings* (20th Session). CIE, Amsterdam.

Rubinstein, F. 1984. Photoelectric control of equi-illumination lighting systems. *Energy Build.* 6: 97-108.

Seppala, P., and Turunen, M. 1974. Ergonomics aspects of microfilm reading (in Finnish). *Tyoterveyslaitoksen Tuthimuksia* 87; abridged English version: Seppala, P. 1975. Visual fatigue in reading microfilms. *Agressologie* 16: 147-150.

Standards Association of Australia. 1987. *AS 2713 Lighting Screen Based Tasks,* Sydney, Australia.

Wells, B. W. P. 1965. Subjective responses to the lighting installation in a modern office building and their design implications. *Building Science* 1: 57.

Wilson, L. M. 1972. Intensive care delirium: The effect of outside deprivation in a windowless unit. *Arch. Internal. Medicine* 130: 225.

Wineman, J. D. 1982. Office design and evaluation: An overview. *Environ. Behav.* 14: 271-298.

Wurtman, R. J. 1975. The effect of light on man and other mammals. *Ann. Rev. Phys.* 37: 467-483.

7

Acoustic Design

Anita Lawrence

The two primary aspects of acoustics to be considered in most buildings are, first, the provision of acceptable noise levels and, second, the control of reverberant sound. As mentioned earlier, people interact with their acoustic environment, and the acceptability of sound levels depends very much on the type of occupancy of the building (and also, in some cases, on the time of day). In multitenancy buildings, it is important not to overdesign, i.e., not to specify noise levels that are too low, since a very low background level requires very good sound insulation between different parts of the building if intruding sounds are not to be perceived or to cause annoyance. Overdesign can be very costly. First, the building envelope may have to be considerably more massive than required otherwise, adding to the costs of foundations and materials generally. Second, if mechanical ventilation or air-conditioning is used, additional costs will be incurred because of the larger-sized ducts and extra sound attenuators that will be needed. All internal partitions (floors, walls, doors) dividing tenancies (or even dividing different parts of the same tenancy) also will need to be more massive and thus will be more expensive to construct and to support if very good sound insulation is required. As a rough rule of thumb, a doubling of surface density is required to improve sound insulation by 5 dB in the midfrequency range.

In this section, therefore, design criteria for sound levels and reverberation control and the acoustic performance of typical building elements and finishes will be discussed.

ACOUSTIC DESIGN CRITERIA: ACCEPTABLE AMBIENT SOUND LEVELS

There are three methods for determining acceptable sound levels inside buildings: (1) prevention of noise-induced hearing loss, (2) ease of speech communication, and (3) prevention of noise annoyance, including sleep disturbance.

Prevention of Noise-Induced Hearing Loss

Noise-induced hearing loss was discussed in Chap. 4 under Physiological and Psychological Effects of Noise (page 65). The "acceptable" sound level is one that will not cause a significant risk of permanent hearing loss in people regularly exposed to the noise. Usually such high sound levels occur only in industrial-type buildings and in machine and plant rooms of other buildings. However, in some commercial buildings with concentrations of high-speed printers and other equipment, care must be taken that excessively high noise levels are not emitted. Other circumstances in which the limit may be related to hearing conservation include airport buildings exposed to the noise from jet aircraft at very close range.

In several countries, legal limits have been set at 90 dB(A) for an 8-hour exposure in 24 hours; preferably the limit should be lowered to 85 or even 80 dB(A) in order to reduce the amount of hearing damage in the general population and to protect those who are more susceptible to hearing loss. It must be emphasized that such high levels are not related to human comfort, only to the prevention of physiological damage to hearing; it is very difficult to communicate aurally in such an environment.

Ease of Speech Communication

Speech communication conditions are related to the "signal-to-noise ratio," i.e., the sound level of the wanted signal, the speech, compared to the sound level of the ambient noise. The human voice, without electronic amplification, has limited acoustic power. The long-term average conversational voice level at 1 m is about 60 dB(A). Out of doors or in a room with very absorbent surfaces, the sound level of a voice will decrease by about 5 dB for every doubling of distance from the source. In a normal room, reflections from the walls, floor, and ceiling will increase the overall speech level above that which would be received out of doors at distances of a few meters from the speaker. However, if there are too many strong reflections, i.e., if the surfaces are hard and the sound is able to travel many times around the room, the individual speech phonemes will take too long to become inaudible and will act as part of the room "noise."

For good speech communication, therefore, short-delayed reflections with excess path lengths compared to the direct path from speaker to listener of no more than about 10 m are beneficial. Echoes may occur if the excess path lengths exceed about 17 m. Thus for auditoriums, lecture rooms, and so on, it is advantageous to provide sound-reflective surfaces around the speakers' position and to make the rest of the room's surfaces sound absorbent (Fig. 7.1).

In office buildings, because of the necessity for flexibility of layout, usually only the floor and ceiling can be relied on for sound absorption. So-called acoustic screens also should have sound-absorbent surfaces in open-plan buildings.

As a guide, noise levels of 50 to 55 dB(A) are considered to be the maximum acceptable for casual, close-range communication, such as in a drawing office. In general office areas, 40 dB(A) is recommended, and in board and conference rooms and private offices, the level should be about 30 dB(A) (SAA 2107 1987).

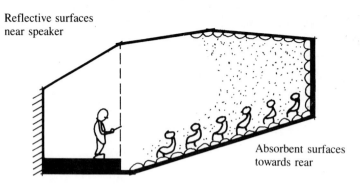

FIGURE 7.1.
Reverberation control in lecture rooms.

Reflective surfaces near speaker

Absorbent surfaces towards rear

Speech privacy is an important consideration in several types of buildings, including offices, consulting rooms, and so on. Lack of speech privacy can cause annoyance and discomfort, and this is a common problem in open-plan schemes. One of the factors that must be considered is the speech level or average vocal effort; this ranges from 60 dB(A) at 1 m from the speaker (as mentioned above) for conversational speech to 66 dB(A) for a raised voice, 72 dB(A) for a "stage" voice, and 78 dB(A) for shouting (SAA 2822 1985). Although the latter may not be expected on a regular basis in commercial buildings, if people are arguing, they may in fact shout, and it could be very embarrassing if this is heard by others. The size of the room or space in which the conversation is conducted together with the amount of absorption provided by furnishings and fittings will affect the overall "source" level; correction factors ranging from +12 to −4 dB(A) may need to be applied. The effective sound reduction between the source and the potential listeners in the adjoining room or space determines the speech levels transmitted, and again, correction factors are necessary, depending on the relative size of the common partition and receiving room floor area. Finally, whether or not privacy is achieved depends on the background sound level in the receiving room. The calculations should be carried out in both directions, reversing the roles of source and receiving rooms (Fig. 7.2).

The acoustic properties required for the dividing partitions and the like may be determined by the speech privacy requirement; if it is not possible to reduce the speech transmitted through such partitions sufficiently, speech privacy may be assisted by raising the background noise level in the receiving room. However, this can cause problems. Studies have shown that office workers are not comfortable if the background sound level is over about 45 to 48 dB(A); in addition, if the background sound levels are raised, speech levels in the area also will be raised to achieve adequate communication. In open-plan offices, it is extremely difficult to design both for good speech communication and acceptable speech privacy.

The required background sound levels may be provided through careful design of the air-conditioning system or, alternatively, by installing loudspeakers in the ceiling space. It is important to ensure an even distribution of background, or "masking," sound if it is artificially produced. If the loudness of the masking sound varies from place to place, people will become aware of it as they move through the room and are likely to object to its presence. There have been several studies regarding the most suitable spectrum for electronically produced masking noise, and there have been some claims made regarding the need for careful spectrum shaping. However, this view is not universally held. An interesting aspect of introduced masking noise is that it tends to be uniform in level and spectrum over time, unlike "natural" background noises from traffic, people, and so on. It has been suggested that some temporal variation in the introduced sound, e.g., such as the noise of waves breaking on a beach, may be more acceptable. However, this also can be challenged, since people may become more conscious of a sound with time-varying levels.

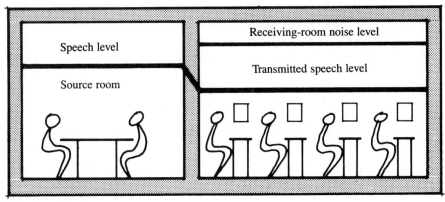

Good speech privacy

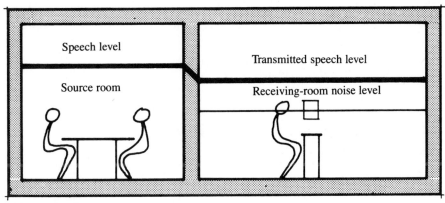

No speech privacy

FIGURE 7.2.
Speech privacy principles.

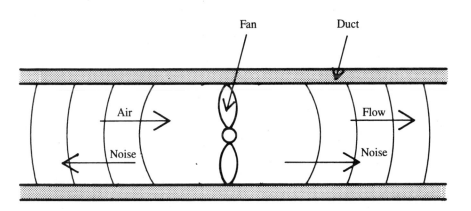

FIGURE 7.3.
Sound energy travels with and against air movement in a duct.

In concert halls, auditoriums, studios, and so on, ambient sound levels are critical, and as well as being specified as overall dB(A) levels, it may be necessary to limit the level in each one-third octave frequency band. This is done to prevent low-frequency problems in recording and replaying performances that have taken place in the room. It is the minimum expected performance level that determines the maximum acceptable ambient sound level, since no extraneous sound should be heard. In critical situations, it is necessary to control vibration levels as well. Major concert halls and auditoriums are usually air-conditioned, and the mechanical services are frequently the source of unacceptably high noise levels, particularly in the lower-frequency bands. It is essential that the building designer take great care in the placement of mechanical plant rooms and provide sufficient space for the plant and duct work, silencers, and so on. (It must also be remembered that sound travels equally well with and against the flow of air in a duct and thus return-air systems also must be as well designed acoustically as the supply side; see Fig. 7.3.)

Ambient sound levels of less than 15 to 25 dB(A) may be required in concert halls, studios, and so on depending on the standard of excellence appropriate to the building. It must be emphasized that the achievement of very low ambient sound levels requires considerable skill in design and execution, and specialist acoustic advice is essential. This should be sought at the preliminary design stage and should continue until the final commissioning of the building.

Prevention of Noise Annoyance, Including Sleep Disturbance

The third aspect of establishing acceptable noise levels in buildings is to prevent annoyance. This is appropriate when speech communication is not the principal requirement. The recommended levels depend on the different room functions. For example, in rooms where noisy activities take place, such as kitchens (both domestic and commercial), the acceptable ambient sound levels are higher than in rooms for relaxation or for sleeping. Levels of 45 to 50 dB(A) are acceptable in commercial kitchens, and levels from 35 to 40 dB(A) are acceptable in domestic kitchens. However, in bedrooms, the levels should preferably not exceed 25 to 30 dB(A), although as discussed in Chap. 4 (page 67) levels of up to 40 dB(A) do not appear to affect sleep significantly, at least for people who are accustomed to noise. In living rooms, levels of 30 to 35 dB(A) are acceptable.

ACOUSTIC DESIGN CRITERIA: REVERBERATION TIME

The control of reverberant sound is important in most buildings, although frequently it is only seriously considered in auditoriums, concert halls, and the like. The deleterious effect of long-delayed reflected sounds on speech communication has already been discussed. Any noise source such as an office machine, a hot-air hand dryer, or a compressor will create higher noise levels in a reverberant room than in one in which the surfaces and fittings are sound absorbent. This occurs because of the additional contributions from the reflected sound waves. Unfortunately, however, noise control through sound absorption is of limited effect. The direct sound components cannot be reduced by room surface treatment. If the surfaces are already absorbent, e.g., if the floor is carpeted and there are thick curtains and "acoustic" ceilings, little noise reduction will be gained by attempting to increase the sound absorption. However, if all surfaces are hard, e.g., tiled floors, masonry walls, and a concrete ceiling, some reduction in noise levels can be obtained by applying sound-absorptive materials. This is not always possible, because of hygiene requirements (Fig. 7.4).

In auditoriums and concert halls, control of reverberation time traditionally has been one of the major design considerations. However, over the last two decades or so it has become apparent that reverberation time is a fairly coarse measure of the quality of a room's acoustics, especially for music rooms. The sequence and relative sound levels of the early reflections, compared to the direct sound from source to listener, are of critical importance. As in the case of speech, reflected musical sounds that arrive within 30 ms of the direct sound are beneficial in increasing the overall loudness of the perceived signal. Those which arrive later than 50 ms after the direct sound may be perceived as undesirable echoes if they are sufficiently loud. However, if the sound decays evenly, the later reflections form part of the overall reverberation process and contribute to the overall musical response of the room (Fig. 7.5).

It is also considered desirable that listeners perceive that they are surrounded by sound. For this to occur, it is necessary that the signals received by the two ears, at least over the first 30 ms or or so, be different. This is achieved

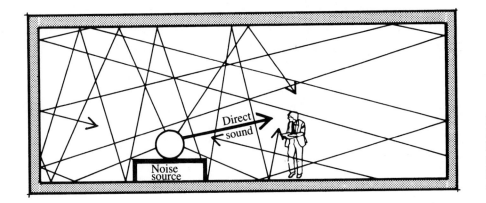

FIGURE 7.4.
Buildup of noise level in a reverberant enclosure. Noise levels will be higher if all surfaces are reflective; however, making surfaces absorbent has limited effect, since only a proportion of the energy is absorbed.

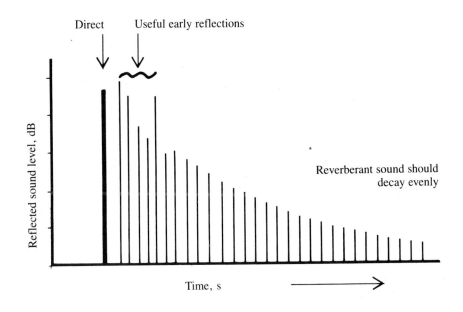

FIGURE 7.5.
Reflected sound sequence in an auditorium.

usually by providing early reflections from lateral directions; this is critically dependent on the overall size and shape of the room, however.

Although there will always be differences between the sound received close to the stage and that received at the middle and rear of an auditorium simply because the direct sound from a large, distributed source such as an orchestra will be incident from a much greater range of angles at the front, it is desirable to obtain an even distribution of the reverberant sound; i.e., a diffuse sound field is required. A "diffuse sound field" is one in which the sound is traveling with equal probability in all directions, and such a sound field can only be achieved by very careful analysis of room shape and distribution of reflective and absorbent surfaces. Computer models can assist in the design of such rooms. However, many assumptions must be made in computer studies, particularly in the low-frequency range, where the sound waves tend to be larger than the dimensions of room features. For important auditoriums, physical acoustic models can be very useful design aids. To be useful, however, they must be carefully constructed, and besides accurately modeling the room shape, they must have surfaces with the same absorbent/reflective properties at model-scale frequencies as the surfaces that will be used in the full-scale building. Sophisticated techniques are now available that will allow subjective assessments of the acoustic quality of speech and music by radiating such sounds into the model, recording them, and then replaying (using multispeed tape recorders to correctly model the scale frequencies) (Els and Blauert 1986).

Typical midfrequency reverberation times recommended for concert halls seating about 2000 people are of the order of 2 s. An increase of up to 150

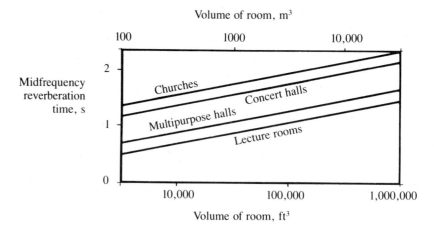

FIGURE 7.6.
Recommended reverberation times.

percent is acceptable in the lower frequencies. For smaller rooms, the reverberation times should be lower. Typically, in rooms for speech, reverberation times should not exceed about 0.8 s. The desirable reverberation times for studios depend very much on their purpose, and specialist advice should be sought (Fig. 7.6).

SELECTION OF INTERIOR BUILDING MATERIALS AND ELEMENTS

Airborne Sound Reduction Between Spaces

Sounds in the audio range of frequencies have wavelengths ranging from about 17 m to 17 mm. It is not surprising, therefore, to find that the acoustic characteristics of building elements are frequency dependent.

At low frequencies, a reasonably homogeneous element such as a masonry wall (or reinforced concrete floor slab) will tend to behave as a panel, and the attenuation of sound will be somewhat dependent on size and stiffness. The panel will have "resonant frequencies," i.e., preferred frequencies of vibration, and sound impinging on it at these frequencies will be transmitted to the other side with little attenuation. At higher frequencies, the panel behaves as if it were comprised of many small, independent elements. Here the attenuation depends on the acoustic impedance ratio between air and the material of which the panel is made. Since impedance is dependent on surface density and the speed of sound in the material (which is itself dependent on density and elasticity), the greater the mass of the material, the greater is the impedance ratio and the higher is the attenuation. This is called the "mass-law region," and an increase in attenuation of 6 dB per doubling of surface density is theoretically possible in this range. In addition, as frequency doubles, there is a theoretical increase in attenuation of 6 dB.

Another effect occurs at higher frequencies for some building elements. This is called the "coincidence effect." A solid will support flexural (bending) and shear waves, as well as the longitudinal sound waves that travel in a gas. Unlike longitudinal waves, which all travel at the same velocity no matter what the frequency, flexural wave velocity is frequency-dependent. At a certain angle of incidence of the airborne sound, it is possible for the flexural wavelength to coincide with the projected airborne wavelength at the same frequency—and, as at resonance, the sound will be transmitted with little attenuation (Fig. 7.7). For materials such as window glass, plywood, and gypsum board, the coincidence effect occurs in the important range between about 1000 and 3000 Hz. [The coincidence effect also occurs in masonry materials, but it is at a low frequency and is difficult to detect among the resonances (Lawrence 1970).]

If a building element consists of two (or more) leaves, such as a double-glazed partition, two other effects can occur. One is a "mass/air-mass resonance,"

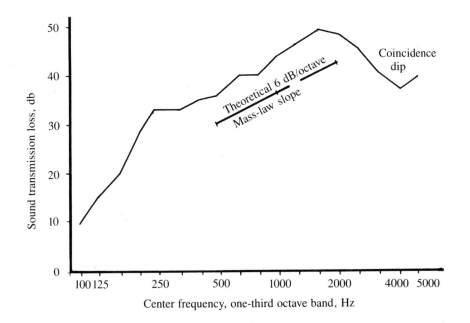

FIGURE 7.7.
Typical variations of airborne sound transmission loss with frequency for dry-wall construction.

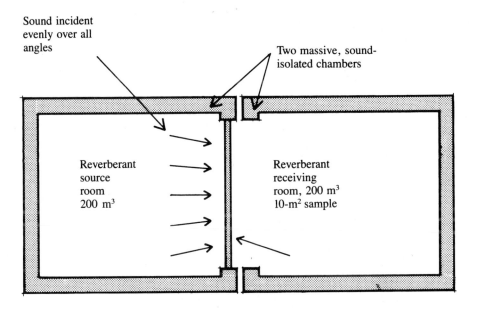

FIGURE 7.8.
Laboratory measurement for sound transmission loss.

in which the air between the leaves acts as a spring (usually a low-frequency effect), and the other is the formation of standing waves in the cavity itself. These effects tend to make the behavior of multileaf elements even more frequency-dependent than single-leaf elements.

It is not surprising, considering all the preceding factors, that it is difficult to predict the sound transmission characteristics of practical building elements from first principles. Many common materials and proprietary products, such as partitions and ceilings, have been measured to determine their sound transmission loss (STL) properties in acoustic laboratories. However, the transfer of these data to real buildings has some problems. For example, elements such as walls, floors, and so on are usually tested as 10-m² samples, and they are placed between two isolated, reverberant rooms, so that the sound energy can be assumed to be incident evenly over all angles (Fig. 7.8). In real buildings, the dimensions of the element may be considerably larger or smaller than the sample tested (which will affect the resonant frequencies), and the sound may be incident only over certain angles.

An even more important problem in translating laboratory performance data to field situations is that a number of building elements will usually be

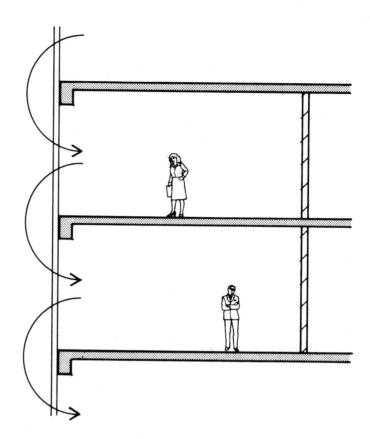

FIGURE 7.9.
Flanking sound transmission through lightweight facades.

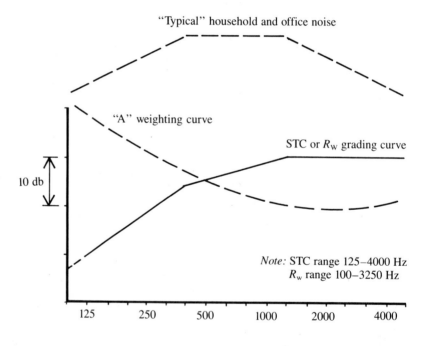

FIGURE 7.10.
Airborne sound transmission loss grading curves.

involved and there may be significant sound transmission paths through joints, around openable windows and doors, or by means of common floors or side walls. In the case of a lightweight facade, sound may travel by means of the facade, short-circuiting the dividing floors and walls between rooms (Fig. 7.9). If possible, acceptance testing of the acoustic performance of the building should be part of the commissioning procedure.

Although detailed information regarding the sound transmission loss for each one-third octave band from 100 or 125 Hz up to 4000 or 5000 Hz is usually available from laboratory tests, this is only of use if the frequency spectrum of the intruding noise is known. For general internal partitions it is usual to

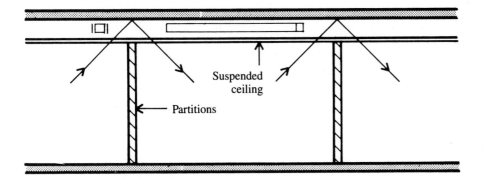

Structural slab

Suspended ceiling

Partitions

FIGURE 7.11.
Flanking sound transmission through suspended ceilings.

average the data, either arithmetically or by using a grading curve, to determine its sound transmission class (STC) or its weighted sound reduction index (R_w) (AS1276-1979, ISO 717/1, 1982). These curves were developed in the context of typical household or commercial noise (with a steeply rising low-frequency spectrum, then flat from about 400 to 1250 Hz, and then declining steeply in the high frequencies). However, with the advent of powerful domestic hifi systems, it may be debated whether the low-level components in the lower frequencies are still appropriate (Fig. 7.10).

Of particular concern in achieving the required acoustic performance in commercial buildings is the effect of suspended ceiling systems. They are frequently chosen to provide sound absorption, and they are usually installed across whole lettable floor areas. Subsequently, the space may be subdivided using partitions that only reach to the underside of the suspended ceiling. In this case, there is a flanking sound transmission path through the ceiling on one side that is reflected from the structural slab above and then down through the ceiling on the other side. It is now possible, in some cases, to obtain data for the total sound transmission reduction of a suspended ceiling system, including this flanking path (Fig. 7.11).

Structure-Borne Sound Transmission and Vibration

Structure-borne sound originates when part of the building is excited by an impact, e.g., a footstep on a concrete floor or a banging door. This type of sound energy may be transmitted for very many meters around the building with very little attenuation; it is sometimes difficult to trace the source. Vibration is usually the result of improper installation of machinery and plant.

As discussed in the preceding section, sound will be reflected if the impedances of adjoining materials are different. In the case of structure-borne sound, therefore, materials that are resilient and of low density will be effective in reducing the transmission of sound in such materials as brickwork and concrete. For machine vibration transmission reduction, careful selection of isolators is required. Vibration isolators must be chosen to have resonant frequencies lower than those of the machine to be isolated; otherwise, the vibration amplitude will be increased. If the machines are supported on suspended floors or similar structures, the resonant frequencies of the supporting structures themselves must be considered. Generally, vibration isolation is a specialized discipline, and advice should be sought. It is important that short-circuiting of vibration isolation does not occur through rigid connections of ancillary pipes, ducts, and so on.

For domestic floors, structure-borne sound isolation can be measured either in a laboratory or in situ (ISO 140, 1978). The resulting one-third octave band data can be compared to a standard grading curve to give the single-number quantity of impact sound insulation rating or the impact sound protection margin (M_i) (ISO 717/2, 1982). However, there has long been controversy regarding the method of measurement of impact sound isolation, since it is claimed that the standard tapping machine, which is used as the source of impact, does not represent footstep impacts adequately. There is some evidence to show that the standard method does not rank order thin floor coverings (such as sheet vinyl or cork) in the same way as people do, but since such coverings tend to be inadequate for impact sound reduction, this is perhaps not very important. Impact noise transmission through walls also can cause annoyance, but as yet there are no standard methods of measuring and rating the effect.

Another source of structure-borne sound in buildings arises from plumbing systems. Typical water velocities in domestic systems are such as to cause turbulent flow, which is inherently noisy. Although the pipes themselves may be small and inefficient radiators of airborne sound, if they are rigidly attached to large surfaces such as walls and floors, these may reradiate the energy. Control valves, pumps, and so on are other potential sources that must be carefully selected.

Sound-Absorbent Materials

Every material has acoustic properties, although only some are commonly considered to be "acoustic materials." As discussed previously, the acoustic properties of the surfaces of a room are important in the control of reflected sound, both the reflected sounds that are wanted, as in auditoriums, and those which are not (to avoid echoes and to reduce the reverberant noise levels).

As would be expected, the acoustic absorption of any material varies considerably with the frequency of the impinging sound. Some systems are excellent absorbers of low-frequency sound and good reflectors of high-frequency sound, and others have the opposite characteristics. Generally, low-density, porous materials are good absorbers of high-frequency sound; they must be very thick if they are also to absorb low- or medium-frequency sounds. The latter are best controlled with special "Helmholtz" resonators, membrane or panel absorbers, or perforated paneling over porous materials (Fig. 7.12).

The sound absorption characteristics of materials and elements may be measured in a laboratory, preferably in a reverberation chamber. Basically, the reverberation time is measured with and without the material or sample being present, and the difference gives the amount of energy absorbed by the sample.

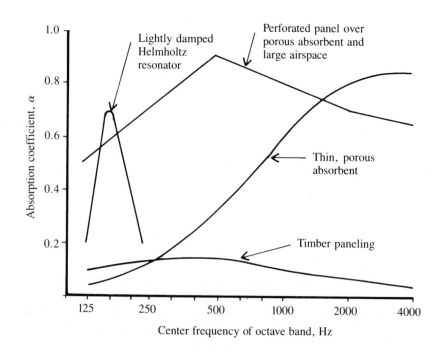

FIGURE 7.12.
Typical absorption characteristics of building materials.

This is usually corrected for the sample size (usually 10 m^2) and presented as an "absorption coefficient" for various frequencies. For individual elements, such as theater chairs, the absorption per element may be provided.

For general commercial applications, where most of the acoustic energy tends to be in the medium- to high-frequency range, materials should be selected to be efficient in this range. One difficulty that should be considered is that most suspended ceiling systems must contain far more than acoustic panels; for example, a considerable percentage of the area may be occupied by luminaires, air-conditioning grilles, loud speakers, and so on. This will tend to reduce the amount of absorption that can be provided by the ceiling.

In auditoriums, it is important to ensure that there is a good balance of absorption over the whole audio range. This means that low-frequency sounds must be considered as well as medium- and higher-frequency ones. It is important to ensure that inadvertent absorption is not installed; for example, large areas of thin timber paneling will act as a low-frequency absorbent, leading to a loss of bass in a room. The audience is a good absorber of higher-frequency sounds too. If possible, the seating chosen should have similar absorption to people, to avoid large changes in the reverberation time of the auditorium when it is only partly occupied.

EVALUATION OF PERFORMANCE

Objective Acoustic Measurements

Both objective and subjective analyses of the acoustic performance of a building may be made. The former relies on acoustic measurements of noise levels and reverberation time. Naturally, for buildings in which the acoustic quality is of prime importance (e.g., concert halls), more sophisticated techniques are required. Subjective analyses may take the form of questionnaires in which people are asked either general or specific questions about their acoustic comfort.

There now exists a considerable number of international and national standard methods for measurement of the acoustic characteristics of buildings and building elements. As discussed in previous sections, it is not easy to translate data obtained in a laboratory into a real building situation; thus it is always prudent to include acceptance testing as part of the commissioning of a building. For field measurements, it is necessary to differentiate between measurement of the sound transmission loss of an individual building element and that of the overall noise reduction between two spaces. Sometimes it is difficult to measure sound transmission loss because it is uncertain through which element the sound is traveling—although with the increasing availability of sound intensity measurement techniques, this problem should be overcome. If noise reduction is measured, all that can be ascertained is whether or not the total system meets the requirements. If the requirements are not met, it may be very difficult to decide where the problem lies. Sophisticated acoustical instrumentation and experienced personnel are required to carry out such measurements and analyses.

Reverberation times also can be measured in completed buildings and compared with the specifications. In a noncritical situation, the overall A-weighted reverberation time may be sufficient, but for auditoriums, studios, and so on, it is essential to measure the variation in reverberation time with frequency. Again, specialized acoustical instrumentation and analyzing equipment is necessary. For auditoriums, concert halls, and so on, the fine details of the spatial and temporal patterns of reflections may be required.

Subjective Acoustic Measurements

For commercial and residential buildings, subjective assessments of acoustic performance may well form part of a general assessment of the building's comfort. Questionnaire design is of importance if sensible results are to be obtained. It is normal to find quite significant individual variations in people's responses to their acoustic environment. This is partly due to the modifying influences of nonacoustic factors. For detailed descriptions of some of the subjective assessments that have been conducted in each area, Kryter's book is recommended (Kryter 1985).

For auditoriums, concert halls, and so on, it may be an advantage if special test performances can be arranged at which expert panels of listeners can make subjective judgments about the acoustic quality of the space. This is also useful if the room has been provided with "variable" acoustics, i.e., with panels, banners, and reflectors that can be adjusted after the building has been completed to suit different types of performances and different audience sizes. It should be remembered that this type of acoustic judgment is affected by aesthetic and emotional responses and that a significant range of opinions could be expressed.

For information regarding standard methods of measurement of the acoustic properties of buildings, refer to either the catalog of the International Standards Organization (ISO) or to that of the national standards authority in your own country.

SUMMARY

The selection and construction of interior elements and surfacing materials have a pronounced effect on the acoustic comfort of any building. There are many interactions between systems in a real building, and expensive mistakes can be made if these interactions are not examined. Since the actual acoustic performance of materials and systems depends very much on the details of installation and construction, careful supervision is essential. It is also recommended that compliance with the specifications be assessed using standard acoustic test methods, and for concert halls in particular, it is advisable to hold test concerts so that fine-tuning can be carried out during the commissioning period.

REFERENCES

Els, H., and Blauert, J. 1986. A Measuring System for Acoustic Scale Models. In *Acoustics and Theatre Planning for the Performing Arts*. 12th International Congress on Acoustics, Vancouver, B.C., Ottawa, Canadian Acoustical Association, pp. 65-70.

International Standards Organization (ISO). 1978. ISO 140: *Acoustics—Measurement of Sound Insulation in Buildings and of Building Elements; Part 6: Laboratory Measurements of Impact Sound Insulation of Floors; Part 7: Field Measurements of Impact Sound Insulation of Floors; Part 8: Laboratory Measurements of the Reduction of Transmitted Impact Noise by Floor Coverings on a Standard Floor.*

International Standards Organizations (ISO). 1982. ISO 717/1: *Acoustics—Rating of Sound Insulation in Buildings and of Building Elements; Part 1: Airborne Sound Insulation in Buildings and of Interior Building Elements.*

International Standards Organization (ISO). 1982. ISO 717/2: *Acoustics—Rating of Sound Insulation in Buildings and of Building Elements; Part 2: Impact Sound Insulation.*

Kryter, K. D. 1985. *The Effects of Noise on Man*, 2d Ed. Academic, New York.

Lawrence, A. 1970. *Architectural Acoustics*. Elsevier, London. Pp. 68-80.

Standards Association of Australia (SAA). 1987. AS 2107: *Acoustics—Recommended Design Sound Levels and Reverberation Times for Building Interiors.*

Standards Association of Australia (SAA). 1985. AS 2822: *Acoustics—Methods of Assessing and Predicting Speech Privacy and Speech Intelligibility.*

Standards Association of Australia (SAA). 1979. AS 1276: *Methods for Determination of Sound Transmission Class and Noise Isolation Class of Building Partitions.*

P A R T III

The Dynamic Role of the Building Envelope

Part II discussed how the design of a building's interior is related to the requirements of the occupants and the interior's function. However, it is the building envelope over which the architect or building designer has the most control, and design of this envelope has an important impact not only on energy savings, but also on human health and well-being.

The building envelope can be considered as a filter with its component parts designed as functions of the external climate and human requirements. There is some evidence, as demonstrated in Part I, that uniform conditions are not as desirable as an interior environment that mimics the natural cycles of climatic change. A building envelope that responds dynamically to the climate not only acts as a human stimulus, but can also reduce the quantity of energy required for environmental control.

In Chapter 8, Peter Smith introduces strategies for modifying building envelope properties in different climates in order to optimize the envelope's thermal performance and reduce energy consumption. The relationships between climate, the building occupants, and the envelope design are discussed in some detail.

In Chapter 9, which should be read in conjunction with Chapter 8, George Cunningham describes the role of air-conditioning in buildings, the load components of the air-conditioning system, and their performance and energy consumption. Methods are proposed by which variable temperatures can be achieved.

As envelope components, window systems play a variety of roles, but it is generally believed that priority should go to providing a comfortable luminous environment and hence a productive workplace. In Chapter 10, Nancy Ruck describes both passive and active strategies in the use of daylight in building interiors for the well being of the occupants.

In addition to ensuring visual and thermal comfort, it is necessary to use the building envelope as a filter between a noisy external environment and the building occupants. To ensure good sound attenuation, the building envelope must be assessed as an overall system. Anita Lawrence describes the characteristics of external noise levels and the selection of materials to reduce sound transmission.

Thermal Performance

8

Peter Smith

FUNCTIONS OF THE BUILDING ENVELOPE

The building envelope serves as a filter between the exterior environment and the interior of a building. The external climate is a property of the locality, and it can be modified only slightly by landscaping and by adjacent buildings. The specific requirements of the building envelope depend on both the climate and the expectations of the occupants. A harsh climate places greater demands on envelope performance than a mild one, whereas occupants who need a closely controlled environment will be harder to satisfy than those who allow a great deal of tolerance in the physical conditions (Fig. 8.1).

In an extreme case, we may want to isolate the interior from the outside so completely that the occupants of the building are completely unaware of the exterior, such as happens, for example, in a theater, which is totally enclosed whatever the climate. At the other extreme, a resort on a tropical island may need only a roof and no walls. However, a bank or office building on the same island would require walls, to provide a small amount of climate control as well as for other practical reasons.

Although the main function of the building envelope is to *protect* the interior from external conditions, it cannot avoid contributing to the interior conditions. In admitting radiant energy and allowing the transmission of conducted heat one way or the other, the building envelope plays a part in establishing the thermal environment, particularly for occupants near the external walls or in the story immediately under the roof. Similarly, daylight

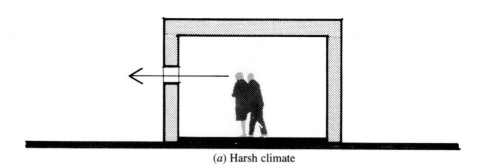

(a) Harsh climate

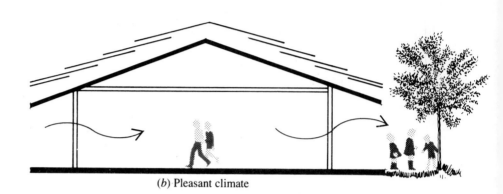

(b) Pleasant climate

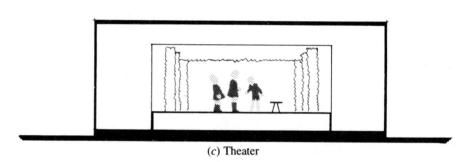

(c) Theater

FIGURE 8.1.
(a) In a harsh climate, the building protects the occupants from the exterior with minimal contact. (b) In a pleasant climate, little protection is needed. (c) When the function of a building demands it, the envelope isolates the occupants from the exterior, whatever the climate.

transmitted through the windows makes a major contribution to the visual environment, and in a noisy locality, the transmitted noise can be significant in the whole acoustic environment (Fig. 8.2). The design of the envelope also has a significant effect on the energy requirements of the building's environmental control systems.

As we have mentioned previously, most occupants of most types of buildings prefer to be aware of the outside world, and it is probably desirable for the interior conditions to fluctuate a little to mimic the daily changes in the real world. The design of the building envelope can contribute to all these requirements.

We saw in Chap. 5 that the preferred value for indoor temperature can be related to the mean monthly outdoor temperature, the two coinciding about 25.5°C. The average temperature inside a house that has no heating or cooling will generally be 3 to 5°C above the outdoor temperature, because the occupants, the lights and appliances, and the solar gain through the windows will all add heat to the interior and there are no corresponding mechanisms to remove such heat. In such instances, mean thermal neutrality inside the building can be achieved when the mean monthly external temperature is in the range 18 to 21°C. That is why "heating degree days" are usually measured relative to a mean daily temperature of about 18°C.

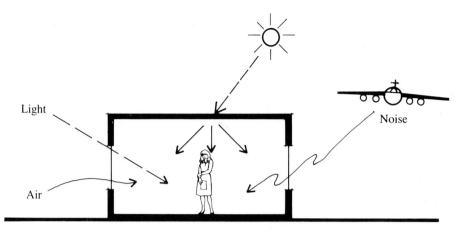

The actual difference between internal and external temperatures depends on many factors, including the extent of heat generated within the building, the surface-to-volume ratio, and the transmittance of the building envelope. For a building of given geometric shape, the surface area increases with the square of the linear dimensions, whereas the volume increases with their cube. Therefore, a large, compact-shaped, heavily populated, and well-insulated building heats up more (or reaches thermal balance at a lower external temperature) than one that is smaller, more extended in shape, less densely occupied, and less well insulated.

The designer can exercise further influence over a building's performance, if necessary. By designing for solar gain through correctly oriented windows, with the remaining surfaces well insulated against subsequent heat loss, mean interior temperatures about 10°C above the exterior should be readily achieved. However, by reducing solar gain to a minimum and manipulating the ventilation to make use of the cool night air, the internal temperature can be kept at or even a few degrees below that of the exterior. Natural cooling is less effective than natural heating, and it works best when there is a large diurnal temperature range to work with. A building can be designed to take advantage of both passive solar heating in winter and natural cooling in summer.

While the mean internal temperature depends largely on the amount of solar gain, insulation, and ventilation, the actual temperature will vary throughout the day and night. This variation is controlled mainly by the amount of material in the building (the mass). Often a massive building (and hence a small diurnal range inside the building) is desirable, but depending on the hours of occupation and the characteristics of the climate, in some cases a lightweight building that responds more rapidly to external changes may be preferable.

The building envelope serves many other functions. It often has a structural purpose, e.g., as a load-bearing wall or structural concrete roof slab; the size and mass of the structural elements have an effect on the thermal performance. The need for security and acoustic isolation may limit both the size and placement of windows. The requirements for cleaning the outsides of windows and for painting and maintaining parts of the exterior may have an effect on the form of the exterior and the placement of external sunshading devices. The facade also plays a major role in the aesthetics of a building. Decisions about the "image" that the building should present can preempt decisions about performance.

PRINCIPLES OF HEAT TRANSFER

Steady-State Heat Transfer

Heat transfer is dealt with in detail in many books, some of them directed particularly toward buildings. Reference should be made to one of the specialist texts, such as Threlkeld (1970) or Szokolay (1980). The following is a very brief resumé of the subject.

Heat is transferred by three modes: conduction, convection, and radiation. "Conduction" is movement of heat energy through a medium which itself does not move. "Convection" is the carriage of heat energy by the movement of a fluid (liquid or gas). "Radiation" is the transmittance of heat energy through a transparent medium or through a vacuum; the energy only has an effect when it strikes an opaque surface.

Heat flows from a hotter to a cooler body. There is no conductive or convective heat exchange in regions of equal temperature. All surfaces emit radiant heat, but if two surfaces are at the same temperature, the *net* radiant heat exchange between them is zero.

In the study of heat transfer in buildings, it is common to refer to "air-to-air transmittance" when there is a difference in air temperature between the two sides of a wall or roof. Air-to-air transmittance is mainly a *conduction* phenomenon, but the other two modes may be involved as well. Heat is exchanged between the air and the outer surface by convection, through the material by conduction, across any internal air cavities by convection and radiation, and from the inner surface to the air by convection again. At the same time, radiant heat exchanges are taking place between each surface of the partition and its surroundings. This heat flow is combined with, and indistinguishable from, the air-to-air heat flow while it is within the partition, but it is *calculated* separately for convenience.

When the temperature on each side of a partition remains constant over a long time (many hours), the heat transfer settles down to a steady state and can be calculated simply from the transmittance and the temperature difference between the two sides. If conditions on one or both sides change more rapidly, then the thermal inertia of the construction also becomes important. This is discussed later.

The principal source of radiant heat loads is solar radiation. In the absence of the sun, most surfaces tend approximately toward the temperature of the air that surrounds them, and the radiant effects can often be ignored or combined with the air-to-air effects. Solar radiation is of such magnitude that if it strikes a surface, it must be considered in its own right. The magnitude of the solar heat load on a surface at any time is the product of the actual intensity of the solar beam, the cosine of the angle of incidence, and the "absorptance" of the surface. The absorptance of a surface to solar radiation is seen by the eye as the "lightness" or "darkness" of its color: white absorbs 20 to 30 percent of the energy falling on it, black 90 to 95 percent, and other colors amounts in between.

Radiant Transmission Through Glass

Clear window glass transmits almost 90 percent of the solar radiation falling on it. It reflects about 8 percent and absorbs a little, varying with the chemical composition of the glass. However, the transmitted energy strikes the interior of the room, which is warmed up and reradiates some of its energy as long-wave radiation, to which the glass is opaque. The energy is thus largely trapped inside the building (Fig. 8.3). This is known as the "greenhouse effect"; it is very useful if the objective is the collection of solar heat, but not if the objective is to keep the interior cool.

Variable External Conditions

Once a steady state of heat transfer becomes established, each part of the construction reaches a temperature that is determined by the requirement of equilibrium, namely, that the same amount of heat that flows into each part must flow out of it on the other side. Once the temperature on one or both sides

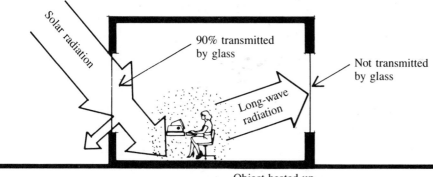

90% transmitted
by glass

Not transmitted
by glass

Solar radiation

Long-wave
radiation

Object heated up

FIGURE 8.3.
The greenhouse effect. Solar radiation is readily transmitted by glass, but the long-wave radiation from the hot surfaces in the room is unable to pass back out through the glass.

of the partition changes, however, a new regime must again be set up to satisfy the new equilibrium condition. While the temperature of any part is changing, the amount of heat flowing out of it will be more or less than that flowing in.

After a change in conditions, all the parts tend toward the new equilibrium condition, getting ever closer but (theoretically) never exactly reaching it. In practical terms, the time delay varies from a fraction of an hour for lightweight construction (such as framed and sheeted walls) to a fraction of a day for heavy masonry walls. After this time, the system is close enough to its equilibrium condition that the steady-state assumption can be used.

A wall or roof usually contains several layers; heavy materials such as masonry have a high thermal capacity but a relatively low insulating value, whereas the best insulating materials are very light and therefore have a low thermal capacity. Under varying heat loads, these two types of materials behave differently, and the building behaves differently depending on which one is on the inside or outside face. Under steady-state conditions, the order is not important.

Many heat loads fluctuate on a 24-hour cycle because they are initiated by solar radiation or by the daily use of a building. A building with high thermal inertia is usually recommended to even out the differences between a hot day and a cool night. When there is little variation between daytime and nighttime conditions, thermal inertia is of less benefit. This is discussed further in the section entitled Mass (page 144).

Variable Interior Conditions

A dense surface such as masonry has high "thermal diffusivity." This term implies that heat striking the surface is readily "diffused" into the body of the material so that the surface adapts very slowly to changes in the temperature of its surroundings. A light, insulating surface such as lightweight plaster or insulating ceiling board has the opposite effect—a low thermal diffusivity and the ability for the surface to come quickly to the temperature of the surroundings.

The properties of the internal surfaces have little effect on the thermal performance of a room under steady-state conditions, since either dense or lightweight surfaces will have come to equilibrium anyway. However, these properties are important when the heat input to the room varies rapidly. This occurs mainly with intermittent heating and with the admission of solar radiation.

With intermittent heating (such as a house that is unoccupied during the day and has the heaters turned on at night), dense surfaces cause much of the initial heat to be absorbed by the walls before they are warmed up, whereas lightweight surfaces will quickly warm up and allow the benefit of the heating to be felt in a short time. When solar heat enters through the windows, dense surfaces allow much of the energy to be stored in the surfaces before the interior becomes too hot; lightweight surfaces will allow the interior to heat up rapidly, with the result that the occupants will either close the curtains or open the windows to regain comfort conditions. Little of the solar heat will be stored for later use.

RELATIONSHIPS BETWEEN CLIMATE, OCCUPANTS, AND ENVELOPE DESIGN

Some of the requirements of building envelope performance are, if not mutually exclusive, at least incompatible. For example, allowing ventilation is in conflict with excluding noise, and allowing daylight and view makes thermal insulation more difficult. Several factors will influence the designer in giving preference to certain aspects of performance over others in any particular case.

What is expected of the building envelope varies from one building *type* to another. For example, a distinction is commonly made between those buildings which are air-conditioned and those which are not, since in the latter case the envelope has to provide controllable ventilation along with its other functions. As indicated in Fig. 8.4, such a distinction may affect not only the *details* of envelope construction, but the form of the building as a whole.

However, different expectations also arise on the basis of the way a building is to be used. In an informal setting, such as a residence, one can move about, in or out of the sun for thermal comfort, near the window for better light, and so forth, whereas in a formal working situation, the workstations are more or less fixed and the environment must be adjusted to suit them. A "passive solar" building presupposes that the occupants are sufficiently involved in maintaining their thermal comfort by natural means that they will operate curtains, windows, and the like at the appropriate times, whereas the occupants of commercial buildings are more concerned with their work and in any case do not occupy the building in the evening and early morning when adjustments are most likely to be required.

The thermal performance of the building envelope is important both in terms of limiting the cost of maintaining comfortable conditions when this is done by the use of fuel and in terms of maintaining the best comfort conditions that can be achieved whether by natural or powered means. It is usually not possible to compensate by means of air-conditioning for the discomfort of being close to an ill-designed wall or roof.

Insulation

If a wall or roof is poorly insulated, this will have two effects. First, the heat flow through it will be high, and this usually leads to excessive energy consumption.

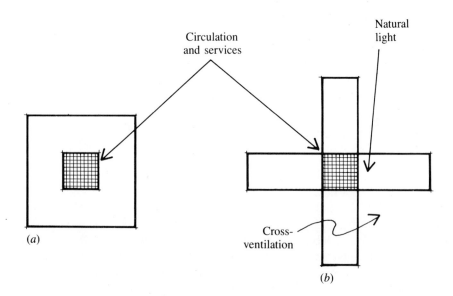

FIGURE 8.4.
The form of a building may be determined partly by the function of the envelope. Plan (*a*) with less wall area would suit an air-conditioned building, whereas plan (*b*) would provide better cross-ventilation and natural light.

This is discussed further in the section entitled Envelope Design and Energy Consumption (page 139). Second, the inside surface will tend toward the temperature of the outside surface. If this is much different from the general inside air temperature, it will affect the "mean radiant temperature" experienced by people close to it and therefore alter the thermal comfort of those people. The roof is a major radiant source for all the occupants of the top story of a building, whereas a window or a wall is only a significant radiant source for persons relatively close to it, and even then, such persons may be partially shielded by furniture.

When the radiant effects of one part of a room depart *a little* from the comfort zone, compensation can be made by adjusting the air temperature. This solution becomes less satisfactory if the radiant effect is highly directional or if it departs *much* from comfort. This becomes more difficult to achieve if people in close proximity to each other receive significantly different thermal radiation, since it is difficult to vary the air temperature over short distances.

Infiltration

A building is never perfectly sealed against the passage of air in or out through small, inadvertent cracks and through the doors or windows when they are opened. In severe climates, both hot and cold, a significant improvement can be made in the energy consumption of a building by reducing infiltration to a minimum. Weatherstripping and storm shutters can be added to help seal cracks around windows that are only opened in fair weather, and revolving doors or airlocks with two sets of doors reduce the inrush of air when doors are opened.

Infiltration is increased by pressure differences caused by wind, but even in still air some exchange will occur by diffusion. An air-conditioning system recirculates much of the air from the space, but some fresh air must always be added to maintain freshness and to replace the oxygen used by the occupants, and a similar amount of used air is spilled. The advantage of *recirculation* is intended to be reduced energy consumption, although this is not always achieved. This is discussed further in Chap. 9.

In some parts of buildings, such as kitchens, toilets, and car parks, the air is deliberately exhausted to the atmosphere to avoid recirculating undesirable contaminants. If insufficient air is lost through these spaces to allow for the fresh-air makeup, then a little additional outward leakage through the construction would not do any harm to the energy consumption, but *excessive* leakage would mean that more fresh air would have to be processed by the air-conditioning plant to make up for the loss, with consequent additional energy requirements. The more severe the climate (either hot or cold), the greater is the energy consequence of unwanted air leakage.

In any case, if air can escape, it also can enter under certain wind conditions or when the air-conditioning fans are not operating. Air-conditioned buildings are operated slightly above atmospheric pressure, to discourage the entry of untreated air, not only because of its different temperature, but also because it contains dust, it can be felt as a draft by persons close to the point of entry, and the air leakage may produce a noise. For all these reasons, the aim should always be to produce an airtight building envelope. Particularly in high-rise or otherwise exposed buildings, where the combination of wind and rain can cause water to enter through unsealed gaps even against the effect of gravity, it is also important to avoid these gaps for the sake of watertightness.

Solar Radiation Control

The energy admitted to a building by the passage of solar radiation through the windows is of greater intensity than that admitted as a result of sunlight striking an opaque wall, and it is felt selectively, having a great and immediate effect on people directly in its path but a lesser and delayed effect on people who are shaded from the direct rays. Furthermore, the intensity of light from the sun is so high that it causes extreme visual contrast between sunlit surfaces in a room and surfaces that are not sunlit. Contrast of this magnitude is unacceptable if it occurs in the field of view of someone performing a visual task such as reading.

In a building with fixed workstations, therefore, it is almost impossible to create comfortable conditions for people in the path of direct solar penetration. The heat load also causes the refrigeration plant to work harder and consume more energy, except under those cold conditions where there is a net loss of heat from the building. For all these reasons, it is highly desirable to control the sunlight, either by external shading, internal shading, or the use of solar-control glass. These methods are discussed in more detail in the section entitled Windows (page 141).

View

The psychological need for a view out is discussed in Chap. 10. While it is impossible to quantify the value of a view in physical terms, the real estate industry is well aware of its monetary value. Offices and apartments with a good view command substantially more rent than those without. If the direction of the view happens to be undesirable in terms of solar heat or glare, so that a tradeoff is necessary, the value of the view almost always exceeds the cost of dealing with the solar problem, but the result is seldom completely satisfactory.

One solution is to use adjustable shades, either inside or outside. This can control the solar problem, but it restricts the view for at least part of the time. Another is to use solar-control glass, which partially controls the sun but also prevents any view out at night, when the glass looks reflective from the inside. Yet another solution is to deal with the solar load by air-conditioning alone, which, as we have already said, provides less than ideal thermal comfort for the occupants. This is also discussed in greater detail in the section entitled Windows (page 141).

Natural Light

Natural light is dealt with in detail in Chaps. 6 and 10. From the thermal point of view, the significance of using natural light is first that the luminous efficacy of skylight is quite high, in the range 100 to 200 lm/W (Hopkinson, Petherbridge, and Longmore 1966), compared with normal fluorescent lighting, which is 60 to 90 lm/W. Therefore, for a given amount of light, skylight produces significantly less heat than artificial light from the usual lamp sources. However, as mentioned in Chap. 10, the luminous efficacy of daylight is measured outdoors, and this comparison can be misleading, because the preceding values may be modified when measured indoors.

Second, however, it is difficult to get only the recommended value of illuminance from natural light because of the variations in its spatial and temporal distribution. In a side-lit room, and particularly toward the back of the room, the proportion of the daylight flux that ends up as useful illuminance on the working plane is likely to be less than the proportion from a well-designed set of overhead luminaires. Therefore, if the aim is to satisfy a requirement for illuminance in the working plane, the solution is likely to be as follows:

1. Allow higher levels of natural light for most of the time and for most of the room in order to satisfy the minimum requirements in the worst case.
2. Add a supplementary lighting system with the chance that it will be turned on most of the time whether it is needed or not.
3. Add a supplementary lighting system with automatic controls to add just enough artificial light to "top up" the deficit from the windows.

Either of the first two solutions is likely to negate any benefit of reduction in the cooling load that would result from the higher luminous efficacy of the daylight. The third solution offers the most likely benefits, at the expense of a more costly installation. This is discussed in Chap. 10.

In practice, skylights (located in the ceiling and admitting light from the sky, but excluding excessive solar heat) have about the same impact on the cooling load as would an equivalent fluorescent lighting installation; sidelighting (from windows only) is likely to incur a greater cooling load, because in order to have enough light at the back of the room, there will normally be excessively high levels of illumination close to the windows. Innovative methods of getting more light into the interiors of deep spaces are discussed in Chap. 10.

Even though the cooling load may not be reduced, the use of daylight can confer substantial savings in the energy used for the lighting itself. The heat gain and loss through the glazed areas under different climatic conditions interacts with the effects just mentioned, and the full impact on energy consumption can only be established by considering all these factors together and by considering them over the full range of conditions throughout the year. This argument is developed in Part IV.

The discussion of daylighting so far has been based on one of the assumptions often made in lighting codes, namely, that the illuminance on the working plane should be uniform throughout the building and throughout the day. Since abundant artificial light has been available for less than a century, it is apparent that humans have made do with much less uniform illumination for most of their existence; there are good reasons to suppose that temporal variation, at least, can produce beneficial stimulation. The purpose of lighting a building interior is for the benefit of the occupants. This is discussed further in Chap. 6.

In addition to the benefit of variability of daylight, the directional component of light entering through windows may give better modeling to objects and people than uniform overhead lighting, but there are obvious difficulties, especially when specific tasks need to be carried out. Unwanted shadows, reflections, and rapid changes can detract from the comfort and convenience of the occupants, probably more than unwanted uniformity.

ENVELOPE DESIGN AND ENERGY CONSUMPTION

Whatever the building type, the envelope can be designed to limit the need for additional energy for controlling comfort conditions inside the building. It also can reduce the requirement for artificial lighting by admitting controlled daylight. The building envelope is the part over which the *building designer* has the most control at the time of the architectural design. Normally, the envelope will remain for the life of the building, even if the mechanical plant and partitioning layouts are updated from time to time. Consequently, the envelope deserves a good deal of consideration at the design stage.

The plant "capacity" is determined by the worst combination of conditions it is expected to cope with, while the "energy consumption" is the sum of the energy used on each day throughout the year (Fig. 8.5). The use of daylight and passive-solar energy can significantly reduce energy consumption by contributing light and heat for a large proportion of the year, but they are unlikely to reduce the size of the plant if it has to be able to cope with occasional periods of overcast and cold weather. Other measures such as improved insulation and sunshading can reduce both plant size and energy use.

Generally, heat is produced within a building by the lights, people, and equipment. Solar radiation, when it strikes the building, adds heat to the interior. Conduction through the envelope causes either heat gain or loss depending on the outside temperature.

The relative contributions of the envelope and the internally generated loads to the required plant capacity and the energy consumption vary greatly

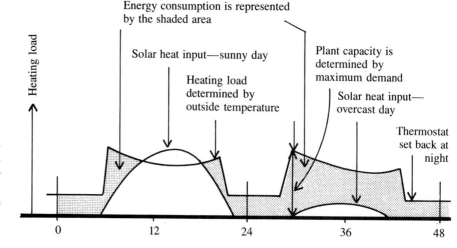

FIGURE 8.5.
The heating load for a building is determined by the heat losses through the envelope and is plotted on a diagram over several days. The solar heat gains are superimposed on this diagram. The difference between the two lines represents the heat that must be supplied from other sources. The annual energy use takes into account all the sunny days and all the overcast days. The plant capacity is determined by the maximum demand.

with the details of the building and the climate in which it is located. Using "typical" rather than very energy conscious buildings, the envelope might account for a third of the energy used for environmental control in a mild climate and two-thirds in a severe climate. Since the internally generated loads are fairly constant throughout the year, while the envelope-related loads fluctuate with the weather, the proportion of *plant capacity* attributable to the envelope would be expected to be greater.

The requirements of the environmental control plant depend on the net combination of these effects, which may result in a need for heating, cooling, and lighting. The main variables in envelope design are its color and insulating value; the proportion, shape, and shading of windows; and the mass of the construction. The insulating value of windows, even multiple-glazed windows, can be less than that of a well-insulated wall. However, reference is made in Chap. 10 to advanced glazing materials now becoming available with properties equivalent to those of a well-insulated wall. Small windows can improve the overall thermal insulation of a wall, but the reduced availability of daylight and the reduced amenity for the occupants usually outweigh the thermal advantages of excessively small windows.

A perfectly insulated building would never need heating except for the initial warmup after a period of nonuse. In practice, the interior zones of large buildings behave like this and are frequently designed with a provision only for cooling, although the air-handling plant may contain provision for warming the outside makeup air. The perimeter zones and the story immediately under the roof usually require some heating in a cold climate, but better insulation means less heating energy. Provision for solar gain will reduce the heating energy consumption, but if it also means larger windows or reduced sunshading, the benefit must be weighed against the greater heat gain or loss through the windows by conduction.

In hot weather, there is a need for cooling to remove both the internally generated heat and the heat conducted through the envelope. Good insulation of the opaque surfaces, together with insulating windows or small windows and sunshading, will help to reduce both the plant size and the energy consumption.

At those times when the external air is *a little* below the internal air temperature, a poorly insulated construction would allow the heat to escape without the intervention of a cooling system, and this would be beneficial. However, at other times it would either allow external heat to enter, placing an additional load on the cooling system, or allow too much heat to escape, requiring the addition of heating near the wall or roof concerned. Even under the ideal condition of external temperatures just below those inside, a poorly insulated construction is not really a great advantage, since the internally generated heat can be removed by simple ventilation (either by opening the windows or by using an air-conditioning system without turning on the heating or cooling) more uniformly and with more opportunity for control.

The material on the inside surfaces of the rooms has a marked effect on the way in which those rooms respond to changes in the internal environment, such as the entry of solar heat or the turning-on of a heater or air-conditioner. In those cases where heating or cooling is required only intermittently, this effect is significant. It was discussed in the section entitled Variable Internal Conditions (page 134).

STRATEGIES FOR MODIFYING ENVELOPE PROPERTIES

Opaque Construction

All the heat transfer except radiation through the windows can be reduced by reducing the conductance of the actual construction. This is achieved by adding insulating material to one face or the other or into a cavity or by adding additional separate air cavities within the construction. The effectiveness of air cavities can be further improved by lining one or both sides of them with reflective material. This restricts the radiant transfer across the cavity, although convective transfer still occurs.

The transmittance values (U-values) of typical constructions can be found in specific texts, e.g., ASHRAE (1985). For uninsulated walls or roofs, values on the order of 1 to 2 W/m^{2}K are common. With the addition of 50 mm of insulating material, the value drops to less than 0.5 W/m^{2}K, which is a very significant improvement. In severe climates, 100 mm or more of insulation may be used, with correspondingly lower U-values.

The solar heat load on an external surface can be reduced by using a light color that reflects much of the solar energy. The color makes no difference to the air-to-air conductance in either direction.

In any multilayer wall or roof construction, the layers will have different resistances to heat flow and also different resistances to the passage of water vapor. Since warm air can hold much more water vapor than cool air, condensation is likely to occur when warm, moist air meets a cool surface. In a cold climate, condensation can occur on the inside face of poorly insulated constructions, particularly on single-glazed windows.

When the construction is insulated, the moist air is likely to penetrate into the insulating material and cause condensation at the point where the temperature falls to *dew point*. To avoid this, a vapor barrier should be placed on the warmer side of the insulating material. Vapor barriers are usually polyethylene film, aluminum foil, or tar paper. Some insulating materials are supplied with a vapor barrier already attached. For further discussion of vapor barriers, dew point, and psychrometry, reference should be made to a building materials text such as Cowan and Smith (1988) or an air-conditioning handbook such as ASHRAE (1985).

Windows

A single-glazed window has a U-value of about 5 W/m^{2}K. The actual value depends mainly on the degree of exposure, which determines the convective transfer between the outside air and the glass surface. The glass itself, although not a good conductor of heat, is too thin to contribute much to the insulating value. (An *open* window must be considered on the basis of the convective exchange contained in the ventilation rate it provides. To consider it as having infinite transmittance obviously gives the wrong answer.)

Since most of the thermal insulating value of a window comes from the *surfaces* where heat is transferred by convection from air to glass, a double-glazed window has about twice and a triple-glazed window about three times the insulating value of a single-glazed one when *air-to-air* transmittance is concerned. When the window is open, the use of multiple panes has no advantage.

A heavy curtain fitted with a pelmet to prevent free flow of air over the top adds virtually another airspace and considerably improves the performance of a window *when it is tightly closed* (Fig. 8.6). The curtain is therefore useful in preventing nighttime heat loss through windows in those buildings which have someone to close the curtains at the appropriate times. Curtains are of less use during the day. The occupants must be prepared to close them at the appropriate times, and then the advantages of natural light and view are lost.

Internal venetian blinds are quite useful in reducing the effect of direct solar radiation, provided someone is prepared to adjust them to keep out the sun but admit daylight and allow some view. Venetian blinds are not as effective as external adjustable louvers, but they are cheaper to install and maintain. Because they are not airtight, they are of limited use in reducing nighttime heat loss.

Depending on the glazing material, windows are often considered a weak link in the thermal performance of a wall. Therefore, if a wall has a large proportion of glass, adding thick insulation to the opaque parts will have relatively little effect on the heat transfer through the wall as a whole (Fig. 8.7).

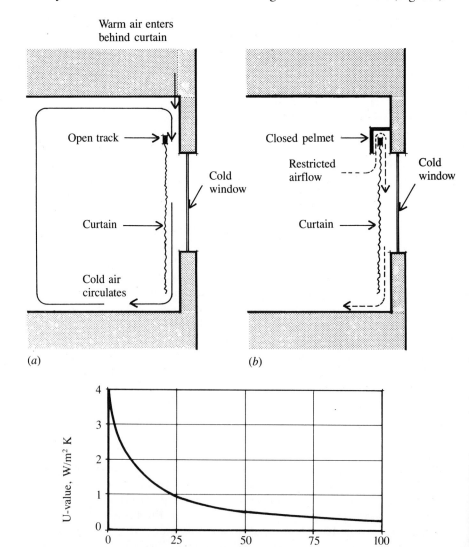

FIGURE 8.6.
A heavy curtain reduces the radiant cooling effect of a cold window, but it is of limited help in reducing airflow over the window unless there is a closed pelmet at the top as in (b).

FIGURE 8.7.
Adding insulation to a poorly insulated wall reduces its U-value rapidly at first, but as the thickness of the insulation increases, the effect of each additional increment is less noticeable.

The design of windows having both high values of solar transmittance and high insulating values is discussed in Chap. 10 and Part IV. There is also little point in raising the thermal performance of either the wall or the windows at great expense if the occupants are likely to insist on leaving the windows open for ventilation.

External nonventilating shutters perform the same function as internal curtains and in addition provide extra protection against air infiltration. They also can be used to keep out unwanted sunlight, but the same objection can be raised as with curtains. External ventilating shutters are used in many hot climates to exclude the sun, but they allow in ventilation and some filtered light (Fig. 8.8).

Improving the performance of a window against *direct solar gain* requires different strategies. For example, double glazing with clear glass might reduce

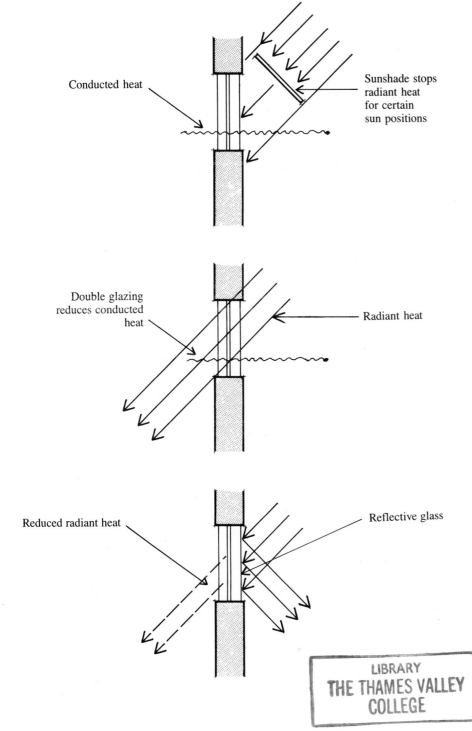

Conducted heat

Sunshade stops radiant heat for certain sun positions

Double glazing reduces conducted heat

Radiant heat

Reduced radiant heat

Reflective glass

FIGURE 8.8.
Sunshading is effective against direct solar heat from certain directions, while multiple glazing is effective against air-to-air conducted heat. Reflective glazing reduces radiant heat loads from all directions. A combination of these techniques may be needed.

the transmittance from 90 to 80 percent, an insignificant improvement compared to the halving of its air-to-air transmittance. On the other hand, the use of reflective, heat-absorbing, or low-emissivity glazing can reduce the solar transmittance to half or a third while making no difference at all to the air-to-air transmittance.

The greatest reduction in direct solar gain is obtained by external sunshading that prevents any sunlight from reaching the glass. On a facade facing the equator (the south face in the Northern Hemisphere, or vice versa), a simple overhang or horizontal louver system will achieve this aim, at least in summer. Other aspects are more difficult to shade, with east and west faces being quite difficult because of the low-altitude sun that shines directly on them. The design of solar control devices is further discussed in Chap. 10, and there are many books on the techniques of sunshading, such as Harkness and Mehta (1978).

Infiltration Control

Air leaks occur particularly around *opening* windows and doors because operating clearances are required. To prevent this leakage, windows and doors must be sealed with flexible gaskets that press against the surfaces when closed but do not interfere with their movement. The problems of air infiltration suggest that "nonopenable" windows have advantages.

Nonopenable windows have the advantages mentioned above, as well as preventing accidents related to falling objects or falling people. However, it is necessary to devise a system of window cleaning from the outside. This is easily done from a gondola if the facade is unencumbered by sunshading devices, but solar radiation is an important source of cooling load and external sunshading is one of the preferred options for dealing with it. External sunshades usually make external window cleaning difficult, unless the sunshades themselves are designed to accommodate walkways for the cleaners.

Mass

As we saw earlier, the mass and thermal capacity of a construction become important when heat flow fluctuates, such as, for example, when the sun shines on a wall for part of the day or during the normal day-night variation in air temperature. When such a transient heat load strikes the exterior of a wall or roof, the rate of heat flow to the interior is *less* than it would be through a lightweight wall of the same U-value, and it is *delayed* (Fig. 8.9). The delay is of great benefit in some circumstances because it may cause the greatest heat gain to occur either at night, when it is needed, or after the occupants have left at the end of the day, when it causes no discomfort. The reduction in intensity of the heat transfer is usually desirable; in those cases where solar heat would be welcomed immediately and undiminished, it should be admitted through windows instead.

The insulation value of a wall subject to continuous heat gain or loss can be improved by the addition of insulation anywhere in its thickness. The resistance of the wall is simply the sum of the resistances of all its layers. A wall subject to fluctuating external loads would benefit most from the addition of insulation to the outside face, with heavy material on the inside (Fig. 8.10). This is sometimes done, but it offers practical difficulties. The heavy masonry materials are durable and resistant to damage when used externally. Insulation material is soft, whether in the form of batts or blankets that cannot be exposed at all or as soft board materials that can be used as internal linings. If a decision is made to use insulation externally, it must be covered with a weatherproof layer such as sheet metal or cement rendered on metal lath.

Dense internal finishes are a disadvantage where there is intermittent heating, since they delay the response of a room to heating input. They cause no problem in a room that is continuously heated or cooled and are preferred in

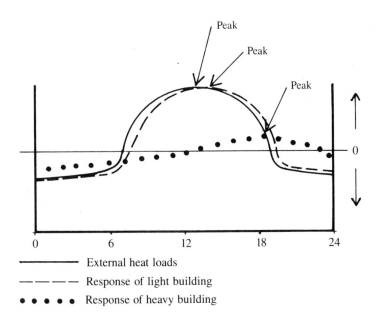

External heat loads

– – – – Response of light building

• • • • Response of heavy building

FIGURE 8.9.
A heavyweight building responds slowly to external heat loads. The peak effect is less noticeable and occurs later than in a lightweight building.

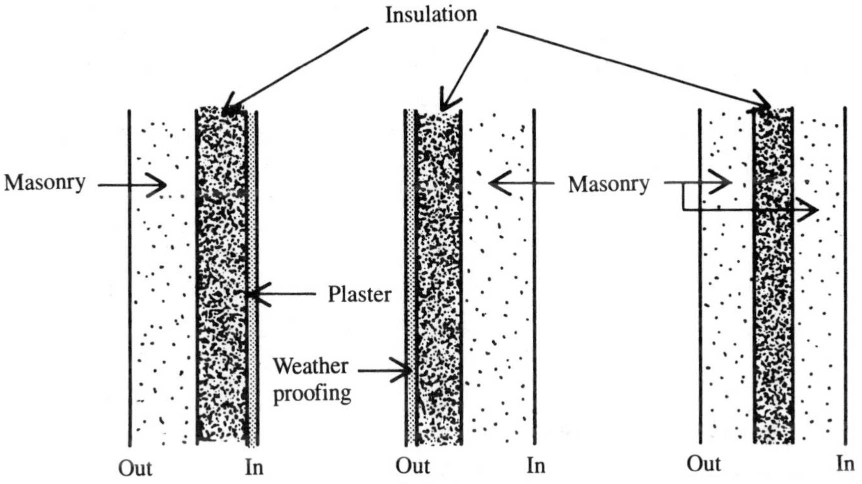

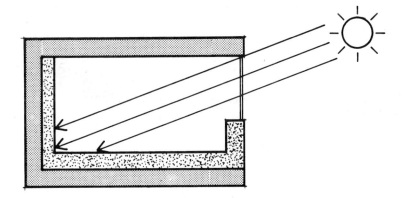

FIGURE 8.10.
Insulation may be included on either face or in the middle of a wall. The best insulators are very soft materials that need a protective layer. An alternative is to use lightweight plaster or blockwork that combines reasonable insulation value with sufficient durability. In all cases, the location of a vapor barrier requires careful consideration.

FIGURE 8.11.
A building designed to collect solar energy by direct gain relies on sufficient mass to absorb the gain slowly through the day and release it slowly at night. The surfaces not used for this process should be insulated to reduce heat losses.

specialized rooms such as cool stores or instrument rooms to discourage rapid fluctuations in internal conditions.

Thermal mass may be added to the interior of a room by means of heavy internal walls, concrete floor slabs, decorative brick walls, or fireplaces. This extra mass is of advantage in a passive-solar building, both to limit the temperature rise during the day while the sun is being admitted and to store and release the heat later in the evening when the sun is not present (Fig. 8.11). To thermally couple a heavy floor to a room, the floor should have hard, noninsulating finishes such as ceramic tiles rather than carpet.

STRATEGIES FOR MODIFYING BUILDING PERFORMANCE

The specific performance requirements of a building's envelope will depend on the climate and the building characteristics referred to earlier. Each element (such as the roof, solid and glass portions of the walls, and shading devices) makes a contribution to the thermal transmittance, the thermal storage, and the admission of daylight, as well as to the appearance, cost, ease of construction and maintenance, security, and many other features of the building as a whole.

A Building That Is Heated but Not Cooled

The detailed requirements of a building envelope's performance depend on the climate. There is a need for controllable natural ventilation. The envelope should be well insulated to save on heat loss, but solar radiation may be quite welcome to reduce the cost of heating.

To increase the solar heat gain in winter, the sun-facing wall can have a large proportion of glass, with curtains or shutters to reduce the heat loss through it at night. The rooms adjacent to this glass should have a large thermal mass, especially in the floor. These windows usually require sunshading to prevent summer heat gain. Windows in the remaining walls can be reduced to the minimum required for daylighting and cross-ventilation. The remaining walls, particularly those facing the coldest winter winds and the hottest summer winds, should be well insulated or protected by plantings or earth berming.

In a single-story building, some roof glazing can be added to increase the solar gain, but roof glazing (depending on the glazing type) is likely to admit more solar heat in summer, when the sun is higher in the sky, than in winter. A vertical clearstory window facing in the same direction as the main windows may be more effective for admitting solar heat to the side of the building away from the sun.

The ceiling of the top story is a significant path for both heat gain and loss. Since the roof-ceiling combination does not require any penetrations, good insulation can be provided. A light-colored roof surface reflects summer heat away. There is less solar heat on the roof in winter than in summer, and therefore, it is not the preferred location for direct gain. It is, however, a convenient place for active solar collectors, both because the roof provides an unencumbered and largely unseen location and also because it is less likely to be overshadowed than the walls. Solar collectors are often inclined at a steeper angle than the roof would otherwise be.

If intermittent heating is proposed, a compromise is needed in the interior surface finishes. A high thermal capacity is preferable for efficiently collecting solar heat through the windows, but a low thermal capacity (light or insulating interior finishes) is preferable to allow the heating to be effective quickly after it is turned on.

Since there is no cooling plant to remove unwanted heat, the sunshine must be controlled to prevent its entry when the building is warm enough without it. There is still a chance of the building overheating in summer, however, and the other means of maintaining comfort include allowing large amounts of ventilation to cool people by airflow or relying on the mass of the building to remain relatively cool throughout the day, combined with allowing cool night air to flow through it to remove the heat gained during the day. These two options require different strategies, and they are discussed further later.

A Passive-Solar Building

In a passive-solar building, the envelope (usually the walls) form a major part of the heat absorption and storage system, as well as preventing unwanted heat gains or losses by having sufficient insulation. The wall exposed to the sun may be of masonry and may be dark-colored without necessarily having added insulation. Other walls should be well insulated. Additional internal mass, such as a concrete floor without carpet, can be added.

The strategies listed earlier apply equally in this case. More emphasis is placed on the use of the envelope to maintain comfort conditions in all climates. It is assumed in a "solar" building that the occupants are prepared to take a more active role in maintaining the climatic controls; therefore, opening shutters or blinds can be used with more confidence that they will be operated at the right times. A part of the total system should be a manual of instructions on the correct operation of these devices and the opening of windows to admit cooling breezes in summer while excluding air that is too hot or too cold.

A Humid Tropical Climate

In a building in a humid tropical climate there is no need for heating, but the emphasis is on large amounts of ventilation and the exclusion of solar heat from the roof and walls. A lightweight construction with large, shaded windows will allow the interior to adapt quickly to the small reduction in temperature that occurs at night.

If there is a need for security, metal or concrete grilles or closely spaced louvers may be appropriate. Ventilation extending close to floor level must be left open at night for the comfort of the occupants.

Since solar heat must be excluded as much as possible, sunshades, louvers, and overhangs are frequently used. Orientation is important, since the sun is easily kept away from both the north and the south walls with simple overhangs, but the east and west walls are much more difficult to shade. Because of the large area of windows required for ventilation, and because narrow-plan shapes are preferred for cross-ventilation, daylighting is more easily achieved in this type of building. Furthermore, the length of day does not vary greatly from one season to the next in the tropics, so the hours of daylight availability are more uniform than is the case further from the Equator. However, heavy cloud cover is common, and the sky luminance available may be lower than in a drier climate.

An Arid Tropical Climate

With hot days and cool nights, the exclusion of solar heat is essential in an arid tropical climate. Heavy construction with small windows forces the interior to respond slowly to the extremes of both heat and cold, thereby achieving a more uniform environment inside. Insulation is essential in all surfaces, but particularly in the roof.

The intense sunlight and relatively low luminance of the clear sky, combined with small window areas, make daylighting of interiors without excessive glare quite difficult in this climate. One solution is to use windows set high in the walls under an overhanging roof. In this way, light from the surrounding sunlit ground is reflected onto the ceiling. Additional small windows would be necessary to provide a view out, possibly using reflective or low-transmittance glass to reduce the glare.

An Air-Conditioned Building

In an air-conditioned building it is desirable for the envelope to be highly insulated to prevent infiltration or escape of air and to prevent the direct transmission of solar radiation. In most air-conditioned buildings, the occupants and the lights provide enough heat except under very cold conditions or during startup in the morning. In these two situations, significant solar heat is unlikely to be available anyway, and at other times, solar gain would be undesired.

Air-conditioned buildings can be planned with greater depth from the windows to the interior than naturally ventilated and lit buildings. Not only is this possible, but the economics of obtaining good site coverage in downtown areas and of efficient floor planning without excessive corridors or excessive distances to fire stairs lead toward deep rather than narrow plans. The air-conditioning of the center zone is easier to design and install and cheaper to run than that in perimeter zones. As a result, natural lighting is seldom able to serve the whole floor, and the contribution it is able to make in the perimeter is often overlooked.

The envelope design can contribute to the thermal performance of the whole building by the control of solar penetration, by the controlled admission of daylight, and by restricting the thermal transmittance of both the glass and the opaque surfaces. Although the air-to-air transmittance may account for a relatively small proportion of the total load on the air-conditioning system, it gives rise to local radiant heating or cooling surfaces that cause thermal discomfort for people close to them.

SUMMARY

The design of the interior environment of a building should be related to the requirements of the occupants and the functions they have to perform in the building. These requirements are partly a function of human physiology, but they are also modified by the climate and other factors relating to locality.

The means of achieving the interior conditions include the active building services and the active and passive role of the building envelope. The envelope acts passively when it separates the interior from the external environment by insulation against heat and sound, by its opacity, and by providing shelter from wind and rain.

The active role of the envelope occurs when it allows light and ventilation to enter in a controlled way and when it interacts with changing heat loads to provide a reduced and delayed heat input. The ceiling and walls become part of the thermal environment, and the windows become a part of the lighting system of the internal spaces.

There is some evidence that uniform conditions are not as desirable as an environment that mimics in some way the natural cycles of climatic change. A building's mechanical systems tend toward uniformity; controlled variation can only be achieved with quite sophisticated control systems. A completely uncontrolled environment, even in the best of climates, is impractical for most purposes. The well-designed building envelope can provide the basis for a set of conditions between these two extremes. The interior usually needs fine-tuning by the use of active service systems and perhaps adjustable devices in the envelope itself; if the function of these is seen as secondary to that of the envelope, however, then the cost of operating the plant should be reduced and the satisfaction of the occupants should be increased.

REFERENCES

American Society of Heating, Refrigeration and Air Conditioning Engineers (ASHRAE). 1985. *ASHRAE Handbook 1985: Fundamentals.* Atlanta, Ga.

Cowan, H. J., and Smith, P. R. 1988. *The Science and Technology of Building Materials.* Van Nostrand Reinhold, New York, N.Y.

Harkness, E. L., and Mehta, M. L. 1978. *Solar Radiation Control in Buildings.* Applied Science Publishers, London. England.

Hopkinson, R. G., Petherbridge, P., and Longmore, J. 1966. *Daylighting.* Heinemann, London, England.

Szokolay, S. V. 1980. *Environmental Science Handbook.* Construction Press, Lancaster, England.

Threlkeld, J. L. 1970. *Thermal Environmental Engineering.* Prentice-Hall, Englewood Cliffs, N.J.

9

Air-Conditioning Performance

George Cunningham

Air-conditioning is an important component in providing a comfortable environment for human occupation. In small buildings, particularly residential buildings, in temperate climates, air-conditioning could still be considered a luxury. In major commercial buildings, it is an essential ingredient in their commercial success.

There are conflicting pressures on the designers of air-conditioning systems to maximize the amenity for occupants of the building and minimize its initial cost. Because it is an active system, thought also must be given to its ongoing performance and its operating costs.

A difficulty in defining the performance of air-conditioning is that its primary goal, human comfort, is a very subjective criterion. This is reflected in the many attempts to define human comfort in reasonable terms and the seeming conflicts among various research activities in this area (see Chap. 1). Against this background of fuzzy goals, commercial considerations, and complex interactions with other building elements, the air-conditioning designer must produce a satisfactory system.

This chapter is not intended to be a definitive text on air-conditioning design. The American Society of Heating, Refrigerating and Air Conditioning Engineers (ASHRAE) provides this in four volumes which are continually updated. Rather, this chapter is an attempt to address contemporary issues of design as they relate to current thinking on human comfort and the emphasis on energy efficiency.

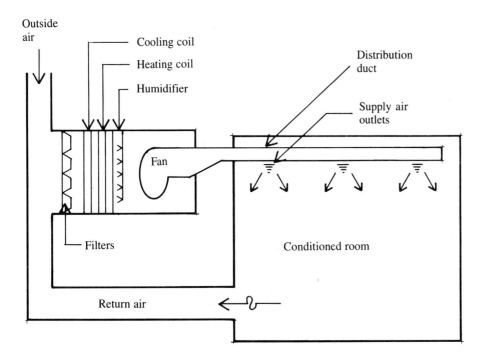

FIGURE 9.1.
A basic air-conditioning system.

THE ROLE OF AIR-CONDITIONING

To be able to fully appreciate the contribution of air-conditioning to human comfort in a building, it is helpful to understand what air-conditioning is, how it works, and some of the terms used in describing it.

By definition, "air-conditioning" is the control, within a space, of temperature, humidity, air cleanliness/quality, and air movement. An air-conditioning system, then, has elements that perform these four functions. A basic air-conditioning system is shown in Fig. 9.1. In this system,

- Temperature is controlled by a cooling coil to lower its value or by a heating coil or heating element to raise its value.
- Humidity can be lowered (dehumidification) by the cooling coil, which condenses moisture from the airstream on its cold surface. If required, humidity can be raised (humidification) by injection of steam or by water sprays. Humidification is normally required for comfort applications only in very cold climates.
- Air cleanliness and quality are controlled by two means. First, the air is filtered to remove solid particles. Second, outdoor air is introduced in sufficient quantity primarily to minimize odor buildup.
- Air movement is effected by a fan, which pushes air down the supply duct to outlets in the room. Ducting and outlets are arranged to provide good distribution with minimal disturbance from draughts.

You will see that not all the air circulated is outdoor air. Some air from the room is reused. Use of this "return air" saves on running costs.

THE INFLUENCE OF THE BUILDING DESIGN

In order to effectively allocate design effort and construction costs among the various facade elements, it is important to recognize the relative influence of these facade elements on air-conditioning and heating plant capacity and energy usage. Generally, energy usage in air-conditioning and heating systems is directly related to plant capacity. Since plant capacity is more easily and

TABLE 9.1 Major Components of Cooling Load in a Building in Sydney, Australia

Component	Percent of Total
Solar gain through windows	30
Transmission through walls and windows resulting from temperature differences between inside and outside	15
Transmission through roof	0
Internal loads (people and lights)	30
Miscellaneous	10
Outside air	15

TABLE 9.2 Major Components of Cooling Load for a Two-Story Shopping Center in Sydney, Australia

Component	Percent of Total
Solar gain through windows	5
Transmission through walls and windows	5
Transmission through roof	20
Internal loads	40
Miscellaneous	10
Outside air	20

TABLE 9.3 Peak Solar Heat Gain (W/m^2)

Latitude (south)	Window Orientation			
	North	West	South	East
0°	260	530	190	530
20°	470	550	88	550
30°	510	550	68	550
40°	520	550	57	550

quickly estimated, a first-pass assessment of facades based on their effect on plant capacity is valuable. If necessary, more detailed analyses can be performed using rigorous computer programs to establish energy usage and life-cycle costs.

In a typical high-rise office building in Sydney, Australia, the cooling load on the building (which must be matched by plant capacity) consists of the major components listed in Table 9.1. The table shows the percentage each component represents in the total building load.

For a two-story shopping center of the same floor area, the breakdown is quite different, as Table 9.2 shows.

The internal loads are a major contributor in both styles of building. About half the internal load is attributable to lighting, and the remainder is attributable to people and equipment. The designers have little influence on the load attributable to people and equipment. However, they can have a significant influence on the lighting load by using more efficient light fittings and lamps or by the use of daylighting. As a rule, for every kilowatt of lighting energy saved, there is a savings of half a kilowatt in air-conditioning plant energy. Saving this energy is not such a priority in cold climates, since the lighting energy will contribute directly to heating the building.

In shopping centers or other low-height buildings with large floor areas, the load attributable to solar and transmission gain through the roof is more significant. There are opportunities for building design to reduce this load. For example, direct solar radiation can be excluded if a covered car park is located on the uppermost level. If direct sunlight cannot be excluded, adequate insulation should be provided. In high-rise office buildings or other buildings with significant areas of external glazing, there are many opportunities for innovative design to reduce the load attributable to this source.

Solar radiation is best handled by preventing its penetration through glass. From a purely technical point of view, this is best handled by minimizing the glass area. Obviously, this is not always the best solution from an aesthetic point of view or from the viewpoint of the occupants. If there is an opportunity to orient the building to face its shortest side or, more particularly, its side with the smallest glass area in a particular direction, then preferably this should be west. A second choice is probably east, closely followed in the Southern Hemisphere by north or in the Northern Hemisphere by south. The predominantly shaded side of the building (south in the Southern Hemisphere) is not exposed to direct solar radiation, and thus cooling load is much less sensitive to the extent of glazing on this face.

A single sheet of plain glass will let almost 100 percent of the solar radiation falling on it through into the building interior. Thus an estimate of the solar radiation falling on a vertical surface of a particular orientation will give an indication of the amount of solar energy passing through a window with that orientation.

Table 9.3 gives an indication of the solar gain through plain-glass windows for the various orientations and for various latitudes. This table is for the Southern Hemisphere. In the Northern Hemisphere, it is, of course, the south-facing glass that is sunlit and the north-facing glass that is shaded.

Conversion of this information to peak cooling load is complex because of the heat-storage effect of the building structure and the fact that solar heat gain varies with time. Calculation of the hourly energy usage is even more complex (see Chap. 13) and is discussed in detail in the *ASHRAE Handbook* (1985).

Where glazing is exposed to direct solar radiation, the most effective treatment is to provide external shading devices (see Chaps. 8 and 10). These will minimize the adverse effects of direct radiation. Glass is quite transparent to solar radiation at its predominant wavelengths. However, when radiation passes into space and strikes a solid surface, the reradiated heat is of a different wavelength, and glass is more opaque to it. (This property of glass is exploited in hothouses for growing plants in cold climates and in solar collectors for water heaters.) There are further disadvantages of direct solar radiation. It can result in disability and discomfort glare and uncontrolled, widely variable lighting levels. It also can produce thermal discomfort for the occupants, despite the temperature in the air-conditioned space. External shading devices can, however, be expensive if they are not coordinated with the architecture of the facade.

It should be noted that as the latitude decreases (i.e., moving toward the tropics), effective shading is provided by even modest overhangs. In more temperate climates, overhangs can be designed to be wide enough to effectively shade the glass in summer but admit the winter sun to provide for some heating. Exploitation of this effect in commercial buildings requires considerable judgment. In many cases, because commercial buildings have significant internal heat sources, solar heat gain in winter may still be an unwanted load.

As mentioned in Chap. 8, the selection of glazing also can significantly affect the heat gain into a space. Air-conditioning engineers measure the effectiveness of glazing combinations by the concept "shading coefficient." A shading coefficient of 1 means that all the solar radiation striking the outside of a window results in heat load, whereas a value of 0.5 means that half the radiation results in heat load. A single pane of clear glass has a shading coefficient of about 1. Double glazing has little effect on the value of the shading coefficient unless the outer glass pane is tinted or reflective. The actual value depends heavily on the properties of the particular glass, and shading coefficients from 0.3 to 0.8 have been experienced with double-glazed units with outer panes of various tints.

Double-glazed windows with venetian blinds between the glass panels typically have a value of 0.3, whereas the same window with venetian blinds inside the room may have a shading coefficient of about 0.6. Problems are sometimes experienced when venetian blinds and curtains are taken into account when calculating cooling load and occupants later insist on keeping them open to admire the view.

Double-glazed windows perform much better than single-glazed windows in the reduction of heat gain from transmission (i.e., the difference in temperature between inside and outside). This is of particular importance in cold climates. An added advantage is that double-glazed windows provide a much more effective acoustic barrier than single-glazed units (an important factor in buildings on busy city streets).

If substantial design attention is applied to lighting and glazing, the other sources of heat gain need much less attention (see Chap. 10). Provided walls are reasonably well insulated (either masonry or insulated lightweight panels) and the facade is substantially airtight to prevent wind pressure from forcing unwanted outside air into the space, the building's thermal performance will usually be satisfactory.

AIR-CONDITIONING LOAD COMPONENTS

In many parts of the world, weather conditions are such that the primary requirement for an air-conditioning system is cooling. Cooling may be thought of as removing heat loads imposed on a building in order to maintain a steady temperature.

Figure 9.2 shows a simplistic building with the commonly encountered heat loads shown diagrammatically. Loads originating external to the building include solar radiation through glass, solar radiation through walls and roof, conduction through windows, walls, roof, and floor, and outside air heat. Loads originating inside the building include lighting, people, and machines and equipment. Of these loads, only conduction and outside air heat are a function of outside air temperature, and in many instances, these represent only a small fraction of the overall cooling requirement.

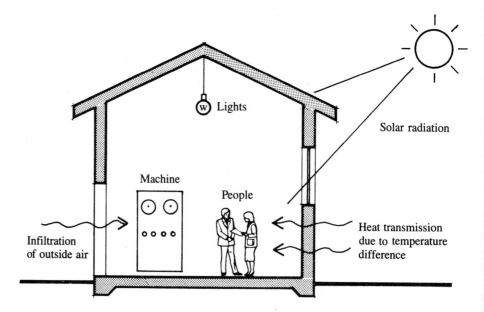

FIGURE 9.2.
Sources of building heat load.

Indeed, in many buildings, even in midwinter, it is necessary to provide a substantial amount of cooling to maintain a comfortable environment. This is particularly true of shopping centers with dense-weight walls and well-insulated roofs.

The importance of each of the load components varies from building to building. For example,

- Solar gain through the roof is a major load component in a single-story building that is insignificant in a high-rise building.
- Solar gain through glass is always a major load unless glass area is minimized or appropriate shading devices are used.
- In shopping centers where the opaque building walls admit little solar gain, lighting and people are major components of the cooling load.

In general, variable loads (e.g., solar radiation) should be minimized. They are most difficult to control if close temperature tolerances are required and in general require a more sophisticated (and expensive) air-conditioning system.

Loads affect not only plant capacity, but also energy consumption. Figure 9.3a shows the percentage of total energy usage for the various components of a commercial office building in Sydney, Australia (which has a very temperate climate). The air-conditioning and heating energy usage accounts for only 40 percent of the total, and heating alone accounts for only 10 percent. By comparison, Fig. 9.3b shows the same breakdown for a similar building in New York. The air-conditioning and heating energy usage has increased to 76 percent, and in particular, the heating has increased to over 30 percent.

Of course, in tropical areas, the total air-conditioning energy will be higher again (about 70 percent), but the heating energy will be zero. Even in Brisbane, Australia, commercial buildings have been designed in which the heating energy use is negligible.

The importance of the heating, ventilating, and air-conditioning (HVAC) element is that it is the one that can be most heavily influenced by building design. For a particular building configuration, energy used by elevators is fixed within a narrow band. Lighting energy usage can be influenced to some extent by the daylighting efficiency of the facade design, but this opportunity diminishes in buildings with large floor areas where core areas are unaffected by daylighting.

The changing importance of air-conditioning and heating energy consumption as climate changes also must be recognized. Because a good deal of

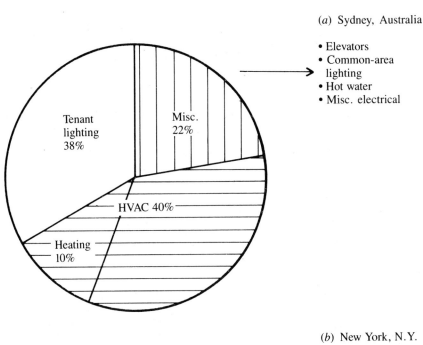

(a) Sydney, Australia

• Elevators
• Common-area lighting
• Hot water
• Misc. electrical

Tenant lighting 38%

Misc. 22%

HVAC 40%

Heating 10%

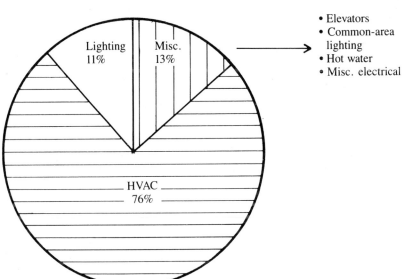

(b) New York, N.Y.

• Elevators
• Common-area lighting
• Hot water
• Misc. electrical

Lighting 11%

Misc. 13%

HVAC 76%

FIGURE 9.3.
Comparison of energy usage distribution in a typical commercial building. (a) Sydney, Australia. (b) New York, N.Y.

research into building energy efficiency has been carried out in North America, there has been a tendency to transfer their good designs of both building elements and engineering systems to other locations (e.g., Australia). The folly of transferring such a design, which heats a building efficiently, to a climate such as, for example, Brisbane, Australia, is obvious.

Many opportunities exist for reducing heat gains by thoughtful architectural and building treatment. In many instances, this approach results in lower first costs and/or lower life-cycle costs than alternatives that may require larger air-conditioning plants. Many of these approaches are discussed in detail in Chap. 8.

SYSTEM DESIGN FOR HUMAN COMFORT

Why Air-Condition at All

In small buildings in temperate climates, the need for air-conditioning is small. A combination of good envelope design and natural ventilation will ensure that indoor temperatures will closely follow outdoor temperatures, humidity levels will be satisfactory, indoor air quality can closely approach outdoor air quality, and air movement can be varied by adjusting windows and doors.

Even in larger buildings, air-conditioning may not be required. In single-story buildings with very large floor areas, natural ventilation can be provided far from the facade by roof ventilators. This is common in factories, workshops, and warehouses. In multistory buildings, natural ventilation can be provided if the depth of the floor from the facade is kept small.

However, natural ventilation cannot maintain temperatures that are lower than outside temperatures. The loads on the building, other than those imposed by temperature difference (lights, machines, people, and solar radiation), will raise the temperature of the ventilation airstream and hence the space.

Thus, in large multistory developments, those which are commercially leased in downtown city locations, air-conditioning is essential in removing heat and in maintaining the temperature in the comfort zone. Because much of such buildings is far from the perimeter and the opportunity for natural ventilation, air-conditioning also provides the means for maintaining humidity, air quality, and air movement within acceptable limits.

Even in smaller buildings in temperate climates, the variation in temperature with time may be too great for maintaining acceptable human comfort. Air-conditioning can provide the means of maintaining temperatures within limits that can be controlled as tightly or loosely as required or as permitted by the initial cost and ongoing operations budgets.

In colder climates, where the primary comfort requirement can be met by heating, this is often achieved by a combination of space heating and natural ventilation without the complexity of a total air-conditioning system. However, in such systems, in an effort to minimize energy use, natural ventilation has been minimized with adverse effects on air quality (see Chap. 2).

The technical design of air-conditioning systems (i.e., calculating the load and determining the air quantity and air temperatures) is well documented (ASHRAE 1984) and well understood by design professionals. There is, however, little documentation or guidance on selecting the appropriate system for a particular application.

Air-Conditioning Systems

Although all air-conditioning systems basically perform the functions and need the hardware shown in Fig. 9.1, there are many variations on this scheme to suit different applications.

The need for different systems arises from more complex load patterns, the degree of quality required, and first-cost versus running-cost tradeoffs. For example, in an international standard hotel, bedrooms could be air-conditioned at a low initial cost by a window unit in each room. Considerations of noise, life, running costs, maintenance costs, and convenience (not to mention aesthetics) dictate that the lowest-initial-cost solution is not appropriate for this building.

Some of the better known systems are packaged units, all air zoned units, all-air single-zone and reheat units, multizone systems, dual-duct systems, induction units, fan-coil units, and variable-volume units. All these systems have the basic elements and perform the basic functions of the system described earlier. Further details of these systems are available in the ASHRAE handbook (1984).

Such a wide range of systems results first from a need to provide the right combination of first cost and operating cost and second from a need to adequately address the zoning problem. The zoning problem arises from the fact that the time variance of air-conditioning loads changes from one area, or "zone," of a building to another. Typical zone profiles for an office building are shown in Fig. 9.4.

Figure 9.4 shows that through the occupied hours, the center zone is relatively constant because it is isolated from the influence of solar radiation

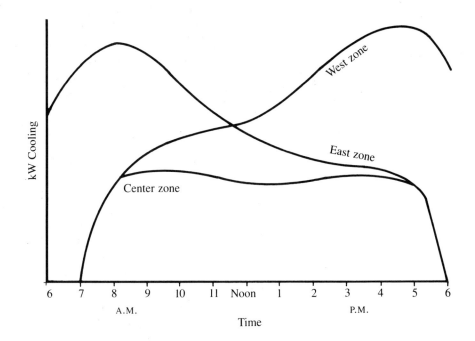

FIGURE 9.4.
Air-conditioning load: typical zone profiles.

and external temperature. The east zone typically has a high load in the morning resulting from solar load, and the west zone has a similar load peak in the afternoon. Some common types of air-conditioning systems are discussed in more detail in the following paragraphs.

Packaged Units

These units are used on smaller jobs and are built to a budget. As a result, they have poor thermal performance, their power requirements are high, and they provide minimum filtration (i.e., not a high-efficiency air filter). They have a limited life (12 to 14 years). One drawback is that access for maintenance could disturb occupants.

A particular application of packaged units is in multiroom buildings such as hotels and apartment buildings. In this application, a unit is supplied for each room and is connected to a common condenser water-reticulation system. In some applications, the individual units are used for both cooling and heating and the water-reticulation system used as both a heat sink and a heat source. These systems are fully discussed in the section entitled Heat Pump Systems (page 165).

All Air Zoned Systems

Each zone has a separate conditioner. Often a whole building facade is one zone; i.e., there is no floor-to-floor control and no room-to-room control.

These systems are relatively inexpensive, but they have limitations where buildings with significant window areas are subject to moving shade patterns from, say, adjacent buildings. They are very appropriate in landscaped offices where intermixing of air from different areas minimizes the influence of localized load variations.

All-Air Single-Zone Reheat Systems

In this system, each zone is supplied by sufficient constant-temperature cold air to satisfy its maximum load. Zones not requiring maximum cooling are reheated by a duct-mounted electric or hot-water heater.

To satisfy varying zone load requirements, simultaneous cooling and heating occurs, but it can be kept to a minimum by providing good solar protection.

Air-Conditioning Performance **157**

This system is very economical and simple to install, but it is usually uneconomical to operate, especially if there are large single-glazed windows or other design load variations. This system is most appropriate in buildings where zone-to-zone loads vary sufficiently to require individual control of temperature but not sufficiently to justify the cost of a separate conditioner for each zone.

Multizone Systems

In these systems, a single conditioner serves a number of zones, typically up to 10. In the conditioner, after the supply air fan, the airflow is divided. Part of the air flows through a heating coil located in a "hot deck" and part flows through a cooling coil in a "cold deck." Air ducts to each of the individual zones are located so that they can draw air from each of the hot and cold decks. At the inlet to these ducts, mixing dampers control the proportion of hot and cold air distributed to each individual zone in response to the demands of that zone.

In early years, many multizone systems were controlled to maintain constant hot- and cold-deck temperatures. In such systems, simultaneous heating and cooling could not be avoided, and thus they were not energy efficient. More modern control systems have enabled hot and cold decks to be controlled in response to zone demands; e.g., the cold-deck temperature is just cold enough to satisfy the demand of the zone with greatest cooling demand when all the air to that zone is drawn from the cold deck. Such control strategies minimize the need for simultaneous heating and cooling inherent in the system.

This system is appropriate for small, mixed-use buildings with a small number of discrete zones. It has the advantage, from an operational and maintenance point of view, that all mechanical equipment is confined to the plantroom. It can be quite energy efficient if thought is given to control strategies.

Dual-Duct System

From a thermal performance point of view, the dual-duct system is identical to the multizone system. However, to allow a larger number of zones to be accommodated, the hot and cold decks are extended to hot and cold ducts running throughout the building. Mixing of hot and cold air is done locally at the zone by means of a "mixing box" that draws air from either or both ducts and distributes it to its zone.

This system can be used for quite large buildings with large numbers of zones. The system was very popular during the 1960s and 1970s, but it has given way to the variable-volume system. It has some advantages over the variable-volume system in its ability to cope with winter heating requirements, however.

Induction-Unit System

This system was extremely popular in major office buildings in the 1960s and 1970s. In this system, each perimeter-zone office has an induction unit. Only a small air quantity necessary for ventilation is treated in a central conditioner and ducted to each induction unit, where it is discharged through small nozzles at high velocity. Room air is induced through a small coil which is part of the unit by virtue of the high-velocity primary air. The amount of cooling is regulated by chilled water flow through this coil (usually about 11°C, since colder water would cause condensation on the induction-unit coil) and controlled by a room thermostat and automatic valve.

The ducted air is dehumidified at the central conditioner to absorb room latent load and to avoid condensation on the induction-unit chilled water coil, and its temperature is regulated to balance outside conditions, i.e., hot when cold outside, and vice versa.

In cold climates, hot water is provided to the secondary coils to provide additional heating capacity. The center zone is handled by a separate single-zone conditioner.

The system is relatively expensive, but it gives individual control to each office or group of offices. It is flexible, since units can be changed or relocated to accommodate varying occupancy or usage patterns.

Since primary air risers are sized for high velocities and part of the cooling is done with chilled water in the space, riser sizes are small, thus minimizing the use of floor space. The system has some requirement for simultaneous heating and cooling, however, although this can be minimized by good system design.

Fan-Coil Units

Each fan-coil unit has its own fan, cooling/heating coil, and air filter. Fan-coil units come in many sizes. In the extreme case, a commercial building could be provided with a fan coil in each individual perimeter office. The fan-coil units are usually supplied with chilled or heated water from a central plant.

Outside air can be pretreated and ducted from a central plant, or it can be brought in through the facade. Where a large number of units are necessary, the system can be very expensive to install. The system provides good control of individual zones, is low on energy cost, but is high in maintenance cost, since there are individual fans and filters in each unit.

Variable Air-Volume System

Variable air-volume systems have received a great deal of attention since the focus on energy consumption. In major buildings, such systems provide good zone control while minimizing energy usage. To operate successfully, some technical aspects must be carefully considered. These include provision of good air movement over a range of flow rates, variation in air quantity in one zone affecting supply to adjacent zones, and accommodating winter heating in what is essentially a variable-cooling-capacity system.

SYSTEM DESIGN FOR ENERGY CONSERVATION

Air-Distribution System

Considerable energy is used by the fan systems that distribute air around a building. Many energy-efficient design strategies may be adopted that are applicable to both constant and variable air-volume systems. The theoretical fan energy is a function of air quantity and the pressure at which this air must be delivered. It is therefore evident that the two strategies for minimizing fan energy are to minimize air quantity and to minimize distribution pressure loss. These will be examined separately.

Sensible cooling capacity is proportional to air quantity and the temperature difference between the supply air and the room. The air quantity required in an air-conditioning system is generally determined by knowing the load and using an "acceptable" difference between supply-air and room-air temperature. Thus the lowest air quantity will result when the lowest possible supply-air temperature is chosen.

There are, of course, constraints on the choice. The most significant of these is the requirement to keep room humidity at a comfortable level. A look at the psychrometric chart in Fig. 9.5 will show the effect of choosing too low a supply-air temperature.

Assuming the ratio of sensible to total load in the room to be 0.85 and the presence of a cooling coil that will cool the air to within 0.5°C of its dew point, then one can see the effect on room relative humidity of supply-air temperatures of 12 and 5°C are chosen. When the supply-air temperature is 12°C, the resultant relative humidity in the room will be 51 percent. With a supply-air temperature of 5°C, the resultant relative humidity in the room is only 37 percent, a condition considered by most to be unacceptable.

As supply-air temperature is reduced, the refrigeration plant providing the cooling effect will require more energy to provide the same amount of cooling capacity. Thus, in selecting a supply-air temperature, the designer must balance the reduced fan energy against the increased refrigeration plant energy.

Another factor that works against choosing a low supply-air temperature is the effect of freezing of condensate on the cooling-coil surface. If the supply-air temperature is lowered to a point where the coil surface temperature falls below freezing, then the condensate forming on the coil will freeze. This will effectively block the passage of air, which in some cases can result in severe problems for the refrigeration plant, not to mention the obvious adverse effect on air distribution.

One must always bear in mind that one of the aims of air-conditioning is to provide satisfactory levels of air movement within a space, i.e., with neither excessive drafts nor dead spots. Too low an air quantity may lead to this result. It is the belief of some that unacceptable air movement will occur when the supply-air quantity falls below 6 air changes per hour, and indeed, this is often so if square or round ceiling diffusers are used. Full-scale tests both in Australia and elsewhere have demonstrated that satisfactory air movement can be achieved using air quantities as low as 3 air changes per hour distributed through carefully selected slot-type diffusers.

It is of interest that currently the city of Melbourne requires a minimum air-supply rate of 6 air changes per hour, as does the Australian Department of Housing and Construction for buildings occupied by commonwealth government departments. On the basis of tests and real-life demonstrations, the Department of Housing and Construction has accepted some installations with air-change rates as low as 3 air changes per hour as meeting the intent of their requirement.

There are a number of approaches to minimizing the pressure loss in air-distribution systems. The pressure losses in a system can be broken into three distinct groups: (1) the air-handling unit, (2) the ductwork system, and (3) the terminal distribution unit.

In the air-handling unit, pressure drop is essentially a function of velocity. Reduction in velocity across filter banks and cooling and heating coils results in an increased face area of these items and a corresponding increase in cost. For many years, a "rule of thumb" velocity of about 3 m/s was used. This was before the energy crisis of 1975. In the late 1970s, the School of Mechanical Engineering at the University of New South Wales investigated the optimum velocity for

S1 = Supply-air temperature—12°C
R1 = Corresponding room condition
S2 = Supply-air temperature—5°C
R2 = Corresponding room condition
(Room sensible heat factor = 0.85)

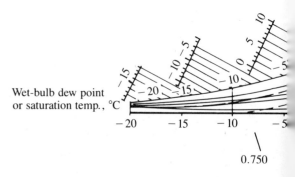

Wet-bulb dew point or saturation temp., °C

0.750

FIGURE 9.5.
Effect of air temperature on room humidity.

a conditioner (Chan 1978). A computer program was developed that took into account not only the costs of the components and fan energy, but also the costs of bigger equipment housings. Based on material and energy costs in 1978, the optimal velocity for a constant-volume system was found to be around 1.6 m/s. Since variable-volume systems run for much of their operating time at less than the design air quantity, it is likely that a design velocity of around 3 m/s is close to the economic optimum for these systems. Since parameters change from time to time, it would be wise to check the optimal conditioner velocities, particularly in large projects.

The pressure loss in the ductwork system is also a function of velocity. Larger ducts run at lower velocity and lower pressure loss. They also use more valuable building space. The effect of higher velocities can be offset somewhat by sizing ductwork using the "static regain" method. This method requires considerable computation if done by hand, but today many computer programs are available to take the tedium out of static-regain calculations. An added benefit of this method of sizing is that it can result in similar pressures being developed at each branch in a system and hence it can save considerable time during the final balancing of the air-distribution system.

Another cause of pressure losses in ductwork systems is the fittings used to change the direction and velocity of the airstream or the shape of the ductwork to avoid obstructions. Design of such fittings is well documented

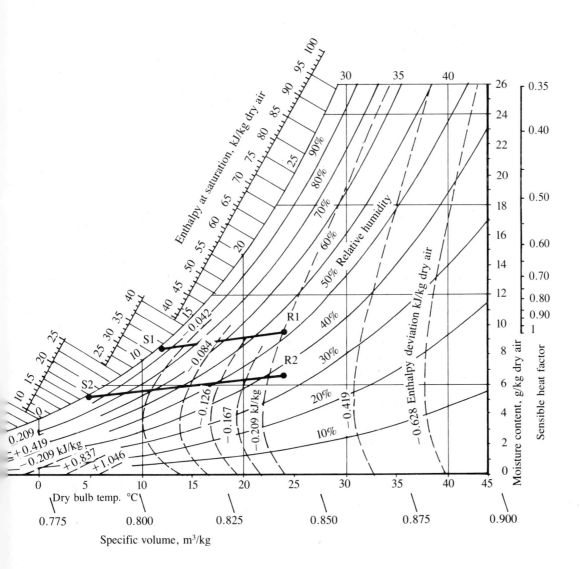

(ASHRAE 1985), and attention to detail in their design will minimize pressure losses. Of course, the best approach is to minimize the use of such fittings at all. This can often be achieved by careful coordination of the air-distribution system and architectural design so that ducts can be accommodated with the least complex layout and minimal changes in direction. Size, shape, and location of riser shafts can all have a significant bearing on fan energy.

One decision that requires close coordination of services and architectural design is the number and size of plantrooms. There has been a recent trend toward providing air-handling plants on individual floors rather than in a central plantroom serving many floors.

The advantages of this "distributed" plant approach are as follows:

- Reduced length of air-distribution ductwork resulting in reduced pressure loss and in turn reduced fan energy
- The opportunity to use lower-cost, factory-assembled air-handling units
- The opportunity to bypass some of the stringent fire code requirements for air-handling systems serving more than one floor (This opportunity may be short-lived as fire codes become increasingly stringent.)

The disadvantages of distributed air-handling systems are as follows:

- The cost of providing multiple "outside air economizer cycles" (see page 166)
- More items of plant result in more maintenance and hence higher maintenance costs
- Factory-assembled air-handling plants may require more maintenance and have shorter working lives than "builtup" central conditioners
- Maintenance must be carried out on occupied floors

The selection of terminal air-distribution equipment (e.g., variable-air-volume boxes, induction units, and so on) also should be done with a view to minimizing pressure losses. Until recently, this required choosing "oversized" and hence "overpriced" units, and the selection was a tradeoff between first costs and running costs. Since the impact of the energy crisis, equipment suppliers have undertaken considerable research to provide low-cost devices that operate with low pressure losses. Careful attention to detail during the selection process will result in selection of the most appropriate equipment.

Variable-Air-Volume System Fan Control

There is no doubt that the single greatest impact on air-conditioning design in the 1980s has been the proliferation of variable-air-volume systems. Part of the energy savings results from their ability to minimize the amount of simultaneous heating and cooling, which is a penalty imposed by many other systems. Another reduction in energy is the result of the reduced fan power required when the fan is supplying a lower air quantity at reduced cooling load. Care must be taken in selecting the appropriate air-volume control for the application.

In small systems, a fan with a forward-curved impeller is often used. As air quantity to the variable-volume boxes is reduced, the air pressure in the system "backs up," reducing the supply air quantity. On these particular fans, there will be an accompanying reduction in power. This is a very economical method of fan control in both first cost and running costs. However, there is a limit to how far this strategy can be exploited before the fan stalls.

In larger systems, the three major methods of fan volume control are variable-inlet guide vanes, variable-speed fans, and variable blade pitch on axial fans. For many years, the high cost of the latter two methods resulted in variable inlet guide vane control being the most commonly used in commercial applications. There were problems in long-term reliability and/or high maintenance costs and a limitation on the reduction achievable before the fan stalled.

This method did, however, produce significant reductions in power at reduced air quantities.

Driven by the energy crisis and developments in electronic control systems, variable-speed motors now represent an effective economical approach to fan volume control. However, one must take care to select a variable-speed control that is designed to save power. Some, designed for process-control applications, save little power but have very accurate speed control. This, of course, is not the selection criterion in air-conditioning systems.

Variable-pitch axial fans provide results similar to those of variable-speed motors. Many curves are published comparing the relative merits of various fan volume controls. They seem to vary depending on the particular interests of the provider. To be sure, it is suggested that suppliers provide power versus air-quantity curves for each major application and final selection be made on the basis of this information.

COOLING AND HEATING PLANT

Having decided on the most appropriate air-conditioning system for a building, it is necessary to consider the various options for cooling and heating plant. While the design of such plants is well understood and documented (ASHRAE 1984), there is little guidance as to the appropriateness of a particular system to a particular building. Systems need to be evaluated on both a technical and commercial basis.

The Refrigeration Cycle

Cooling for the majority of air-conditioning systems is provided by the mechanical vapor-compression cycle, which functions in principle in the same way as a domestic refrigerator. Understanding the refrigeration cycle is made easier if it is considered not as the provision of cooling but rather as the removal of heat. Whether it be a domestic refrigerator or an air-conditioning application, the problem is the same—there is a need to remove heat from a cold space and reject that heat to warmer surroundings. This, of course, defies the rules governing simple heat transfer, and thus a more complex mechanism (the refrigeration cycle) is required.

In the basic refrigeration cycle, a cooling coil containing liquid refrigerant at low temperature and pressure is placed in thermal contact with the load (e.g., food in a refrigerator or air in an air-conditioned building). Heat flows from the load to the cooling coil, where the heat evaporates the liquid refrigerant. Thus this cooling coil is also known as the "evaporator."

The outlet of the evaporator is connected to the suction of a compressor, which keeps the evaporator at a low pressure and subsequently maintains the evaporation process at low temperature. The compressor raises the pressure of the refrigerant vapor and in doing so raises the temperature at which it will condense back to a liquid. This condensing temperature will be higher than that of the fluid used to cool the refrigerant. This cooling fluid is usually ambient air or water from a pond or cooling tower. Thus the heat extracted from the cold space is rejected to a warmer environment.

To complete the refrigeration cycle, the hot, high-pressure liquid refrigerant passes through a throttling device and becomes a cold, low-pressure liquid entering the evaporator.

In some applications, the mechanical compressor is replaced by a chemical absorption/desorption process. Such refrigeration plants are known as "absorption systems." They are inherently far less efficient than mechanical compression systems, but they are powered directly from a heat source (e.g., steam or gas flame) rather than an electric-powered motor. The main application for absorption systems is where the heat source is much less expensive than electric power.

Primary versus Secondary Refrigerants

In many air-conditioning applications, a cooling coil that evaporates refrigerant within its tubes is used to directly cool the air. This is common in room air-conditioners, packaged units, and systems with one or only a small number of zones.

There are physical difficulties in transporting refrigerant over large distances. Thus, in systems with many cooling coils (e.g., fan-coil units systems and induction-unit systems), an easily distributed secondary refrigerant is used. The most common secondary refrigerant is water because it is easy to distribute and inexpensive should a leak occur. In these systems, the water is chilled by flowing over tubes in an evaporator. The chilled water is circulated to the cooling coils in individual air-conditioning units, where it picks up heat from the air. The water is then returned to the evaporator, where it is recooled.

In large systems, the losses in efficiency inherent in using a secondary refrigerant are a small penalty to pay for the improved flexibility and trouble-free operation.

Heat-Rejection Options

Heat from the condenser must be rejected to a suitably large heat sink. In the case of a domestic refrigerator, the heat sink is the air in the kitchen. In the case of room air-conditioners, it is the air outside the room. The condensers in many large systems are also cooled by ambient air. This is common practice in packaged air-conditioners or where refrigeration plants can be located close to an outside wall or roof. Limitations on the use of ambient air as a heat sink occur when long runs of refrigerant pipe are required between the refrigeration plant and a suitable condenser location. In larger plants, the sheer size of the condensers and the required quantities of cooling air make their use difficult.

Another possible heat sink is water. This is particularly attractive if the temperature of the water is lower than that of ambient air. A lower water temperature can result in a lower condensing temperature, which will result in higher efficiency. Sometimes nearby lakes or harbors can be used as a heat sink. This option may attract operational costs in keeping the heat-exchanger surfaces free of marine growth. There also may be problems of thermal pollution of the waterways.

The majority of water-cooled systems make use of a water cooling tower. These devices bring the condenser water into contact with ambient air. This is usually done by spraying or splashing water onto thin, closely spaced solid surfaces such as timber slats or corrugated plastic sheets. Ambient air is drawn or blown through the spaces between these surfaces. By evaporating a small percentage of the water into the air, the remaining water is cooled significantly.

Except in very humid climates, it could be expected that condensing temperatures, when using water-cooled condensers and cooling towers, would be about 5 to 10°C lower than those achieved using air-cooled condensers. This would result in refrigeration system efficiency increases of some 5 to 10 percent.

In some areas, even though they offer technical and space saving advantages, cooling towers may not be possible because of cost or even the lack of availability of suitable water. For example, in Hong Kong, the law prohibits the use of cooling towers for commercial air-conditioning applications.

Evaporative Cooling

In climatic regions where large differences between ambient dry-bulb and ambient wet-bulb temperatures consistently occur, evaporative cooling may provide an effective alternative to mechanical refrigeration. Evaporative coolers draw air over wetted pads, where some of the water evaporates. The heat required to evaporate the water is extracted from the airstream, thus lowering the temperature of the airstream. However, the humidity of the airstream will increase. Provided the end result is a more comfortable combination of temperature and humidity (see Chap. 5) than would otherwise exist, evaporative cooling can be effective.

Heating

In cold climates, it is necessary to heat the air to achieve comfort conditions. This may be done directly by means of an air furnace or indirectly by heating water or steam and distributing this secondary fluid to heating coils in individual conditioners. Where oil- or gas-fired heaters are used, indirect heating has the advantage of flexibility and ease of distribution of the heat to multiple locations from a single central boiler. Direct heating is more appropriate in applications where only a single heating zone is required. When electric heating is used, this can be easily and more economically distributed without the need for a secondary fluid.

A major decision in the choice of a heating system is the choice of fuel, which is commonly oil, gas, or electricity. The parameters that must be considered when making this choice include availability of fuel (including reliability of supply), initial cost of installation, fuel cost (and its stability), maintenance costs, safety aspects, and potential pollution problems.

Heat Pump Systems

Heat pump systems take advantage of the principles of the refrigeration cycle to increase the useful heat generated for a given electric power input. Even though outside air is cold during winter, because it is above absolute zero temperature, it still contains some heat. The most common heat pumps take advantage of this fact. The cold outside air is passed over a refrigeration system evaporator, where heat is extracted from it and transferred to the evaporating refrigerant. This refrigerant vapor is then pumped from a low-temperature, low-pressure state to a high-temperature, high-pressure state. The vapor is then condensed at a high temperature in a condenser over which air from the conditioned space is passed. The heat given up by the condensing vapor raises the temperature of this air. This process is essentially the same as the refrigeration cycle described later.

Sometimes ambient air is unsuitable as a heat source for heat pump systems. Other sources include water from lakes, bores, harbors, and so forth which are suitable, provided there is no risk of freezing the water on the evaporator.

In some climates, solar heating is used to keep the air heat source at a more suitable condition. A most effective heat source is the building itself, particularly if one part of the building is being cooled while another requires heating. The heat extracted from the area requiring cooling will provide a heat source for the evaporator, while the heat being rejected from the condenser is used to heat the air in the other part of the building.

An extension of this principle is used in multiroom buildings such as hotels and apartment blocks. In this application, an air-conditioning unit serves each room. In each unit, a valve is used to redirect refrigerant so that the unit can act either as a conventional cooling unit or as a heat pump heating unit. The units are all connected to a common water-reticulation system. This enables heat generated in parts of the building that require cooling to be transferred to the water-reticulation system to provide the heat source for areas requiring heating, and vice versa.

Should the majority of units be operating in the cooling mode, then the net requirement for heat rejection is accommodated by a cooling tower. If the majority of the units are operating in the heating mode, then it may be necessary to provide additional heating to the water-reticulation system via a boiler.

Heat pumps usually provide about 3 to 4 kW of heating for each kilowatt of electrical power input. In many areas of the world, the peak electrical demand placed on the generating and distribution facility occurs in winter. Thus the use of heat pumps is encouraged by electric utilities because the heat pumps reduce the requirement for a large plant that is operating at less than peak load for much of the year. There are obvious financial benefits to the user in that the convenience of electric heating is available at a cost competitive with that of other fuels. This is particularly so where the need for summer cooling and winter heating are of the same order of magnitude. The most popular application of heat pumps is in room air-conditioners.

FANCY PLANTS

A major influence on the energy efficiency of a building is the selection of the air-conditioning and heating system design. Before the escalation of energy prices in the 1970s, little emphasis was placed on energy efficiency of air-conditioning systems and much more emphasis was placed on low initial costs. The result was that during that period, many simple, inexpensive, but quite energy-inefficient plants were designed and installed. The major drawback of these systems was that they encouraged simultaneous heating and cooling in order to control space temperatures. In the late 1970s, there was a strong swing toward more energy-efficient systems. The most popular of these is the variable-volume system. In general, this type of system delivers varying quantities of cold air to the different zones within a building to match the prevailing heat gain. In winter, when the heat transfer at the perimeter of an office building becomes a loss rather than a gain, a simple variable-air-volume system cannot match the load. There are many innovative ways in which the variable-air-volume system is adapted to cope with this problem. The final design is heavily influenced by the local climate and by commercial considerations.

In temperate climates, another popular energy-saving technique, particularly in larger projects, is the use of outdoor air for cooling. This system is often called the "economizer cycle." It works on the principle that whenever the outdoor air contains less heat than the return air from the space, the conditioner will draw in outside air and cool it to the required temperature. If conditions are such that net heating of the space is required, the outside air will be shut down to a minimum. The cost of maintaining the air dampers that control the flow of outside and return air is very significant. Thus the system is often not considered in very cold climates where the outside air dampers would be shut for much of the year. The economizer cycle also makes no sense in tropical climates.

Some systems take advantage of the storage capacity of the building itself. This can be effective where there is a large daily range of ambient temperature, that is, where there is a large difference between the minimum and maximum temperature over a 24-hour period. This phenomenon is typical of inland desert areas. Thus it may be advantageous in Australia in places such as Alice Springs, but it is not applicable to the capital cities, which are coastal. These systems use a "night-purge cycle," in which cold night air is circulated past the massive structural elements of the building (e.g., concrete floor slabs) to precool them without refrigeration to start the new day.

In areas where the daily range of temperatures is small, the cooling effect is minimal and the savings in refrigeration energy are more than offset by the energy to run the circulation fan during the night. The system becomes more attractive if fans can use electric power during the night at off-peak rates. This is not currently available in Australian capital cities.

Where all the favorable conditions apply, this strategy has energy-savings potential. It is currently being used in an energy-efficiency project in California state office buildings, particularly in Sacramento, where most favorable conditions occur (Davis 1981). It is anticipated that some measurable results of the value of this system will be forthcoming from this project.

Many fashionable energy-saving systems are considered from time to time. While many of these are very cost-efficient in particular types of buildings, by and large their costs cannot be justified in normal commercial buildings.

One of the most publicized of these is "heat recovery," in which heat generated by the lights in a building or the computers or as a byproduct of some energy process is transferred to areas that require heating. This is no doubt a valid proposition in the cold conditions experienced in a North American winter. In the temperate conditions experienced in major Australian capital cities, the amount of heating required is minimal and the high cost of collecting and redistributing this waste heat usually outweighs the benefit.

Heat-recovery systems could be of value in some buildings in temperate climates. The initial questions to ask before considering a heat-recovery system are

Is heating required?
At the time heating is required, is waste heat available?
Can this heat be economically transported from its source to where it can be used?

Example 1

In winter in a Sydney commercial building, heating is required on the perimeter while cooling is required in the interior. In this case, some heat-recovery methods that could be considered are as follows:

1. Draw hot air from light fittings and redistribute it to the perimeter.

   The problem with this proposal is that heat is available at relatively low temperatures, so that large quantities of air must be moved to transfer the required amount of heat. The fan energy to move this heat makes the cost of its recovery ineffective.

2. Use the condenser-water heat rejection from the center or core zone cooling plant to heat the perimeter.

   The problem with this proposal is that is will usually be more economical to cool the center zone via an outside-air economizer cycle, in which case the refrigeration plant would not be operating.

Example 2

A major computer center has an attached office building with perimeter glazing. The office needs heating in winter, while the computer facility still needs cooling.

Because of the requirement in computer rooms to achieve close tolerances on temperature and humidity, outside-air cycles are rarely used for such facilities. The condenser-water heat rejection from the computer facility refrigeration plant would therefore be available to provide an economical heat source for an office block.

Example 3

An office block is associated with an industrial plant that produces abundant waste (e.g., in steam condensate). This heat could be used directly to heat the office block.

Solar heating and cooling is also well publicized. However, at today's costs of solar collectors and energy, solar heating can be marginally cost-effective for domestic hot water but usually not cost-effective for space heating and even less effective for cooling.

Storage systems for cooling do not usually save energy in a commercial building. In fact, they may use more energy when cooling the stored water to lower than normal temperatures. Their advantage is in saving energy costs by doing cooling work at times of lower electricity costs. The cost of the storage tanks is high. The risk that tariffs will change in the future and possibly render the system nonviable also must be considered. The viability could be improved in the future if specifically developed latent heat storage can be incorporated into structural elements (Rosenfeld and de la Moriniere 1985).

Storage systems do have an application where high peak loads occur for very short periods of time. Some examples of this might be in heating or cooling a sporting or entertainment venue that is used for only a few hours per week. If the storage is located in an inexpensive area of the building, in inexpensive real estate, so much the better. In general, simple, well-engineered systems that avoid simultaneous heating and cooling of the space provide the most cost-effective solutions.

ONGOING MANAGEMENT

Both the energy effectiveness of the design of a building and its services system and its ability to provide the conditions within the space can be significantly eroded if ongoing system maintenance and energy management are not performed well. The various energy-management systems available can be categorized on the basis of the types of energy-management strategies they can perform.

Switch It Off

If equipment is off, it cannot use energy. Ill-considered applications of this strategy also may result in an unacceptable environment for building occupants. Thus we need an energy-management system that will switch equipment off only when that equipment is not needed.

The simplest device for performing this function is the humble time clock. Provided the time programs are simple and accurate, then a simple time clock is adequate. Simple time clocks are limited in their flexibility. They need to be reset to recognize daylight savings or holidays. If this resetting is not carried out, the time clocks could waste energy by, say, running lights and plant during the holiday period.

Recently, advances in microprocessor technology have made available the multiprogrammable time clock. This device can be programmed ahead typically for 365 days from a simple keypad. Thus account can be taken of public holidays and daylight savings. Moreover, the units can accommodate a number of different time schedules, enabling different time programs to be used for lighting, air-conditioning, and/or different areas of a complex (e.g., office and retail). Multiprogrammable time clocks are effective and relatively inexpensive, representing good value for money for most applications and for many the highest level of sophistication that can be commercially justified.

In some applications, particularly conference rooms and other areas of intermittent usage, an effective means of switching lights or other plant services on and off is a motion detector. Such devices, using infrared or low-intensity microwave sensors, detect motion within a space and operate a switch. The sensor can be tuned to respond to humans, and time delays can be provided to keep plant and lights on for a reasonable time should the occupants remain motionless for a significant period of time. These devices switch energy virtually on demand.

Time-Based Control Strategies

A number of systems are available which act as a time clock and which have enhancements to switch some plant services on or off in a way that more closely approximates demand. They are essentially microprocessor-based time clocks which adjust the start and stop times in response to other measured conditions.

Optimal Start Time

One such enhancement is the "optimum start time technique," which can be applied to the starting (and in some cases the stopping) of air-conditioning and heating plants. Such a facility should have a means of measuring outside and inside air temperatures and a method of establishing the thermal inertia of the building. By using this information, the microprocessor is able to calculate the start time required to just achieve an acceptable space temperature at the scheduled occupancy time for the building.

The method used for establishing the optimal start time should be as effective for cooling applications as it is for winter heating. Many programs developed in North America have concentrated on winter heating requirements, with much less attention paid to the cooling requirements more prevalent in Australian cities.

Duty Cycling

Another sophisticated time-switching system is known as "duty cycling." This strategy also can be applied to air-conditioning plants and works on the basis that, provided temperatures are satisfactory in the building, the plant can be switched off entirely for a short period each hour, thus saving energy. While this strategy may save energy, it has some shortcomings in the area of occupant reaction. Air-conditioning controls not only temperature and humidity within a space, but also air quality and air movement. Duty cycling, by stopping the supply air fans, effectively stops the introduction of outside air into a building. This affects the "quality" of the air, and it is doubtful whether it complies with local health codes.

Stopping the supply airflow also stops air movement, and this can lead to occupant complaints. Occupants notice the supply air stopping not only by its effect on air movement, but also audibly, and these air changes throughout the day lead to complaints from building occupants. It is also possible that indiscriminate duty cycling, while appearing to save energy, could increase peak demand and thus increase energy costs. Any attempt to implement duty-cycling strategies should be approached with caution.

Load Shedding

Another strategy is known as "load shedding." This strategy sheds predetermined loads as electricity demand increases. Some of these loads may be nonessential, such as decorative fountains, and some may require a good deal of judgment, such as, for example, part or all of the refrigeration plant. The aim of the strategy is to reduce peak electricity demand and thus decrease the energy-cost component associated with electrical demand. The strategy is effective only where the electricity tariff for the building has a demand component. Care also should be taken when switching off other than nonessential loads.

Other Control Strategies

All the foregoing strategies are involved in switching on and off. Other strategies can be adopted that make use of available "free" energy sources or make better use of the energy available.

Outside-Air Economizer Cycle

One of the most common is the "outside-air economizer cycle," which uses outside air for cooling whenever conditions are suitable. Such a strategy has been implemented using traditional controls for over 20 years. Microprocessor controls merely make it easier to implement.

Chilled-Water Reset

Another strategy raises the air-conditioning system's chilled-water temperature where this can be tolerated and can select the best combination of chiller plant under given circumstances.

Variable Temperature Control

While air-conditioning systems allegedly control temperature in a space to a single value within a small tolerance band, in practice this is not the case. Because there must be some temperature difference to cause a change in plant operation, typically buildings run a little colder in winter than in summer. For example, cooling may be initiated on a temperature rise to, say, 23.5°C and terminated when it falls to 23°C. As temperature falls further to 22°C, heating may be initiated, and heating is then terminated as temperature rises to 22.5°C. The interval between termination of cooling and termination of heating is known as the "dead band," where neither heating nor cooling occurs.

To conserve energy, this dead band can be extended so that heating is initiated only when temperature falls to 21°C and cooling is initiated when the temperature rises to 25°C. This would undoubtedly result in energy savings. It would also result in indoor temperatures tending to follow outdoor temperatures in which case, if Auliciem's proposal in Chap. 5 is adopted, human comfort also would be improved.

Direct Digital Control

A recent development in microprocessor-based control is known as "direct digital control" or "distributed digital control" (DDC). This system replaces traditional analog control with digital control. As well as its other advantages, DDC can be put to use to save energy. Its ability to control to closer tolerances (particularly if proportional plus integral control is used) enables wider dead bands (where neither heating nor cooling energy is used) to be maintained. It also enables plant services to be more easily controlled directly from space demand than are earlier systems. Direct-digital-control systems usually include many sophisticated time-switching functions.

Keeping It in Tune

Second in importance to switching it off is keeping it in tune. Plants that are operating correctly and efficiently usually use less energy than those which are not. The essential requirement for keeping plants in tune is knowing that they are out of tune, preferably when they are starting to go out of tune.

Many energy-management systems rely heavily on this monitoring capability. The hardware can vary from specialist microcomputers to personal computers, minicomputers, and mainframes. The number of points monitored can vary from tens to tens of thousands. Some systems offer monitoring only, and these can run in parallel with conventional control systems.

Some systems offer monitoring with central control. In these systems, the traditional hardware logic is replaced with software-based strategies in the central computer. This software can include specific energy-management strategies already discussed. The disadvantage of a central control facility is that if the central computer fails or the communication lines fail, the whole plant will be disabled.

More recently introduced are systems that can provide central monitoring and can interface with direct-digital-control units distributed around individual plant systems (e.g., each air-conditioning unit). In new buildings, such a system provides the advantage of modern solid-state control without the dependence on a single central computer. Some energy-management routines can be implemented at the local level, or where appropriate, they can reside in the central computer for "global" application.

When assessing information appearing on a monitor screen, it is important not to believe it with blind faith. It should be recognized that many circumstances affect the reported data. These include such minor problems as calibration drift of sensors and corruption of data during transfer by stray electric current. Good engineering practice during installation can minimize these problems.

A common problem in low-cost installations is that the real result is not monitored at all. For example, it may be necessary to open an air damper. A control signal may be given from the computer to do this and a signal may be sent back confirming that the damper is closed. However, the wire from the computer to the control motor may be broken, the damper motor may be burnt out, or the mechanical linkage from the motor to the damper may be inoperative. In any of these circumstances, despite what the monitor shows, the damper will probably not be closed. A greater degree of certainty is achievable if the monitored message is initiated by a microswitch which senses that the damper blade is actually physically closed.

It is essential that the person carrying out the monitoring be aware of the details of the plant being monitored and the limitations of the monitoring system to ensure that data are correctly interpreted. Whichever of the monitoring systems is selected, it is important to recognize that knowing that equipment is out of tune is just the first step. Monitoring more and more points does not get problems fixed. Thus it is important to have the personnel backup to monitor data, diagnose problems, and have them fixed.

To this end, it is important that monitoring systems provide the best possible human interface. Data should preferably be presented in English and in an easy-to-read format rather than a string of meaningless numbers or code. Graphic representation of plant data is very useful for diagnosing problems. It gives the operator an instant picture of what the plant is doing.

The number of points monitored should be the minimum required to make diagnosis possible. When deciding whether or not to monitor a point, ask the question "If I obtain information about this point, will I be able to react to it?" If the point is on the list just because it looks interesting, delete it. Another technique to relieve the data jam that can occur with some monitoring systems is for the data to be presented only when they are abnormal. When things are going well, the screen should be blank. The operator can, of course, interrogate any particular point or system on demand.

Integrated Systems

There is a trend in the marketplace toward integrated systems. One approach is to use a single computer to handle data for many different systems—energy management, plant control and monitoring, fire safety, and security access. This computer can bring all the information to a single location to either let one person know everything about the building or to require a number of people to gather around a single data facility.

An alternative is to use individual computers to perform individual (or perhaps two or three related) functions. These individual computers can be connected by means of a shared data network to send data, only as required, from one system to another. Each person has access to all data relevant to his or her role.

The Most Cost-Effective System

On a recent tour in search of energy-efficient buildings, I visited an energy consultant who is a leader in his particular field. In his office I saw what is a sharp reminder of what energy management is all about. Mounted on a plaque was a screwdriver above the words

> The most cost-effective energy-management tool is a five-dollar screwdriver in the hands of an experienced operator.

SUMMARY

To extract the greatest benefit from an air-conditioning system, its design must be considered in conjunction with the design of the building envelope and other services systems, particularly lighting. The skills and knowledge of the air-conditioning designer should be exploited by the architect for input into selection of envelope components, particularly glazing.

The design of air-conditioning systems must take into account not only the human comfort factors and technical issues, but also the complex interaction between initial and ongoing costs. Design criteria vary from location to location and from application to application so that while past experience can be useful, design techniques should always be judged for applicability to the project in hand.

The success of an air-conditioning system does not rest solely with the design function. Because the system is active, it must be maintained and operated in such a way as to achieve its original design goals and to accommodate changes in occupancy and use of a building throughout its life.

REFERENCES

American Society of Heating, Refrigerating and Air Conditioning Engineers (ASHRAE). 1984. *ASHRAE Handbook 1984—Systems.* Atlanta, Ga.

ASHRAE. 1985. *ASHRAE Handbook 1985—Fundamentals.* Atlanta, Ga.

Chan, D. 1978. Optimisation of Energy Costs of Air Handling Unit. Thesis, School of Mechanical Engineering, University of New South Wales.

Davis, S. 1981. *Designing for Energy Efficiency: A Study of Eight California State Office Buildings.* Department of Architecture, University of California, Berkeley.

Rosenfeld, A., and de la Moriniere, O. 1985. The high cost-effectiveness of cool storage in new commercial buildings. *ASHRAE Trans.* Vol. 91, pp. 818-831.

Daylight Performance

10

Nancy Ruck

WINDOW DESIGN

Windows play a variety of roles as components of a building's envelope, whereas envelope design is primarily dependent on the environmental control strategy utilized. Not only do windows fulfill nonvisual functions (i.e., satisfy psychological and biological human needs), they also provide natural light and heat to an interior and thereby influence the energy performance of a building.

The standard environmental-control solution of the past 30 years, in which climate-imposed loads are diminished by sealed envelopes and tinted glazing and the functions of lighting and heating are met through artificial and mechanical means, has resulted in visual and thermal comfort being considered in such terms as balancing the contrast between the windows and the electric lighting and removing the heat from the lights, people, and machines. As a result of the energy crisis of the 1970s, this form of environmental control is now being modified to lessen its high energy consumption. Climate-adapting solutions in which the positive and negative influences of climate are selectively filtered and balanced at the building envelope with daylight are being utilized to conserve energy. In this approach, windows play a very important and dynamic role.

Utilizing daylight for interior illumination not only has a major effect on an occupant's visual (and thermal) comfort, as outlined in Chaps. 3 and 6, but it also can influence energy use in buildings. It has been shown that using daylight to reduce electric lighting requirements is a most effective energy-conservation strategy in building envelope design that can result in reduced thermal effects by providing an energy balance between heat gain and loss (Johnson et al. 1984).

Window performance depends on both the optical and thermal properties of the glazing material. These properties control the transfer of both light and heat into buildings. The daylight illuminances achieved in the interior will depend on the available daylight outdoors and hence the sky conditions, the orientation of the windows, glazing transmittance, the presence of external obstructions, and the internal surface reflectances.

Daylighting design also depends on building type and occupancy. For example, in residences, heat loss and sun control are primary concerns, but in commercial buildings, the daylight provided by windows may be the major interest because of its energy-saving potential, provided that solar gain is suitably controlled to reduce cooling loads. The effective use of daylight in buildings, therefore, requires specific and detailed attention to the function of the interior space relative to the quantity and quality of the light required.

Conventionally, the natural lighting of buildings relies on side windows, clerestories, and lights to admit daylight into interior spaces, and the depth to which daylight can penetrate and provide the required illuminance will depend on the method of entry. In addition, by articulating the building walls and using courtyards and light wells, one can open up more interior space to daylight.

Other methods for improving light penetration have been researched, and these depend mostly on the redirection of sunlight by active or passive means using mirrors, light shelves, louvers, or lenses by means of overhead and other glare-free paths. In most cases, tracking or some form of automatic adjustment is required to obtain an efficient system. Some effects are spectacular, but their success relies heavily on good maintenance and the availability of light direct from the sun.

If natural light is to be relayed to deep interiors from the sky and the sun passively (i.e., without tracking), then it is necessary to collect it from as much of the sky hemisphere as possible and condense its volume for transmission. In both active and passive systems, it is also necessary to minimize losses during transmission and to release the light efficiently in the target space. The overall transmission of daylight to deep spaces in terms of potential methods and their limitations is outlined in the section entitled Patterns of Innovation and Change (page 195).

In addition to these architectural solutions to increase the transmission of natural light, new advanced optical and thermal technologies for controlling energy flows through glazing are being researched. Some are now being marketed, such as, for example, low-emissivity window coatings. These coatings, static in the sense that their properties remain unchanged under varying daylight conditions, are now being improved and extended into the field of dynamics. Switchable glazing materials are being developed which take daylight variability into account and hence reduce energy consumption further.

This chapter reviews the daylight available outdoors on a worldwide basis, factors influencing daylight availability, traditional and new methods of introducing daylight into buildings, advanced glazing technologies and their potential, and user response to these technologies.

SKYLIGHT AND SUNLIGHT AS LIGHT SOURCES

The sun is the origin of all daylight, but its light is diffused and scattered by the atmosphere and it is the resulting skylight that forms the basis of daylighting design in buildings. Direct sunlight, unless suitably controlled, should be avoided in working interiors because it produces visual and thermal discomfort. In this book, daylight refers to sunlight and skylight together.

There are significant differences between using the sun and the sky as a source of light and using electric light sources. The luminous efficacy of a light source (defined in lumens per Watt) is generally used to compare the efficiency of light sources. As mentioned in Chap. 8, skylight and sunlight both have higher luminous efficacies than either fluorescent or incandescent light sources. Values given for sunlight (Moon 1940) show a range of 102 to 116 lm/W depending on solar altitude and atmospheric conditions. Measured values for skylight range from 100 to 200 lm/W. Variations depend on solar altitude, atmospheric clarity, and type and amount of clouds. In contrast, a typical value for a fluorescent lamp (based on a 1500-mm, 65-W tube and assuming a 10-W ballast loss) is 63 to 64 lm/W. Incandescent lamps vary between 10 and 22 lm/W. Thus a given amount of daylight is approximately 50 percent more effective than the comparable energy input of electric light sources traditionally used in buildings. However, the luminous efficacy of daylight is measured outdoors, and comparing it to the luminous efficacies of lamps can be misleading because the distribution of daylight within an interior space can modify the preceding figures, e.g., in a sidelit room where daylight levels fall off rapidly with distance from the window wall.

A primary difference between natural and electric light is the inherent variability of daylight and its unpredictability. Therefore, in determining the role of daylight in buildings, one of the crucial problems is the need for reliable localized data on skylight availability for the amount of natural light at a point in an interior, and this depends on the luminance or brightness of the sky "seen" through the window which produces this illuminance.

"Skylight availability" can be expressed as the external skylight illuminance (which is dependent on the sky luminance) available on an unobstructed horizontal plane for a certain percentage of daytime working hours. It also can be defined as the average amount of skylight and sunlight available during typical periods (day, month, season, or year). Both these concepts can be related to a specific locality or climatic zone and are usually expressed as changing hourly levels of external horizontal illuminance.

Although a number of researchers have reported data from measured availability studies in different climate zones in Australia, the United States, Japan, South Africa, and several European countries (Ruck 1984, Navvab et al. 1984, Nakamura and Oki 1979, Kittler 1975), data for some countries or regions are scarce or limited.

PREDICTION OF SKYLIGHT AVAILABILITY

The daylight climate of a region expressed in terms of the sky luminance pattern and its resultant external horizontal illuminance is closely related to the meteorologic climate and in particular to cloud type and amount and the clarity of the atmosphere or turbidity. The "turbidity" can be described as the amount of precipitable water vapor (w) in the atmosphere, which varies with the elevation and by the particle scattering in the atmosphere, represented by the Angstrom coefficient of turbidity (B). The latter can be related to the specific microclimate of the site according to its use (rural area, urban area, or industrial area).

For prediction purposes, the amount of water vapor can be deduced from the water vapor pressure at ground level by Hann's formula, or it can be calculated from atmospheric pressure and dew point using an ASHRAE algorithm (ASHRAE 1981). The Angstrom turbidity coefficient can be obtained from an equation using Coulson's data (Coulson 1975). Cloud-cover data can be obtained from local meteorologic stations and are based on subjective observations in which the experience and judgment of the station observer are used.

It is possible to predict illuminances from the sun and sky. In the case of illuminance from direct sunlight, the total horizontal illuminance can be predicted within fairly close limits depending precisely on the sun's altitude and the clarity of the atmosphere. Under a clear (cloudless) sky, the level of skylight depends on the same factors, but it is more variable than direct sunlight. When the sky is partly or wholly cloud-covered, however, the level of available skylight varies widely. In view of such fluctuations, both temporal and seasonal, any attempt to predict variations in skylight illuminance on a theoretical basis is extremely difficult.

The luminance distribution patterns of clear and overcast skies have been standardized and used for some time by the international lighting community. For estimating daylight levels or calculating energy effects, these sky models are now being expanded to include partly cloudy skies and direct-sun effects.

It is also possible to derive external sky illuminances from measured irradiance data available in most countries by using a luminous efficacy conversion factor. "Luminous efficacy" is the ratio of illuminance to irradiance, and it provides a useful link between the quantities of daylight and solar radiation. Luminous efficacy values for clear, partly cloudy, and overcast skies generally fall in the range of 90 to 130 lm/W. However, substantial variations can occur in the conversion process, due to the lack of information on the dependence of efficacy on atmospheric variables, such as water content of the atmosphere and turbidity.

Current Daylighting Data

While values of average illuminance from the sky (monthly or yearly) are of considerable interest, the actual frequency with which any given illuminances occur gives a more accurate indication of the duration of given levels of daylight. Therefore, daylight design is commonly based on the selection of a level of skylight that will be available as a minimum with some specified frequency such as 90 percent of working hours. Illuminance levels other than

TABLE 10.1 Representative Values of Skylight Availability

Location	Representative Diffuse Horizontal Illuminance (lx)	Availability	Reference
Australia:			
Darwin	12,700	90 percent of	Ruck (1984)
Brisbane	7,900	9 A.M. to 5 P.M.	
Broken Hill	5,900	work day (all	
Sydney	8,800	skies)	
France:			
Paris	5,000	94 percent of daylight hours (cloud on sun)	Fournol (1951)
Nantes	10,000	74 percent of daylight hours (all skies)	Perraudeau and Chauvel (1986)
United Kingdom:			
Kew/Bracknell	5,000	90 percent of 9 A.M. to 5 P.M. work day (all skies)	Hunt (1979)

(continued)

minimum are more feasible if integration with electric light is contemplated, e.g., in deep rooms where the rear of the room cannot be adequately lit with daylight alone. In this case, a percentage value can be selected that will permit the provision of a specified illuminance level or daylight factor.

Table 10.1 provides a summary of the representative values of skylight availability in different countries. It can be seen that the values vary over a range of 4 to 1. The differences are partly due to the time over which the given levels are assumed to be available. Variations in the illuminance levels in similar climatic regions and working hour periods are due to differences in cloudiness, atmospheric pollution, and site elevation. A comparison of such data for different geographic locations is shown in the form of frequency curves in Fig. 10.1. The curves, their locations, and the microclimate are given in Table 10.2.

Figure 10.1 clearly demonstrates that the external horizontal illuminance is dependent on cloudiness and turbidity rather than on latitude. Generally, illuminances are the highest in the hot humid climatic areas (e.g., Darwin, Australia, where there are bright overcast skies in the summer) and lowest in the hot arid zone, where the clear blue sky is the predominant sky condition. In general, data on partially cloudy skies show that these sky conditions are likely to produce the highest levels of illuminance, although fully overcast skies with altostratus clouds can produce higher illuminance levels.

The Effect of Turbidity and Cloudiness on Skylight Illuminances

The microclimate characterizes the structure of the atmospheric layer at a particular location. Its influence on skylight (and sunlight) is dependent on variations in the rate of radiation transmission through the atmosphere. The effects of cloudiness, water vapor, and dust (aerosol) content in the atmosphere are interlinked to the sun path and sunshine duration in a very complex manner. They are further influenced by such "civilization" factors as smog, smoke, and other gaseous and chemical pollutants. The expression of all these influences by a single turbidity factor is a very practical solution, and many researchers

TABLE 10.1 *Continued*

Location	Representative Diffuse Horizontal Illuminance (lx)	Availability	Reference
India: (standard)	8,000	85 percent of 8:30 A.M. to 5:30 P.M. work day (clear skies)	Narasimhan, Saxena, and Maitreya (1970)
Japan			
Nagoya (standard)	13,500	80 percent of 9:00 A.M. to 5:00 P.M. work day (all skies)	Nakamura and Oki (1979)
South Africa			
Pretoria	10,000	97 percent of 8:00 A.M. to 5:00 P.M. work day (all skies)	Richards and Rennhackkamp (1959)
Cape Town	7,500		
United States			
San Francisco	5,000	100 percent of daylight hours (all skies)	Navvab et al. (1984)

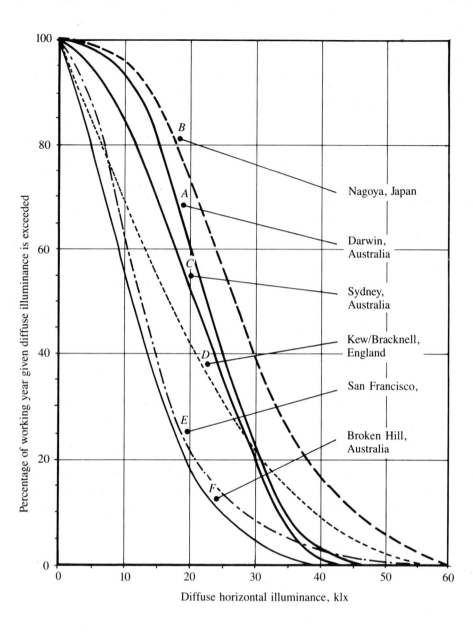

FIGURE 10.1.
The percentage of the working year, for a 9 A.M. to 5 P.M. work day, that the horizontal illuminance from the sky is available for various locations.

TABLE 10.2 Locations of Frequency Curves Given in Fig. 10.1

Curve	Location	Latitude	Time Period	Climatic Region	Reference
A	Darwin, Australia	12°S	9 A.M. to 5 P.M.	Hot, humid	Ruck (1984)
B	Nagoya, Japan	35°N	8 A.M. to 4 P.M.	Temperate	Nakamura and Oki (1979)
C	Sydney, Australia	34°S	9 A.M. to 5 P.M.	Temperate	Ruck (1984)
D	Kew/Bracknell, England	51.5°N	9 A.M. to 5 P.M.	Temperate	Hunt (1979)
E	San Francisco, Calif.	38°N	9 A.M. to 5 P.M.	Temperate	Navvab et al. (1984)
F	Broken Hill, Australia	31°57′S	9 A.M. to 5 P.M.	Hot, arid	Ruck (1984)

have attempted to define the turbidity factor with the assistance of available data from meteorologic records.

According to Bouger's laws, there are two turbidity factors that can describe the momentary state of the atmosphere (i.e., the atmospheric transmittance in the solar radiation spectrum and in the visible part of the spectrum): the turbidity factor according to Linke, T_L, and the illuminance turbidity factor, T_{il}, which relates to the visible spectrum only. Both turbidity concepts are similar and indicate the turbidity in the direction of the sun. While Linke's turbidity factor has been used since 1930, the illuminance turbidity factor, T_{il}, was introduced by Clear sky in 1982 (Navvab et al. 1984). It replaces the irradiance variables in T_L with their illuminance counterparts. Definitions of the two concepts are as follows:

$$T_L = \frac{\ln E_{eo} - \ln E_{esn}}{a_R m_o} \tag{10.1}$$

$$T_{il} = \frac{\ln E_{vo} - \ln E_{vsn}}{a_{il} m_o} \tag{10.2}$$

where T_L = Linke's turbidity factor
    T_{il} = illuminance turbidity factor
    E_{eo} = solar constant (1367 W/m^2) for whole of solar spectrum
    E_{vo} = sunlight constant (klx)
    E_{esn} = Irradiance from direct sun normal to solar beam
    E_{vsn} = sunlight illuminance on a surface element normal to the solar beam
    a_R = radiation extinction coefficient (Rayleigh atmosphere)
    a_{il} = illuminance extinction coefficient
    m = relative optical air mass

While the Linke turbidity factor can be used to calculate irradiances which are then converted to illuminances using the luminous efficacy of radiation, the illuminance turbidity factor is more closely aligned to the illuminance levels associated with all relatively homogeneous turbid skies.

A simple expression for T_{il} has been derived: $T_{il} = 1 + 21.6B$ (Navvab et al. 1984), where B is the Angstrom scattering coefficient.

It has been demonstrated from statistical analysis of measurements in various parts of the world that turbidity and cloud cover have a greater influence on illuminance levels than latitude. For clear skies, this has been shown by a study of measurements, notably in Tashkent, Russia (Lopukhin 1953), Washington, D.C. (Jones and Condit 1948, U.S. Weather Bureau 1953-1956), Uccle, Belgium (Dogniaux 1960), Pretoria, South Africa (Drummond 1956), Roorkee, India (Narisimhan, Saxena, and Maitreya 1970), and Broken Hill, Australia (Ruck 1984). They are summarized here for purposes of comparison. With the exception of the earlier Washington figures, these illumination measurements were all obtained with suitably filtered selenium photocells. Details of these locations are shown in Table 10.3 and in Fig. 10.2.

Figure 10.2 shows variations in horizontal illuminance from clear skies for Pretoria, Roorkee, Brussels, Tashkent, and Broken Hill as a function of solar altitude. The curves are compared with upper and lower limiting values (*dotted lines*) of illuminances defined by Nakamura and Oki (1979) from all types of skies.

The data variations can be attributed primarily to turbidity. The winter in Pretoria is very dusty, whereas the summer is characterized by a very clear atmosphere; therefore, illuminance levels are much lower. Tashkent is also subject to dust haze. Sky illuminance levels for Pretoria (summer) and Broken Hill therefore represent conditions of low turbidity, and those of Pretoria (winter) and Tashkent represent high turbidity. The lower limiting values of diffuse illuminance defined by Nakamura and Oki were for a limited sample measured in India.

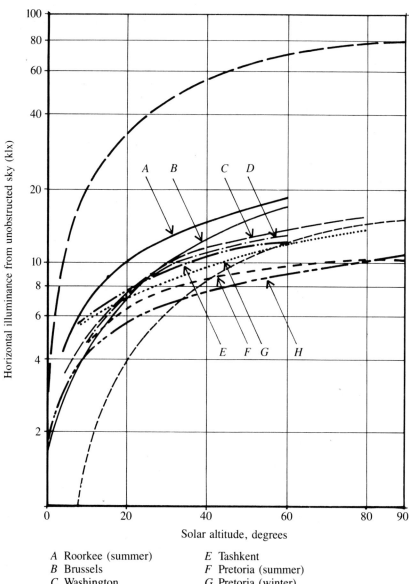

FIGURE 10.2.
Variations of horizontal illuminance from clear skies as a function of solar altitude for Pretoria, South Africa, Roorkee, India, Brussels, Belgium, Tashkent, Russia, and Broken Hill, Australia, Washington, D.C., U.S.A.

A	Roorkee (summer)	E	Tashkent
B	Brussels	F	Pretoria (summer)
C	Washington	G	Pretoria (winter)
D	Roorkee (winter)	H	Broken Hill

TABLE 10.3 Details of Locations for Illuminance Curves

Location	Latitude	Measurement Period
Pretoria, South Africa	25°45′S	1955
Washington, D.C.	38°56′N	1953-1956
Tashkent, Russia	41°30′N	1946-1950
Roorkee, India	22°N	1963
Uccle, Belgium	40°N	1956-1980
Broken Hill, Australia	31°57′S	1979-1980

PREDICTION TECHNIQUES

There are many methods for predicting interior daylight illuminances ranging from simple "rule of thumb" formulae to complex computer programs (see Chap. 12). Understanding how a design method is derived helps one to understand its capabilities and limitations. The technical constraints on the use of a design method (such as accuracy) are based on its technical derivation. Therefore,

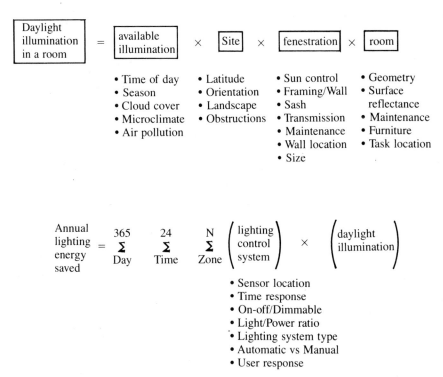

FIGURE 10.3.
Skylight illuminance variables. (After Selkowitz 1984.)

it is important to understand how daylighting design methods have evolved and the technical basis of their predictive capability.

The larger the aperture, the more light is admitted. Therefore, window size and location and the transmission characteristics of the glazing material are major factors influencing interior illuminance levels. Other variables that influence the determination of skylight illuminance in building interiors include skylight availability, the location's latitude, the window orientation, external obstructions, the geometry and reflectances of the interior spaces, and the task location (Fig. 10.3). In addition, the lighting energy saved by the use of daylight is influenced by the skylight availability and the lighting control system used.

This section outlines the mathematical basis for daylighting calculations, taking into account the preceding variables, and briefly describes two methods that can be used to predict and assess daylight in an interior space.

The Derivation of Daylight Calculations

The Three Components of Daylight

The proportion of sky seen through a window from the reference point is the determining factor in the calculation of daylight in interiors, and it is the luminance distribution of this patch of sky that provides the direct-light component of skylight. The external reflecting surfaces seen from the reference point determine the amount of external reflected light, and the light reflectance from internal surfaces of the room determines the amount of internal reflected light.

Direct Light from the Sky

The illuminance at a point in a room resulting from a small element of the sky seen through the window can be defined as follows:

$$dE = Ldw \sin \gamma \qquad (10.3)$$

where L is the luminance of the sky element, dw is the solid angle subtended by the element of sky at the reference point, and γ is the altitude angle of the sky element relative to the plane at the point. This equation can be used to determine the skylight illuminance at a point inside a building.

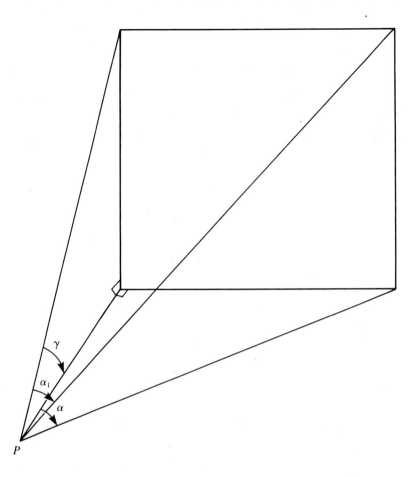

Figure 10.4 shows a simple example with a reference point P on a horizontal plane and a rectangular vertical aperture with its base in the reference plane. The illuminance E at the point resulting from the area of sky visible through the window can be found by integrating Eq. (10.3) over the area of the window.

Consider a small element subtending altitude angles γ', $\gamma' + d\gamma'$ and azimuth angles α and $\alpha' + d\alpha'$ at P. The solid angle subtended by the element is given by

$$dw = \cos \gamma' \, d\gamma' \, d\alpha' \qquad (10.4)$$

and therefore, the illuminance at P resulting from this element is

$$dE = L(\gamma', \alpha') \cos \gamma' \sin \gamma' \, d\gamma' \, d\alpha' \qquad (10.5)$$

Integrating over the aperture, we obtain

$$E = \int_0^\alpha \int_0^{\gamma_n} L(\gamma', \alpha') \cos \gamma' \sin \gamma' \, d\gamma' \, d\alpha' \qquad (10.6)$$

where γ_n is given by $\tan \gamma_n = \tan \gamma \cos \alpha'$ and γ and α are altitude and azimuth angles, respectively, subtended at P by the head and side of the opening.

This equation holds for any sky luminance distribution and is the basis for expressions for illuminance from overcast and uniform skies. For the CIE overcast sky:

$$L(\gamma', \alpha') = L_z (1 + 2 \sin \gamma')/3 \qquad (10.7)$$

Equation (10.6) can then be integrated analytically (Hopkinson et al. 1966, Bensasson and Burgess 1978). These equations are often for unglazed apertures and need to be corrected for glass transmission used in computer programs to calculate daylight in interiors.

A large number of methods have been devised to simplify the evaluation of the basic integrals such as in Eq. (10.6) (Longmore 1968; Bryan and Carlberg 1984; Hopkinson et al. 1958).

Externally Reflected Light

Light reflected from external obstructions can be quantified in a similar way. Equation (10-6) can be used as before for rectangular obstructions, with L being the luminance of the obstruction and the integration taking place over the angular area of the visible obstruction. In this instance, the luminance can be given by the following if the obstruction is uniformly diffusing:

$$L = \frac{\rho_{obs} E_{obs}}{\pi} \qquad \textbf{(10.8)}$$

where E_{obs} is the illuminance on the obstruction and ρ_{obs} is its reflectance. Tables of reflectance of typical building materials are given by Hopkinson et al. (1966), Robbins (1986), and Egan (1983).

 In most cases, the contribution from externally reflected light to interior light is small, but in tropical regions, externally reflected sunlight can be the principal illuminant. Design aids have been produced for this situation (Maitreya 1980).

Internally Reflected Light

The interreflection of light within an interior space is also a contributor to the illuminance at a specified point, but this is more difficult to calculate. The "form-factor approach" is often incorporated in computer programs. In this method, the first reflected flux incident on each element by reflection from all the other elements is calculated. When large computers are unavailable, a simpler method is the "split-flux formula" (Hopkinson et al. 1966). This formula divides the room into two parts and calculates internally reflected light as an average throughout the room. Other simplified methods also have been produced for calculating internally reflected light, including a method for clear skies (Krochmann, 1962; Lynes, 1968).

 These calculation methods are often used in computer programs to determine daylight in sidelit interiors. The general principles involved also can be extended to skylights, whether horizontal or sloping. Hopkinson et al. (1954), Bodman et al. (1985), and McCluney (1984) have all developed methods for illuminance calculations for skylights.

Illuminance Calculations for Special Building Geometries

A special case of illuminance calculations involves using shading devices. The effects of blinds and louvers on daylight flux distribution can be quantified from first principles (Shukuya and Kimura 1983, Kim et al. 1986).

 In principle, and with the aid of computers, daylight calculation techniques can be applied to a wide range of complex building geometries. However, in practice, it is often more worthwhile to construct a scale model of a complex system and record measurements under a real or artificial sky.

The Daylight-Factor Method

Illuminance from the sky is not constant, but rather it varies continuously. In addition there are random variations which are due to the density and movement of cloud cover and turbidity.

 The illuminance on a horizontal surface from the whole sky outdoors generally reaches a maximum in the summer and a minimum in winter; therefore, the duration of "useful" interior daylighting also varies with the season. Because of these variations in sky illuminance, it has been customary in most countries

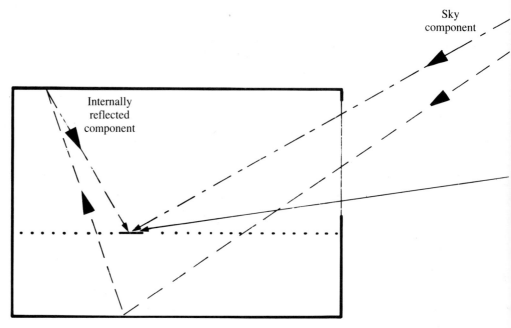

FIGURE 10.5.
The three components of the daylight factor.

other than the United States to express the interior illuminance from skylight (E_i) as a percentage of the illuminance occurring simultaneously on a horizontal surface exposed to the whole sky (E_h). This ratio is known as the "daylight factor" (DF), which can be defined as

$$DF = \frac{E_i}{E_h} \times 100\% \qquad (10.9)$$

Sunlight is excluded from the concept of daylight factor, which includes three components: the direct sky component, the externally reflected component, and the internally reflected component, as shown in Fig. 10.5.

Although daylight prediction methods in existing buildings using the daylight factor have become increasingly complex over the years and enable illuminances to be determined with a high degree of precision, this very precision has worked against practical application of the daylight factor in design and has led to certain basic simplifying assumptions whereby predicted daylight factors which are only strictly valid for overcast skies are not representative of actual sky conditions. Most daylight factor prediction methods use the overcast or uniform standard skies. The daylight factor can then be converted into internal illuminance by multiplication with an appropriate external illuminance.

Under an overcast sky, the daylight factor is assumed to be constant for each specified point, whatever the value of the external horizontal illuminance, but not for other sky types (i.e., clear and partly cloudy skies, which are not uniformly constant in azimuth). This produces extremely conservative results when these types of skies are predominant as the orientation effect is not taken into account. The effects of orientation can be overcome in lighting calculations by using vertical external illuminance (on the window wall) instead of horizontal illuminance. Robbins (1986) has developed an approach based on the lumen method which is valid for overcast and clear skies. Littlefair also has proposed

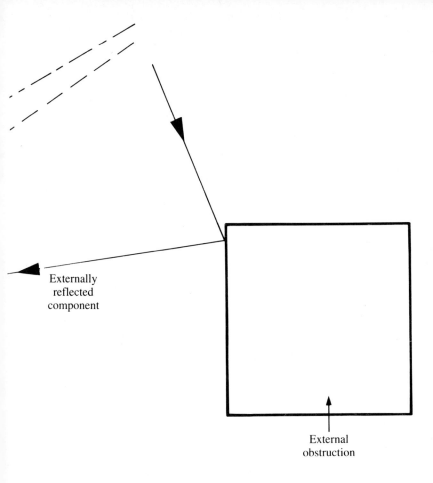

Externally
reflected
component

External
obstruction

a method based on the daylight factor and measured vertical illuminance data (Littlefair 1984).

The usual method of designing a building for daylight using the daylight factor is as follows:

1. Select the recommended illuminance level, e.g., 500 lx, from an appropriate lighting code.
2. Determine a value for the design sky in terms of the total external illuminance available for some percentage of working hours, e.g., 5000 lx.
3. Determine the resulting required daylight factor as a percentage, that is, 500/5000 × 100, or 10 percent.
4. After making allowance for dirt on glass and internal obstructions, select an arrangement of windows that will provide the required daylight factor.

The key factor in this procedure is the selection of a value for the design sky. If a minimum value representative of a fully overcast sky is selected, that is, 5000 lx available for 90 percent of working hours in some countries, most commercial interiors cannot be adequately lit by daylight. A more realistic value would be a higher illuminance available for a shorter period of time, and this would produce an adequate level of illuminance. At times when the external illuminance drops below the required illuminance, interior light levels could be supplemented with electric lights.

The daylight-factor method is generally used to specify a minimum daylight factor. This infers neglect of areas close to windows where the daylighting may be excessive and there is visual and thermal discomfort. Use of a "minimal daylight factor" as a criterion for the adequacy of interior lighting has limitations in that it gives no indication of the adequacy of the daylighting in the room as a whole without a knowledge of window arrangement and the reflectances of surfaces. In view of the increased integration with electric light, it is now necessary to reexamine this minimal daylight factor concept.

Average Daylight Factor

The "average daylight factor" (Longmore 1978), which has been introduced to give a rough approximation of interior daylight availability for different window areas, can be defined as the arithmetic average of all daylight factors on a working plane. For sidelit interiors, this can be calculated using the following formula:

$$\text{Average daylight factor} = \frac{0.85\, W\theta}{A(1 - R^2)} \qquad (10.10)$$

where W = total glazed area of windows
θ = vertical angle subtended in degrees subtended at the center of the window by unobstructed sky
A = total area of ceiling, floor, walls, and windows
R = average reflectance of ceiling, walls, and windows

For skylights, the following formula can be used:

$$\text{Average daylight factor} = \frac{KW}{A_f(1 - R_f R_c)} \qquad (10.11)$$

where W = total area of glazed skylight aperture
A_f = floor area
R_f = average reflectance of floor
K = obstruction coefficient
R_c = average reflectance of ceiling including skylights

This formula is only applicable to a room whose width and length are effectively infinite compared to its height.

The average daylight factor is subject to correction for dirt on glazing and obstructions. With the large variety of glazing materials currently available, it is no longer possible to recommend particular correction factors for tinted glazing or diffusing light panels. These can be calculated from the manufacturers' data.

The Lumen Method

The lumen method was developed by J. W. Griffith (1976) on the basis of model studies under an artificial sky. The method allows a comparison of various window-wall schemes tested and is limited by the parameters set, for example, the head of the window at ceiling level and window sill at desk height. The effects of various glass transmittances, ground reflectances, wall reflectances, diffusing shades, overhangs, and the simulation of clear and overcast skies, and direct solar radiation are included in the calculation.

For sidelighting, the lumen method predicts illuminance levels at three reference points in a room on a center line from the window: 1.8 meters (5 feet) from the window (max), on center of room depth (mid), and 1.8 meters (5 feet) from the wall opposite from the window (min) as shown in Fig. 10.6. The technique was designed to evaluate variable window management systems or fixed controls and to obtain the total benefit of daylight utilization rather than meeting a minimum requirement. A complete description of the method is given by the North American Illuminating Engineering Society (1979). An outline of their procedure is as follows:

1. Select sky conditions and determine the sky and solar illuminance (E_s) from the appropriate charts (Fig. 10.7).
2. Identify the appropriate ground reflectances (see Table 10.4) and multiply these reflectances by the percentages of the 2π sr. of solid angle presented to

TABLE 10.4 Typical Reflectances of Building Materials and Outside Surfaces

Concrete	55%	Vegetation	25%
Asphalt	7%	Snow	74%
Earth	7%	Brick	40%
Grass	6%	Gravel	15%

Distances used in determining X_e-factors

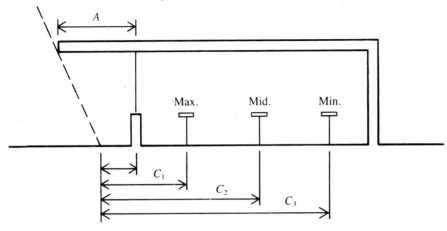

FIGURE 10.6.
Distances used in determining X_e.

the window that are occupied by these reflectances, P_g. Calculate the illuminance incident on the window (E_g) from the ground reflectance.

3. Calculate the net vertical glazing area in square feet (A_w).
4. Determine the light loss factor (F) from Table 10.5.
5. Determine glazing transmittance (T_g) from manufacturer's data.
6. Determine effective transmittance (T_s) of any shading device, if none use 1.0.
7. If there is an overhang, calculate the expanded room dimensions, e.g. for an overhang of width w, room length L parallel to window wall, room width W, and room height H, the expanded room dimensions are:

$$W' = W + w; L' = L(W + w)/W$$

For the sky component, if l is the window length and s the sill height the effective window area A_{ws} is L (H-s) when there is an overhang. For the ground reflected component the effective window area A_{wg} is lH. Use A_{ws} and A_{wg} instead of A_w in calculations if there is an overhang.

8. If direct sunlight reaches the ground under the overhang, determine the width of the illuminated area and lengths B, C1, C2, and C3 as shown in Fig. 10.6. Use the graphs (I.E.S., 1979) to determine the illumination from the sunlighted areas by first deriving X_e from $A - B/C_i + 20$; i = 1, 2, 3 and X_f factors.

Note that the I.E.S., procedures for determining illuminances involve the use of a particular method of calculating the effect of overhangs. With more recently developed sky models, other methods can be used (Moore, 1985).

TABLE 10.5 Average Light-Loss Factors for Various Window Positions (Expressed as a Percentage of Clear-Glass Transmission)

	Window Position				
	Office*	**Factory**[†]			
	Vertical	*Vertical*	*30° from Vertical*	*60° from Vertical*	*Horizontal*
Avg value over 6 mo period	83%	71%	65%	58%	54%
Value end of 3 mo period	82%	69%	62%	54%	50%
Value end of 6 mo period	73%	55%	45%	39%	34%

*Typical clean location.   [†]Typical dirty location.

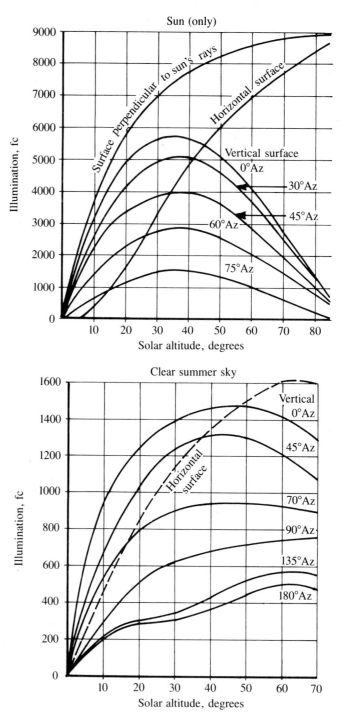

FIGURE 10.7.
Sky illuminance graphs for overcast and clear skies.

9. Look up the sky-contributed and ground-contributed coefficients of utilization C_{max}, C_{mid}, C_{min}, K_{max}, K_{mid}, K_{min} for the room conditions nearest those being designed from the tables (I.E.S., 1979), for each of the three points in the room. Use L' and W' if an overhang is present. If direct sunlight reaches the ground under the overhang, modify these factors with the X-factors from the graphs used in Step 8.

10. Calculate the sky-contributed interior illuminance for max, mid, and min positions:

$$E_{int.\ sky} = E_s\ A_w\ F\ T_g\ T_s\ C_i\ K_i \qquad i = max, mid, min$$

11. Calculate the ground-contributed interior illuminance for max, mid, and min positions:

$$E_{int.\ gnd} = E_g\ A_w\ F\ T_g\ T_s\ C_i\ K_i \qquad i = max, mid, min$$

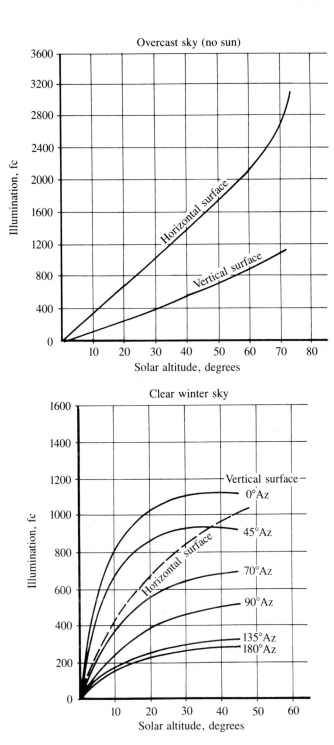

12. Determine the total interior illuminance at max, mid, and min positions by adding the sky-contribution (Step 10) and the ground-contribution (Step 11) components.

$$E_i = E_{int,\,sky} + E_{int,\,gnd}$$

The lumen method for toplighting calculates the average illuminance on the workplane due to horizontal (or near horizontal) roof openings.

The Prediction of Daylight for Energy Calculations

The reduction in electric lighting use during times when daylight can contribute to interior lighting depends on daylight levels and on the type of lighting control system in the building. Photoelectric controls represent the simplest case because they automatically respond to given daylight levels. In its simplest form, the on-off control switches off the lights when a particular daylight illuminance is reached at the control point and switches them on again when the daylight illuminance drops below this control value. Such a control has the disadvantage of being distracting to occupants. Variants of the on-off control have been proposed which incorporate a time delay or an illuminance differential in order to lessen this problem.

Energy savings from the on-off control are relatively easy to predict. If the control switches the lights off when a specific illuminance is reached at the control point, the lights will be off for the fraction of the year this illuminance is exceeded. In order to calculate the lighting use under this system, the fraction of the year that internal daylight illuminances are exceeded must be known.

For the top-up or dimming control system, the fraction of energy saved compared to full lighting use is more complex, but it still depends on the cumulative distribution (Littlefair 1984). Thus for calculations of lighting use, the key quantity is the cumulative distribution of internal illuminances, and various methods have been proposed to calculate this quantity.

One method is to multiply the daylight factor by the external illuminance to obtain the "internal illuminance distribution." This method has the drawback that the daylight factor is only valid for overcast (or uniform) skies and can vary according to orientation. The orientation effects can be allowed by using vertical in lieu of horizontal external illuminance. As mentioned previously, Robbins (1986) has developed such a method, known as the "lumen method," and it is valid for both overcast and clear skies but has some limitations.

An alternative approach is to use a type of sky luminance distribution which automatically allows for the effects of orientation and partly cloudy skies. Aydinli (1983) and Littlefair (1985) have produced data for an average sky but it is difficult to transfer such data to other locations with different weather conditions.

For some applications, such as lighting as a casual gain in large thermal modeling computer programs, a more dynamic approach is needed; that is, the state of lighting for each hour and day of the year is required. Gillette (1983) and Winkleman and Selkowitz (1985) have developed models on this basis in which the momentary sky luminance distribution is assumed to be a linear combination of clear and overcast skies, depending on cloud cover.

Another approach has been adopted by Kittler (1986) whereby a series of representational homogeneous skies (including the clear and overcast skies) are based on actual turbidity values. With this latter approach, absolute hourly values of internal illuminance can be obtained without the utilization of a daylight factor.

TRADITIONAL METHODS USING DAYLIGHT IN BUILDING INTERIORS

Sidelighting

Traditionally, sidelighting and rooflighting are the most widely applied daylighting strategies, and there are a number of conventional methods of introducing daylight into buildings in these contexts, as shown in Fig. 10.8. However, depth of penetration is limited, particularly in sidelit interiors, where skylight illuminance falls off rapidly with increasing distance from the window wall. Figure 10.9 shows the skylight distribution in a sidelit interior using clear or reflective glazing versus distance from the window wall.

The rule of thumb developed many years ago is that useful penetration can rarely exceed two to three times the height of the window. Using this guideline, penetration in a perimeter zone is limited to from 4 to 6 m (15 to 20 ft)

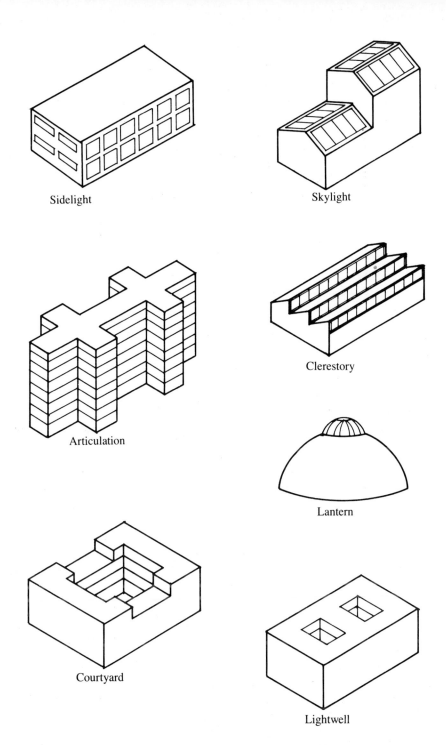

Sidelight

Skylight

Articulation

Clerestory

Lantern

Courtyard

Lightwell

FIGURE 10.8.
Traditional methods of daylighting buildings.

of depth in, for example, a typical office space with a 3-m (10-ft) ceiling height and average daylight conditions. The distance to which daylight remains effective depends on the window shape and the reflectances of the room surfaces. Maximal penetration is favored by high windows and light-colored surfaces.

Established daylighting techniques using the standard overcast sky or clear sky, as mentioned earlier, cannot provide an appropriate level of working illuminance at a reasonable depth into a room without promoting glare and thermal discomfort from the large glazing areas required. These large glazing areas also introduce unnecessary cooling loads in summer.

In addition, when applying these current techniques in regions that have a great deal of sunshine or in which the sky is not predominantly overcast, as in Australia, it should be borne in mind that the conventional use of minimal

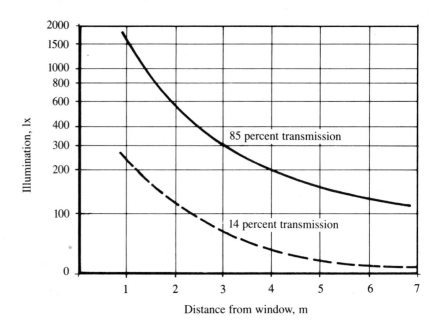

FIGURE 10.9.
Skylight illuminance versus distance from window wall.

conditions (fully overcast sky with stratus-type clouds) will produce extremely conservative results and consequently larger areas of glass than needed with skies producing higher illuminance levels. To avoid glare and solar heat gain in these circumstances, there should be a careful selection of glazing materials and/or shading devices so that solar gain is reduced but daylighting transmittance is maintained. Part IV provides an evaluation of glazing area and transmittance in terms of reduction of cooling loads and gives some recommendations.

Effective Sidelighting Techniques

Because of the increased depth of sidelit high-rise buildings today, it is often impossible to illuminate such buildings exclusively by daylight. To introduce daylight beyond the perimeter zone in most sidelighted interiors requires an unconventional envelope design. Techniques have been developed over the last 10 years to ensure more effective use of daylight while eliminating discomfort glare. One solution, which retains the use of natural light and can maintain good lighting conditions, integrates daylight with electric light (see pages 103–108). Other passive solutions which maintain the input and depth of penetration include the use of simple architectural elements such as light shelves to restrict glare and solar heat gain.

A light reflecting or shading element within the upper portion of the glazing such as a horizontal reflective light shelf (Fig. 10.10) acts as an overhang to control direct sun penetration, while a light-colored diffuse or specular upper surface reflects the intercepted light into the space via a highly reflective ceiling. Modifying the ceiling geometry and texture enhances the contribution of daylight and using a specular surface near the window wall also can reflect light at grazing angles into the space.

An alternative approach using devices such as venetian blinds or louvers provides similar protection from glare and equal light distribution, but it reduces contact with the outside world. Such devices also can be designed to break up and diffuse the sun's rays or redirect them into the interior in such a manner that there is no discomfort glare.

Rooflights

Rooflights are applicable to the interior zones of single-story buildings and to the top floor of multistory buildings, where most or all of the floor area can be

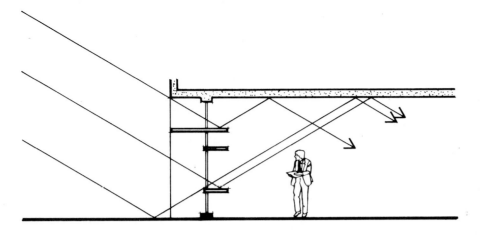

FIGURE 10.10.
Light shelves to enhance daylight penetration.

lighted through the roof. Unlike sidelighting, where the distribution of light places limits on the amount of usable light, a toplighted space can be uniformly daylighted if openings are appropriately sized and spaced over the floor area. There are two major design options: flat glazing used in a horizontal or sloping roof plane or roof monitors incorporating vertical-sloped or horizontal glazing elements. The fundamental difficulty with horizontal or approximately horizontal glazing is that solar transmission peaks in the summer, increasing cooling loads, and diminishes in the winter, which is detrimental to the energy performance of buildings in cool climates. To offset heat transfer, skylights are now available as double-glazed units and in some countries as triple-glazed units. Excessive heat gains also can be limited by orienting the glazing away from the sun or by using shading devices. Diffusing glazings used in rooflights can become glare sources and should therefore be designed carefully.

There is, however, one drawback to rooflight systems. Visual contact with the outdoors is limited to general knowledge of the weather. A view outdoors is virtually nonexistent. This means that, psychologically, rooflights cannot be equivalent to windows on vertical walls at eye height. If rooflights are used, some vertical glazing should be installed to allow visual contact with the outdoors.

Solar Control Solutions

Shading devices are required not only to reduce thermal gain, but also to increase visual comfort. By controlling the luminance of the sky and the admission of sunlight, both thermal heat gain and glare can be limited. Window glare, as discussed in Chap. 6 (page 56), is characterized by the contrast between the high luminance of the sun and the sky and/or external obstructions seen through the window and the adjacent wall surfaces in the room. The visual discomfort from window glare grows in relation to the inflexibility of position of the office worker and the direction of view.

Conventional approaches to solar control have tended to rely largely on tinted and reflective glasses, sometimes in combination with interior shading devices. Window setbacks such that the building envelope itself is a shading element are not commonly used. There is a strong economic incentive to place the glazing at the outermost edge of the building to maximize interior space. Fins and external shading devices have long been important elements of architectural design, but they run counter to the recent architectural emphasis on slick, smooth building surfaces. However, with energy conservation reappearing as an important design consideration, interest is being expressed in the use of external or internal shading devices that are sufficiently flexible to cater to all sky conditions and contribute daylight to the interior as well.

Shading Devices

Exterior shading devices may be fixed or movable, manual or automatic, and they vary widely in cost, durability, appearance, and performance. The general advantage of using external controls is the obstruction of solar radiation before it enters the window. For estimating the effect of a combined system of glazing and external shading devices on a building's cooling load, the concept of "shading coefficient" or "solar gain factor" has been introduced by ASHRAE and the Institution of Heating and Ventilating Engineers (IHVE). As mentioned in Chap. 9, shading coefficients for exterior systems fall into the range of 0.1 to 0.5.

All shading devices, with the exception of those which are fully retractable, reduce the effective light-admitting area. With fixed shading, this reduction can be easily evaluated, and the admission of daylight can be calculated accordingly. With adjustable and retractable shading devices, this is more complicated, and with these shades, calculations should be performed for several settings: the fully opened, the fully closed (which totally obstructs the admission of daylight), and any intermediate positions that are frequently used.

Fixed shading devices such as overhangs, fins, or various types of shade screen materials can be mounted externally (i.e., outside the glazing) and incorporated into the architectural design of the building exterior. These inhibit sun penetration but allow some view of the sky, so that daylight can be admitted. Horizontal overhangs are effective for south-facing windows (north-facing in the Southern Hemisphere), as shown in Fig. 10.11, which is applicable to temperate climates at middle latitudes. An awning can provide shade to a window without heat buildup because of the air circulating between it and the windows. The surface of the awning should be light-colored to minimize solar heat absorption. Vertical planes as sunbreaks can be effective for east- and west-facing windows.

In addition, overhangs can be designed to break up and diffuse the incident solar beam so that diffused rather than direct sunlight also can be introduced into the building. Since they are fixed, however, they invariably represent a compromise between the requirements of sun control, daylight admittance, and glare control. In principle, movable sun-control systems will provide better performance than fixed systems.

These movable systems are more expensive to install, but they allow better daylight management. The most advanced systems control the shading continuously to just block the penetration of direct sunlight. They will automatically open when shading is not required. Such systems also can be connected to automatic dimming systems for the electric lighting to optimize the integration of daylight and electric light. Although these systems are relatively costly compared to interior treatments of reflective glass, they can be economically feasible because they allow for reductions in cooling system size and they make better use of daylight.

The principal advantage of internal shading devices is their accessibility and hence ease of management. The main disadvantage is the absorption of solar heat and its transference to the air in the room. The shading coefficient values for internal shading devices are much higher than for external shading devices.

Venetian blinds with horizontal slats can reflect the summer sun. They also can be designed to redirect daylight to the ceiling for deeper penetration into the room. Tedious cleaning and mechanical problems with the tilt adjustment to accommodate the sun are disadvantages of this type of blind.

Venetian blinds also can be located between panes of glass. If blinds are used with an exhaust-air or airflow window, the air from an interior space is exhausted between the panes of a glazing system over a venetian blind either to the outdoors or to the heating or cooling system. In the winter, this provides an interior glass surface temperature that closely matches the room air temperature,

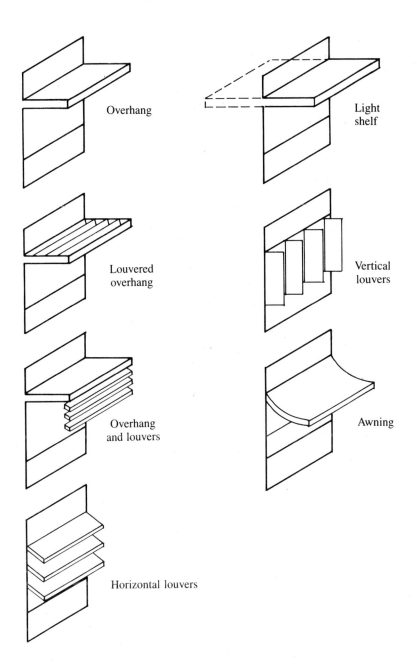

Overhang

Light shelf

Louvered overhang

Vertical louvers

Overhang and louvers

Awning

Horizontal louvers

FIGURE 10.11.
External sunshading devices. (After Selkowitz 1986.)

providing good thermal comfort. In the summer, the blinds can absorb the sun's energy, and the heat collected can be transferred to the moving airstream.

PATTERNS OF INNOVATION AND CHANGE

The growing concern for energy conservation and recognition of the value of the natural qualities of daylight have encouraged research into new optical devices for the daylighting of deep rooms and underground and otherwise obscured building interiors. While significant improvements can be made relative to most window and rooflight systems in current use, these are limited in terms of daylight penetration. Improved methods for lighting interior spaces are now being initiated worldwide.

The use of innovative optical systems to improve daylighting performance is not a new concept. Numerous examples of the use of optical systems to enhance daylight performance in buildings have been recorded in patent literature as far back as the 1880s. These optical systems, whose basic performance has been understood for some time, are now being investigated using advanced technologies incorporating fiber optics and holographic coatings.

Most of these systems depend on the redirection of sunlight using mirrors, louvers, or lenses via overhead or other glare-free paths. In some cases, tracking or some form of automatic adjustment is required. Some effects are spectacular, but their success relies heavily on good maintenance and the availability of light direct from the sun.

The visual and psychological qualities of daylight in deep interiors, together with the lack of maintenance, weigh in favor of the concept of passive utilization of not only sunlight, but also skylight. However, in order to compete with existing systems, the new technologies also must prove viable and economically feasible. If light is to be relayed to deep interiors from the sun and sky without tracking, it is necessary to collect it from as much of the sky as possible, to redirect and/or condense its volume for transmission, to minimize losses during transmission, and to release it efficiently in the target space.

These patterns of innovation and change, together with commentary, are outlined in this section. Technical approaches for introducing skylight and sunlight into deep interior spaces are described, and situations are considered where light transmission is desired over even longer distances and to penetrate in any direction through the interior of a building.

Passive Techniques for Sidelit Buildings

Natural light in sidelit buildings can be effective in areas beyond the perimeter zone, provided there are devices to bend sunlight so that it penetrates further into the interior space. This can be accomplished by redirecting sunlight onto the ceiling and from there onto the workplace. Several optical techniques are available for achieving these objectives, including such light-bending elements as prismatic blocks and lenses and reflecting louvers, as shown in Fig. 10.12.

Glass blocks and linear lenses both have prismatic cross sections so that a section of the window consists effectively of a stack of triangular prisms. Light from outside is generally deflected upward and hence penetrates deeper into the room onto both the horizontal working plane and the ceiling. Such blocks and lenses also can be used to offset the darkening effect of a shaded obstruction, as shown in Fig. 10.13.

In previous years, the use of prismatic glass blocks was discarded because of their thermal storage effects and as a result of the lack of a suitable manufacturing process to produce precise profiles and hence eliminate the scattering or radii and closure surfaces. New advances in production techniques in recent years, however, have enabled the manufacture of more precise prism profiles.

A prismatic panel designed and patented in Australia directs light up to 10 m (Ruck 1985) into the interior of a building by locating the panel above eye height, as shown in the top of Fig. 10.14, and using a highly reflective ceiling. The panel consists of a single linear Fresnel lens of modified Plexiglas, as shown in the bottom of Fig. 10.14, and can be adapted to all seasons of the year and latitudes by a specified degree of tilt. Solar radiation is redirected to the ceiling

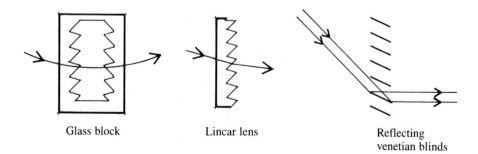

FIGURE 10.12.
Light-bending elements.

Glass block   Lincar lens   Reflecting
venetian blinds

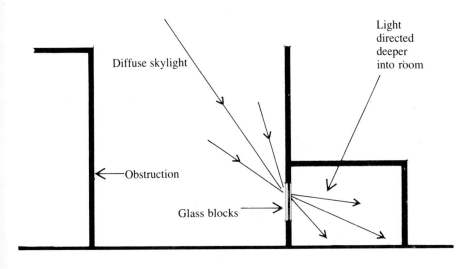

FIGURE 10.13.
The use of glass blocks to offset shaded obstructions.

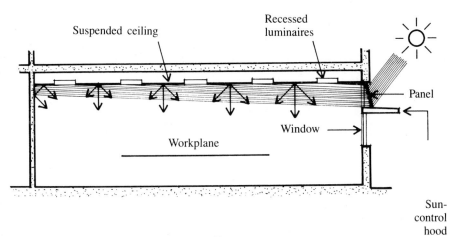

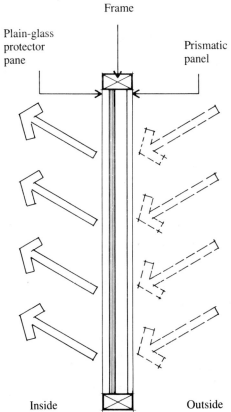

FIGURE 10.14.
The prismatic panel.

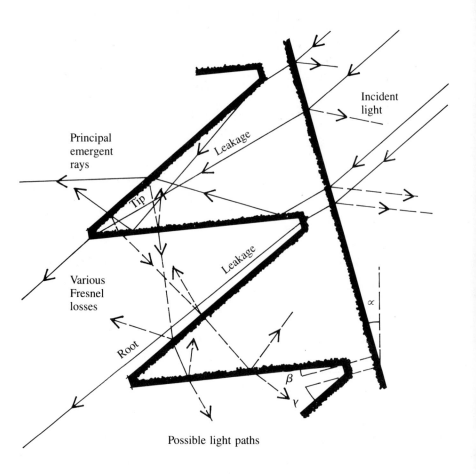

Possible light paths

FIGURE 10.15.
Possible light paths through a prismatic panel.

at a relatively low angle to obtain deep penetration and to contain solar rays within acceptable limits to negate glare.

Only a small area is required to produce 500 lx at a depth of 10 m. The transmission efficiency of the system is 50 to 70 percent depending on the solar altitude. Figure 10.15 shows the possible light paths through a Plexiglas prismatic profile. All refraction in and out of the prisms is necessarily accompanied by Fresnel reflections that reduce transmission and increase misdirected light. By specifically designing the profile to accommodate a maximum range of solar altitudes, and also by tilting the panel, high or grazing angles of the incident solar beam can be avoided, initial reflection losses can be minimized, and angles of incidence leading to misdirected light can be excluded.

Figure 10.16 shows an experimental panel with a 31-degree tilt to the vertical under simulated noon exposure at the summer solstice for latitude 34°. The panel shown is fabricated to a scale four times that envisioned for the final product.

Figure 10.17 gives computation results for lighting in an office 10 m deep facing north (in the Southern Hemisphere) for the months of March, June, September, and December (winter and summer solstices and the equinoxes). It can be seen that the reflected sunlight is mostly sufficient in itself to provide an illuminance level of 500 lx. With the added contribution of skylight, it can produce this illuminance easily. It has been estimated that total electrical energy savings in a climate similar to that of Sydney, Australia, with 57 percent of possible sun hours, would be in the order of 70 to 75 percent.

Light also can be bent by diffraction. Research is currently being carried out on "holography" (Solymer and Cooke 1981) in which laser light patterns are recorded on a thick photographic emulsion forming a three-dimensional structure. Light emerges at set exit angles if the angle of incidence and its wavelength satisfy a particular relationship. Color dispersion of the output light is, however, a potential disadvantage. Efficiencies for such a system have been quoted at 30 to 70 percent.

FIGURE 10.16.
Prototype prismatic panel at 31-degree tilt to the vertical for noon on the summer solstice (latitude 34°S).

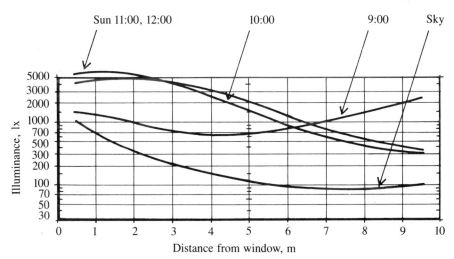

FIGURE 10.17.
Illuminance levels from beamed sunlight and skylight at the equinox (latitude 34°N).

Active Systems and Components

Collectors and Concentrators

Active systems that direct daylight in any direction through the interior of a building and over longer distances require three major elements: a collection system to gather and concentrate available light flux, a light guide system to transmit the light flux to the point of use, and a distribution system consistent with the end use of the lighting in that portion of the building.

In the solar-energy field, much sophisticated study has gone into the capture and concentration of direct sunlight, mostly with devices to track the sun's movement, as shown in Fig. 10.18. However, their success relies heavily on good maintenance and the availability of light direct from the sun.

Figure 10.19 shows a basic system using a heliostat. The simplest form is the polar-axle heliostat; the most complicated is the altitude-azimuth sun tracker. Double-axis polar trackers can introduce a beam of light with approximately constant cross section into an opening on the roof of a building for almost all sun positions.

Simpler systems using only a single mirror rotating in azimuth and changing in altitude are less efficient in intercepting light flux. Skylights with these types of tracking systems are commercially available, although there are little or no definitive performance data at this time. Any system designed to track the sun will suffer from reductions in flux output when clouds or haze obscure the sun's disk. The flux intensity can change in magnitude in a matter of seconds as a cloud moves in front of the sun, which means that the building's lighting system has to be designed to respond to such changes on the time scale with which they occur.

Heliostats are costly and require control and maintenance because they have to track the sun at all times. The size required also may be a problem. For example, a mirror area of 8 m^2 would be needed to light 1000 m^2 of office space even with a 100 percent efficient system. In reality, the efficiency would be

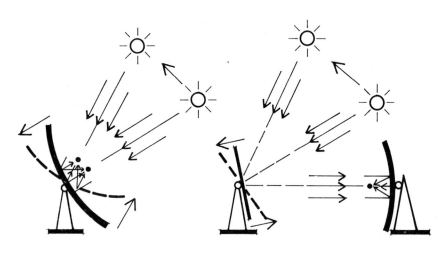

FIGURE 10.18.
Solar collection systems.

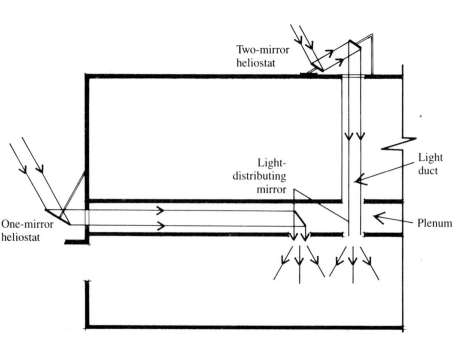

FIGURE 10.19.
Active solar collection systems using a heliostat.

approximately 20 percent, which infers that a heliostat collector of 40 m^2 would be needed.

Collection Systems

There are several alternatives using nontracking systems for collecting both skylight and sunlight, particularly in regions where little sunlight is available. If light is to be relayed to deep interiors from the sun and sky without tracking, then it is necessary to collect it from as much of the sky as possible, to condense its volume for transmission, and to release it efficiently into the target area. Some of these requirements conflict, since the concentration possible with collection systems designed to collect light from a larger solid angle of the sky is limited not only by the acceptance characteristics of the transmission system, but also by the radiance therefrom, which limits the concentration of collected light to an intensity no greater than its source in the same medium (Rabl 1980).

High concentration is associated with a small aperture and a high ratio of depth to width or diameter. For example, consider a deep space that is 25 m by 20 m illuminated at 400 lx using daylight ducted with 50 percent overall efficiency from focons exposed to 20 klx under a diffuse sky. The term "focon" is commonly used to mean a focusing (concentrating) cone. Light for this purpose could be captured over a nominal area of 5 m by 5 m using a square packed array of 25 "Winston compound parabolic focons," each 1 m diameter at the mouth and concentrating 100:1 (equivalent to condensing the daylight onto a 500-mm-diameter spot). However, this would require a depth of 5.5 m, or an assembly of roughly a 5-m cube! Moreover, the acceptance angle would be only 11.5°, or the equivalent of 23 minutes of solar movement either side of midday, and no rays of diffuse or solar radiation outside this angle would be accepted.

A solution to this problem is to refract the light into a denser optic medium during or after concentration or with no concentration at all. The effect is to give skylight a narrower conical distribution and to allow greater concentration. According to electromagnetic theory (Planck 1913), a medium of refractive index n allows a concentration of n^2; therefore, glass with $n = 1.73$ could effectively increase concentration to 4 from 3 (for 120° of sky collected). However, this does not result in much improvement in terms of efficiency.

A fluorescent concentrator in which a fluorescent dye absorbs incident light and then readmits light within a narrow set of wavelengths also has a limited potential. Multiple fluorescent plates would improve overall efficiency. A second option is the use of holographic coatings to collect light from different portions of the sky as the sun moves. It is possible that using angular control for limited wavelengths, light can be redirected in a manner that may be useful in a building. Experimental work is now being carried out in the United States and Australia. Both fluorescent concentrators and holographic coatings are radical departures from the more conventional reflective and refractive systems; neither as yet has demonstrated practically its success as an efficient collection system.

Transmission Systems

Light guide systems are necessary to transmit the light collected by any of the foregoing collection systems to the target area. Five methods are shown in Fig. 10.20:

1. Traditional cleared "lightways" or shafts (Plummer 1980)
2. Lightways with collimating lenses or lens guides (Bennett and Elijadi 1980)
3. Mirrored or reflective metal ducts (Baranov 1966)
4. Prismatic (refractive) ducts (Whitehead et al. 1984)
5. Optic fiber bundles (Allan 1980, Lacey 1982)

The method of simple propagation across cleared airspace or down a shaft is prone to ray scattering off dust in the air and passageways, resulting in

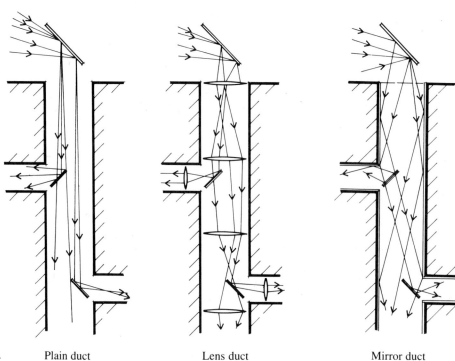

FIGURE 10.20.
Transmission systems for deep building interiors.

Plain duct Lens duct Mirror duct

deviation and loss of efficiency. Collimating lenses also collect dust and suffer from reflections at each interface, also reducing the intensity and efficiency of the light. Some of these losses can be eliminated by the use of antireflection coatings. However, this system is dependent on accurate alignment, so that shifts from environmental causes impair performance.

Hollow reflective light guides have considerable potential. These guides, either circular, rectangular, triangular, or square in cross section, have a highly reflective coating on the interior surface. The application of a stacked dielectric reflector film in place of metal spraying produces reflectances of up to 99 percent and an estimated efficiency for a 50-m-long and 1-m-diameter sealed unit of at least 50 percent. Specularly reflective metal films are now commercially available. Silver has been found to be the most practical material to use in a hollow light guide of small diameter and moderate length. If protected by a thin acrylic coating, the efficiency is high depending on diameter.

Losses due to reflectivity can be limited by confining the light inside a hollow guide with a prismatic cross section that traps light by total internal reflection and redirects it back down the core of the light guide. The light lost in transmission is scattered out into the surrounding space, and therefore, this type of guide can be used as both a transport and distribution device. Solid transparent light guides such as optic fibers also can be used to transmit light. Fiberoptic systems have the potential benefit of the use of flexible cables that can be routed as desired. However, close-packed bundles lose active cross-sectional end area to spacing and cladding by a factor of about 0.7, which detracts from the space savings of a fiber medium, and in addition, the bulk and weight of such a system would be considerable. Much lower ratios between collection and use can be expected with optic fibers because the small-diameter fiber system would require concentrating collectors to introduce luminous flux through a small cross-sectional collection area, and the losses inherent in the system would reduce this by a factor of 2.

Redistribution

Distribution design has been little researched because it is basically influenced by the type of light guide system employed. In an optical fiber system, for

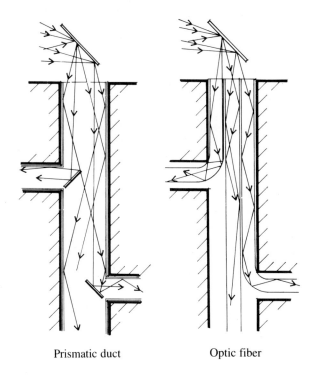

Prismatic duct   Optic fiber

example, simply ending the fibers in the space to be illuminated might well suffice. Thus, while the primary psychological objective is to introduce natural light to deep spaces as a changing background, there also would seem to be advantages in exploiting its architectural and sculptural qualities in a more spectacular manner. Deep interior lit spaces, then, demonstrate a lighting dynamic and special dimensionality akin to those which architects strive to achieve outside in the clear light of day.

Research on advanced systems for the daylighting of deep interiors is at an early stage. If successful economical solutions can be found, it would make possible daylight use in some buildings, particularly in densely built-up areas in which the vertical surfaces of buildings have limited sky view but there is access to the roof.

In addition, consideration must be given to lighting controls to make the best use of daylight. The systems mentioned require that photoelectric switching be appropriate when and as the sun penetrates a space. Photoelectric diming is less obtrusive, but careful planning of photocell sensor positions is required. Timed off/manual controls with a sunlighting system may be inefficient under intermittent sunlight.

ADVANCED GLAZING TECHNOLOGIES

Glazing Materials

There are now a wide variety of glazing materials available to reduce solar gain, including tinted and reflective glazings. Tinted glass uses absorbing materials dispersed throughout the glass, whereas reflective glazings employ a surface coating deposited directly on the glass to reflect and absorb incident energy. Both reduce light transmission.

Heat-absorbing glass is available in gray, bronze, and blue-green. Gray glass will transmit approximately the same percentage of visible light as solar thermal radiation. Bronze glass will generally transmit less visible light, whereas blue-green glass has a much higher visible transmittance than solar thermal transmittance.

Reflective coatings can be deposited on clear or tinted glass or on plastic substrates. Coating performance depends on the materials used. Reflective coatings are located in sealed multiple-glazing units because they are not able to withstand weathering or cleaning operations. Newer versions of these films are more selective in their transmittance and provide higher daylight transmittance relative to total solar transmittance. Some of these films have a low-emittance surface that reduces radiative heat loss.

Previous and existing considerations of window performance have viewed the window as a static device that has optical properties selected to optimize energy consumption in response to climate and building type. This has generally resulted in a compromise solution. Ideally, glazing design should be responsive to the hourly, daily, and seasonal climatic cycles that influence building consumption. Therefore, the properties of a glazing system should be varied in response to climatic conditions, and the net performance of the system should be timed to respond to thermal control and daylighting requirements. This is discussed in the next subsection.

Performance Characteristics

From an energy viewpoint, desirable window performance characteristics can be organized into three broad functional categories: low-conductance, high-transmittance glazing; optical switching materials; and selective-transmittance glazing.

High-Transmittance Glazing

This type of glazing, sometimes known as "low-emittance glazing," has good optical clarity, high solar transmittance to admit sunlight in winter, and low thermal conductance to reduce heating costs, and it is therefore well suited to cold climates. As shown in Fig. 10.21, low-emittance (low-E) coatings have a high visible transmittance, being predominantly transparent over the visible wavelengths (300 to 760 nm), and are reflective in the infrared region (2.0 to 100 μm) (Selkowitz 1985). For the near infrared (0.77 to 2.0 μm), the material can exhibit combined properties depending on design and end use. The coating's high infrared reflectance provides a low-emittance surface. The lower the emittance, the less is the magnitude of radiative transfer in the window (Fig. 10.22).

By using these nearly transparent coatings on a window surface applied directly to the glass in double or triple glazing or as plastic films glued to the glass, the thermal characteristics can be dramatically altered and energy loss can be more efficiently controlled.

In buildings where the heating load dominates and winter solar gains are beneficial, the low-E coating (highest absorptance) should be placed on no. 3 surface in double glazing and on no. 5 surface in triple glazing. The effect of coating placement on the overall thermal conductance, or U-value, is shown in Fig. 10.23. This figure demonstrates the properties of glazing required to reduce cooling loads. The figures at the bottom show the relationship of the solar spectrum to the reflectance and transmission of the glass.

The exterior placement is poor because of potential convection loss. New developments, such as using low-conductance gases and two low-E coated plastic inserts, make it possible to build windows having a solar transmittance of 50 percent. Such windows would outperform insulated walls in any orientation for most climates. The thermal performance of these new advanced glazing materials has been investigated using the computer program WINDOW for a wide range of environmental conditions. It was found that the low-E coatings greatly reduce the radiative loss component in a window.

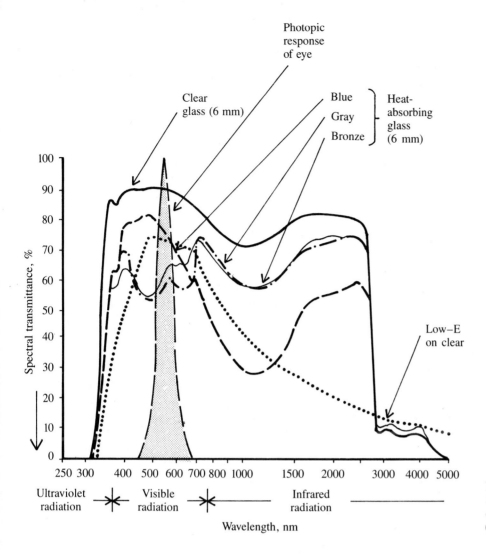

FIGURE 10.21.
Spectral transmittance of glazing materials as compared with the photopic response of the eye. (After CIE 1989).

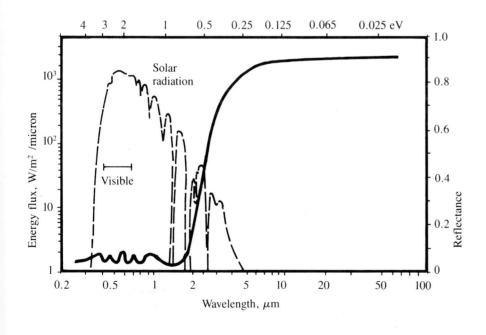

FIGURE 10.22.
Spectrum of solar radiation. Superimposed is the idealized selective reflectance of a low-E coating. (After Lampert 1981.)

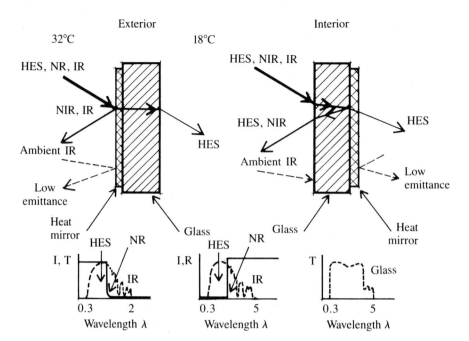

FIGURE 10.23.
The effect of coating placement on single glazing. The figures at the bottom show the relationship of the solar spectrum to the reflectance and transmission of glass. The exterior placement is of poor design because of potential convection losses. High energy solar, including visible (HES), 0.3 to 0.77 μm, near infrared (NIR), 0.77 to 2.0 μm, and infrared, 2.0 to 100 μm. (After Lampert 1982.)

Windows incorporating low-emittance coatings are now commercially available. The commercial products, based on multilayer designs, are not sufficiently durable to withstand the abrasion and atmospheric exposure of unprotected environments, and this limits their applicability to hermetically sealed windows. Research is now being carried out on new types of hard refractory materials to improve the resistance of these windows to weather.

Another solution to glazing that minimizes conducted and convected heat transfer is the use of transparent silica aerogels which can be formed by drying a colloidal gel of silica (Hunt 1984). The resulting material is highly transparent because the silica particles are much smaller than a wavelength of visible light. For use in window systems, aerogel must be protected from moisture, shock, and handling by being located between rigid panes of glass that are sealed at the edges. Research is continuing to develop means to protect this substance from the environment and to improve methods for its production.

Optical Switching Materials

Glazing materials that are responsive to hourly, daily, and seasonal climatic changes are known as "optical switching materials." These coatings can control the flow of light or heat in and out of a building window, thus performing an energy-management function. They inherently provide a change in the glazing's optical properties under the influence of light, heat, or an electrical field or by their combination. Depending on the design, the coatings can control glare, modulate daylight transmittance, limit solar heat gain to reduce cooling loads, and improve thermal comfort.

Current research is involved with thermochromic, photochromic, and electrochromic materials. "Thermochromic materials" change their optical properties with temperature, "electrochromic materials" change by an applied voltage in response to building conditions, and "photochromic materials" change with light intensity. The best-known application of the latter material is the photochromic glass used in eye glasses and goggles.

Window systems in the future may use optical switching materials and coatings to provide much of the solar control that now requires mechanical devices. Advance coating applications to produce glazing materials in which transmittance is a function of solar incidence angle have yet to be researched.

Thermochromic materials undergo a color transition at a specific temperature (Lampert and Li 1983). When the coating is below a specified transition temperature, it will transmit solar energy, but as the coating heats up, it switches to a metallic state and reflects the infrared solar radiation, thus allowing a lower influx of solar energy through the window. The only thermochromic material currently useful for building glazing applications is vanadium dioxide (VO_2). It is not known as yet whether VO_2 will give acceptable visible transmittance and have a pleasing color both in transmittance and from the exterior.

Photochromic materials alter their optical properties with light intensity. Generally, they are energy-absorptive owing to a reversible change of a single chemical species between two energy states each having different absorption spectra. This change is induced by illumination. Further research is needed to develop and utilize commercially available silver halide gases and to establish the most efficient application for absorptive windows (Lampert and Li 1983).

Electrochromic materials change their color reversibly in response to an applied electric potential. Tungsten trioxide (WO_3) is the most widely studied of this class of material. It switches mostly in the solar infrared region and maintains high visible transmittance in both states. Current research has shown that the electrochromic materials have the most potential for solar energy control applications (Lampert 1984).

The development of selective transmittance materials to produce glazing materials that have angular and spectral control to minimize cooling energy requirements and to maximize the efficiency of daylight utilization is currently under investigation. Angle-selective films offer the possibility of modifying the cosine response of specular glazings to reflect or admit solar radiation as a function of solar altitude or angle of incidence. Such films would enhance winter collection with high transmittance and reduce summer transmittance. They also would improve the effectiveness of daylight utilization.

The Effect of Glazing Technology on Human Comfort

Studies of summer conditions in offices (Langdon 1966) have indicated the general dissatisfaction with overheating resulting from solar heat gain. It has been shown that the amount of solar heat transmitted by a window can be controlled by adjusting the glazed area or by using low-transmission glasses or shading devices.

Any method of solar control has a direct effect on the admission of daylight (Markus 1963); therefore, both thermal and lighting conditions should be considered when selecting glazing materials, particularly from the viewpoint of light transmittance and color change.

SUMMARY

Window and rooflights are components of the building envelope that introduce daylight into the building interior as well as satisfying other user needs. The prediction of the daylight available is, therefore, an essential part of building design. Simple and powerful daylight design tools are described that will improve the understanding of the role of daylight in making buildings energy efficient. However it is necessary to recognize the limitations of these prediction tools in order to use them as effectively as possible.

Traditionally, the natural lighting of buildings has relied on side windows and rooflights to admit daylight to interior spaces, and this has brought problems of overheating and glare and the development of new glazing technologies to overcome some of these deficiencies. The dynamic control of window properties, either through the glazing material or with the addition of exterior or interior control devices, provides the most effective solution. Where cooling is the primary design factor, it is possible to combine static and movable shading systems to provide a view and daylight while minimizing the cooling load. The interest in daylighting also has brought interest in new optical techniques to "push" light deeper into buildings without glare or heat problems.

ACKNOWLEDGMENT

The contribution of material on advanced glazing technologies by Stephen Selkowitz is gratefully acknowledged.

REFERENCES

Allan, W. B. 1980. *Fibre Optics.* English Design Guides No. 36. Oxford University Press, Oxford, England.

American Society of Heating, Refrigerating and Air Conditioning Engineers. 1981. *ASHRAE Handbook of Fundamentals.* ASHRAE, Atlanta, Ga.

Aydinli, S. 1983. Availability of Solar Radiation and Daylight. *Proceedings of the 1983 International Daylighting Conference,* Phoenix, Ariz., 15-21.

Baranov, V. 1966. Parabolotoroidal mirrors as elements of solar energy concentrators. *Appl. Solar Energy.* 2: 9-12.

Bennett, D. J., and Elijadi, D. J. 1980. Solar optics: Energy as light. *Underground Space.* 4: 349-354.

Bensasson, S., and Burgess, K. 1978. Computer programs for daylighting in buildings. Design Office Consortium (now CICA), Cambridge.

Blackwell, M. J. 1953. Five years continuous recording of daylight illumination at Kew Observatory. Meteorological Research Committee, Meteorological Research Point no. 831.

Blackwell, M. J., and Powell, D. B. B. 1956. On the development of an improved daylight illumination recorder. Meteorological Research Committee, Meteorological Reporting Point no. 988.

Bodman, H. W., Eberbach, K., and Reuter, P. 1985. Oberlicht und Sonnenschutz. F B Nr 415, Bundesanstalt fur Arbeitsschutz, Dortmund.

Bryan, H. J., and Carlberg, D. 1984. Development of protractors for calculating the effects of daylight from clear skies. Proc. I.E.S. Conf., Atlanta.

Burts, E. 1961. Windowless classrooms: Windows help to promote better classroom learning. *NEA J.* 50: 13-14.

Collins, B. L. 1978. *Windows and People: A Literature Survey.* Institute for Applied Technology, National Bureau of Standards, Washington.

Commission Internationale de L'Eclairage 1989. Guide on Daylighting of Building Interiors. Final Draft.

Coulson, K. L. 1975. *Solar and Terrestrial Radiation.* Academic, New York. pp. 44-50.

Dogniax, R. 1960. Données meteorologiques concernant l'ensoleillement et l'eclairage naturel. Cahiers du Centre Scient. et Tech. du Batiment, Paris.

Drummond, A. J. 1956. Notes on the Measurement of Natural Illumination—Some Characteristics of Illumination Recorders. Archiv. for Meteorologie Geophysik, and Broklimatologie, Series B7, pp. 438-465.

Egan, M. D. 1983. *Concepts in Architectural Lighting.* McGraw-Hill, New York.

Fournol, A. 1951. Resultats français concernant les eclairments naturels. *CIE Proceedings II,* Paper Q.

Gillette, G. 1983. *A Daylighting Model for Building Energy Simulation.* Bldg. Sci. Series 152, NBS, Washington.

Griffith, J. W. 1976. *Predicting Daylight as Interior Illumination.* Libby-Owens-Ford Glass Company, Toledo, Ohio.

Hopkinson, R. G., Longmore, J., and Graham, A. M. 1958. Simplified daylight tables. Nat. Build. Studies Special Report No. 26, H.M.S.O., London.

Hopkinson, R. G., Petherbridge, P., and Longmore, J. 1966. *Daylighting.* Heinemann, London.

Hunt, A. 1984. Advances in Transparent Insulating Aerogels for Windows. In *Proceedings of Passive and Hybrid Solar Energy Update,* LBL Report 18507, Washington.

Hunt, D. R. G. 1979. *Availability of Daylight.* Building Research Establishment, Garston, England.

Illuminating Engineering Society, 1979. *Recommended Practice of Daylighting RP-5,* IES, New York.

Johnson, R., Arasteh, D., Connell, D., and Selkowitz, S. 1985. The Effect of Daylighting Strategies on Building Cooling Loads and Overall Energy Performance. In *Proceedings of the ASHRAE/DOE/BTECC Conference: Thermal Performance of the Exterior Envelopes of Buildings, Clearwater Beach, Florida.*

Jones, L. A., and Condit, H. R. 1948. Sunlight and Skylight as Determinants of Photographic Exposure: Luminous Density as Determined by Solar Altitude and Atmospheric Conditions. *J. Opt. Soc. Am.* 38: 123.

Kaufman, J. 1981. *I.E.S. Lighting Handbook,* 1981 Reference Volume. Illuminating Engineering Society of North America, New York.

Kim, J. J., Papamichael, K. M., Spitzglas, M., and Selkowitz, S. 1986. Determining Daylight Illuminance in Rooms Having Complex Fenestration Systems. *Proceedings of the 1986 International Daylighting Conference, Architecture and Natural Light,* Long Beach, Calif., 204-209.

Kittler, R. 1975. Standardization of Outdoor Conditions for the Calculation of the Daylight Factor with Clear Skies. In *Proceedings of the CIE Intersession Conference, Newcastle-upon-Tyne.*

Kittler, R. 1986. Luminance Models of Homogeneous Skies for Design and Energy Performance Predictions. *Proceedings of the 1986 International Daylighting Conference, Architecture and Natural Light,* Long Beach, Calif., 18-23.

Krochmann, J. 1962. Uber die Bestimmung des Innenreflexionsanteils des Tageslichtquotienten Lichttechnik 14, 3, pp. 105-109.

Lacey, E. A. 1982. *Fiberoptics.* Prentice-Hall, Englewood Cliffs, N.J.

Lampert, C. 1981. Materials chemistry and optical properties of transparent conductive thin films for solar energy utilization. LBL Report LBL-13502, Berkeley, Calif.

Lampert, C. 1982. Solar optical materials for innovative window design. LBL Report LBL-14694, Berkeley, Calif.

Lampert, C. A., and Li, S. M. 1983. *Photochromic and Thermochromic Phenomena for Switching Glazing.* LBL Report 16886, Berkeley, Calif.

Langdon, J. 1966. *Modern Offices: A User Survey.* SO, London.

Littlefair, P. J. 1984. A New Method for Predicting Energy Saving from on/off Photoelectric Controls. IP 14/84 Building Research Establishment, Garston.

Littlefair, P. J. 1985. The Luminous Efficacy of Daylight: A Review. *Light. Res. Technol.* 17: 162-182.

Longmore, J. 1968. BRS Daylight Protractors. H.M.S.O., London.

Longmore, J. 1978. The Engineering of Daylight. In J. A. Lynes (Ed.), *Developments in Lighting—1.* Applied Science, London.

Longmore, J., and Ne'eman, E. 1973. The Availability of Sunshine and Human Requirements for Sunlight in Buildings. Presented at the Conference on Environmental Research in Real Buildings, NIC Committee, TC 3.3, Cardiff.

Lopukhin, E. A. 1953. Natural daylight in Tashkent. *Akademiya nauk SSR izvestiya seriya geofizicheskaya,* 469.

Lynes, J. R. 1968. *Principles of Natural Lighting.* Elsevier, London.

McCluney, W. R. 1984. SKYSIZE—A simple procedure for sizing skylights based on statistical illumination performance. *Energy and Buildings.* 6: 213-219.

Maitreya, V. K. 1980. Evaluation of Reflected Light for Buildings in the Tropics. *Proceedings, Commission Internationale de L'Eclairage, Berlin,* 117-123.

Manning, P. 1965. *Office Design: A Study of Environment.* Pilkington Research Unit, Liverpool University, Liverpool, England.

Markus, T. A. 1967. The Significance of Sunshine and View for Office Workers. In R. G. Hopkinson (Ed.), *Sunlight in Buildings.* Boewcentrum International, Rotterdam.

Markus, T. A. 1963. Changing nature of daylight studies. *Light Lighting.* 56: 119-124.

Moon, P. 1940. Proposed standard solar-radiation curves for engineering use. *Franklin Inst.* 230: 583.

Moore, F. 1985. *Concepts and Practice of Architectural Daylighting.* Van Nostrand Reinhold. New York.

Nakamura, H., and Oki, M. 1979. Study on the statistic estimation of the horizontal illuminance from unobstructed sky. *J. Light Vis. Environ.* 3: 11.

Narasimhan, V., Saxena, B. K., and Maitreya, V. K., 1970. 412-413.

Narasimhan, V., Saxena, B. K. and Maitreya, V. K. 1970. Measurements of luminance and illuminance of cloudy skies at Roorkee. *Indian Journal of Technology* 8, 340-342.

Navvab, M., Karayei, M., Ne'eman, E., and Selkowitz, S. 1984. Daylight availability for San Francisco. *Energy, Buildings.* 6: 273-281.

Navvab, M. Karayel, M., Ne'eman, E., and Selkowitz, S. 1984. Analysis of atmospheric turbidity for daylight calculation. *Energy and Buildings,* 6: 293-303.

Ne'eman, E. 1984. A comprehensive approach to the integration of daylight and electric light in buildings. *Energy, Buildings.* 6: 97-108.

Ne'eman, E., and Hopkinson, R. G. 1970. Critical minimum acceptable window size: A study of window design and provision of view. *Light. Res. Technol.* 2: 17.

Paix, D. 1962. The Design of Buildings for Daylight. Commonwealth Building Station, Bulletin No. 7., Sydney, Australia.

Perraudeau, M., and Chauvel, P. 1986. One Year's Measurement of Luminous Climate in Nantes. In *Proceedings of the 1986 International Daylighting Conference, Architecture and Natural Light, Long Beach, Calif.*

Planck, M. 1913. *The Theory of Heat Radiation,* Dover.

Plant, C. G. H. 1970. The light of day. *Light Lighting.* 63: 292.

Plummer, H. 1980. Lanterns of sun. *Solar Age.*

Rabl, A. 1980. *Concentrating Collectors, Solar Technology Handbook (Pt. A, Engineering Fundamentals).* Eds., Dickinson, W. C., and Cheremisinoff, P. N. Marcel Dekker, New York.

Richards, S. J. and Rennhackkamp, W. M. H. 1959. Measurement of outdoor lighting conditions. *South African Electrical Review* 50, 193, 22-25 and 412-413.

Ruck, N. C. 1984. *Skylight Availability in Australia: Data and Their Application to Design.* Illuminating Engineering Society of Australia, Sydney.

Ruck, N. C.; 1985. Beaming daylight into deep rooms. *Building Research and Practice,* Vol. 13, 3, 144-147.

Selkowitz, S. 1984. Influence of windows on building energy use. LBL Report LBL-18663, Berkeley, Calif.

Selkowitz, S. 1985. Window Performance and Building Energy Use: Some Technical Options for Increasing Energy Efficiency. LBL Report 20213, Berkeley, Calif.

Selkowitz, S. 1986. Smart Windows. *Glass Magazine.* 8, 86-91.

Shukuya, M., and Kimura, K. 1983. Calculation of the work plane illuminance by daylighting including the effect of direct sunlight through windows with horizontal or vertical louvres. *Proceedings, Commission Internationale de l'Eclairage,* 20th Session, Amsterdam. CIE, Paris.

Solymer, L., and Cooke, D. J. 1981. *Volume Holography and Volume Gratings.* Academic, New York.

U.S. Weather Bureau. 1953-1956. Climatological Data. *National Summaries.* Asheville.

Whitehead, Brown, D. A., and Nodwell, R. A. 1984. A new device for distributing concentrated sunlight in building interiors. *Energy and Buildings,* 6: 119-125.

Winkelmann, F. C., and Selkowitz, S. 1985. Daylighting simulation in the DOE-2 building energy analysis program. *Energy and Buildings.* 8: 271-286.

Acoustic Performance

Anita Lawrence

11

In an ideal world, urban planning, site planning, and building design would ensure that every room in every building was located within an appropriate acoustic environment. Unfortunately, this is a rare situation, and it is necessary to use the building envelope as the final filter between noisy surroundings and the building's occupants.

The required acoustic performance of the building envelope depends on two things: (1) the acoustic environment in which the building is situated and (2) the acoustic design criteria inside the building's various areas. The second of these factors has already been discussed in relation to acoustic design in Chap. 7 and will not be repeated here. The external acoustical environment must be carefully assessed; in many cases, the noise levels fluctuate considerably and the question arises as to what percentage of time the internal acoustic criteria should be satisfied. For buildings such as concert halls, the answer is usually 100 percent; however, for commercial buildings, the occasional intrusion of sound from a particularly noisy vehicle or aircraft may be acceptable, and if this is the case, there could be considerable savings in building costs.

EXTERNAL NOISE LEVELS

Transportation in its various forms is the primary source of external noise at most sites. Other common noise sources are industrial processes, entertainment and sporting activities, and general neighborhood noise from appliances, parties, pets, and so on. There are several established methods of assessing each type of noise source, and these will now be discussed.

Road Traffic Noise

Road traffic noise is the most common source of external noise in urban areas. The actual noise levels and the temporal fluctuations depend on the traffic flow rate and the mix of vehicle types, as well as the distance from the road or highway. As a result of many studies of human reactions to traffic noise, it is now generally agreed that for most situations the noise levels should be described either in terms of the equivalent A-weighted sound-pressure levels, $L_{Aeq T}$, or the level exceeded for a certain percentage of time, usually 10 percent, L_{A10T}, or 50 percent, L_{A50T}. (Fortunately, for flow rates over about 1,000 vehicles per hour, the three descriptors are closely related, for example, $L_{A10T} = L_{Aeq T} + 3$ dB; see Fig. 11.1.) The "equivalent A-weighted sound-pressure level" is the level of a nonvarying sound that has the same total energy as the actual fluctuating sound over the same time period. Several methods are available to predict traffic noise levels if the traffic flow rate and composition (percentage of heavy commercial vehicles) is known. Since traffic flow rates generally vary considerably between daytime and nighttime, as well as on weekends, it is necessary to use the relevant data when making the predictions. For example, if the building is only to be used during weekdays, then traffic flow rates at night or during weekends are irrelevant.

Traffic noise prediction methods vary from very simple ones to others that take into account the length of visible road, shielding, gradients, and topography (U.K. Department of the Environment 1976, Burgess 1977). For complicated situations or where the exclusion of road traffic noise is critical, comprehensive measurements at the relevant site locations and at the appropriate times should be made. If this is necessary, meteorologic factors should be taken into account if the road is, say, 30 m or more from the site.

Typical $L_{Aeq.T}$ traffic noise levels near busy roads can exceed 70 dB(A), which means that the building envelope attenuation must be 40 dB(A) or more if quiet interiors are required. This is impossible to achieve with natural ventilation if the opening(s) are in the facade(s) facing the road.

In some situations, particularly for residential buildings, nighttime noise may be the most critical. There may be generally a low flow rate and low traffic noise levels, but a few heavy commercial vehicles passing may cause a large temporary increase in level that may awaken sleepers. It is difficult to quantify these situations, since the equivalent energy level would tend to underestimate the annoyance caused, and if the occurrences extend over less than 10 percent of the time period, $L_{A10.T}$ also would not be a valid descriptor.

Aircraft Noise

Aircraft are an important noise source for buildings located in the vicinity of airports. Again, there has been much research into the effects of aircraft noise on people, particularly in the domestic situation. There is no general agreement

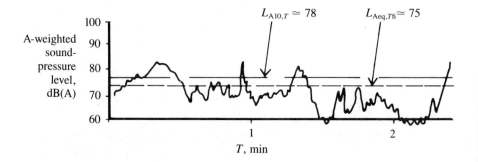

FIGURE 11.1.
Typical chart record of road traffic noise.

on the way in which noise from individual aircraft landing and taking off should be combined to give an overall aircraft noise impact, since each country appears to have developed its own system. In general, however, the need to take into account the number of aircraft movements and the maximum flyover levels at a particular site is recognized.

One aircraft noise-assessment method developed in the United States is the "noise exposure forecast (NEF) system." The noiseprint, or contours of equal sound level on the ground, resulting from the operations of a particular aircraft type are determined; these are then combined for the different aircraft types expected to use the airport (usually at some date a few years ahead). A penalty is applied to night movements. Contours of "equal" aircraft noise impact are drawn, and recommendations are made as to the suitability of the land for various building types (Fig. 11.2). Typically, the land may be "acceptable," "conditionally acceptable," or "not acceptable." For the conditional cases, it is necessary to carry out a detailed acoustical analysis of the attenuation of the building envelope. A similar method has been adopted in Australia, although aircraft movements in the evening as well as at night are penalized (but using a lower penalty than for the US NEF system for night only). Guidelines for the selection of building-envelope elements for buildings of different types whose sites lie within the "conditionally acceptable" contours are given (SAA 2021 1985).

It should be noted that although it may be possible to achieve acoustic design criteria inside an aircraft noise-affected building, external areas associated with the building's use, such as gardens, balconies, or outdoor play areas, cannot be protected, and this may adversely affect people's comfort. In addition, if the building envelope has sufficient sound attenuation to reduce very high levels of aircraft noise, it will tend to reduce other noises to inaudibility. People usually do not like living within a sound-isolated shell without *any* aural contact with the world around them.

Less information is available regarding noise from helicopters or light and ultralight aircraft and its effect on people. If such sources are likely to have an impact on the building site, details of maximal noise levels may need to be acquired through measurement, although there are many difficulties involved in obtaining reliable measurements of the noise emitted by such sources.

Railway Noise

Railway noise can affect buildings in two ways. First, surface trains will emit airborne sound, which can be at a high level, particularly for high-speed passenger trains and freight trains. Ground vibration often may be associated with these passbys. Second, underground railway vibration may be transmitted through the ground to the building's footings. If the train passbys are frequent, then the equivalent A-weighted sound-pressure level is probably the most

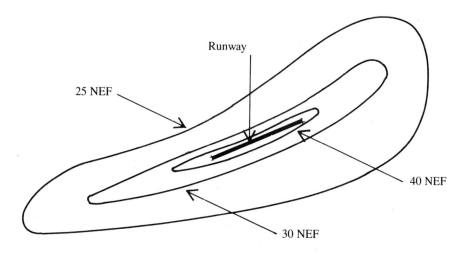

FIGURE 11.2.
Aircraft noise-exposure forecast contours for use with land-use compatibility assessment.

appropriate descriptor. If there are only a few movements per day, it may be more reasonable to use some combination of maximum passby level and number of noise events.

There are few prediction methods available for railway noise, and measurement of the actual levels may be the most accurate procedure. However, care must be taken that the rolling stock, track, and supporting structures are similar to those which will affect the building, since there is a wide variation in the noise emission of different types. As with road traffic noise, if the distance between source and site is 30 m or more, meteorologic factors must be taken into account and the measurements must be made when appropriate weather conditions prevail.

Other Noise Sources

Other external noise sources include industry, shipping activities, entertainment, sport, and neighborhood noise. Since there is so much variation, it is usual to conduct environmental noise measurements to determine the design noise levels for important buildings. Care must be taken with the sampling for such surveys, both with respect to location and time.

SELECTION OF BUILDING-ENVELOPE ELEMENTS

A similar principle applies for the selection of elements and materials to divide spaces within buildings. However, it is even more difficult to translate laboratory data to the field situation, since the external sound will frequently impinge at near-grazing incidence. The transmission of sound through an element varies with angle of incidence, being lowest at a normal incidence and greatest at grazing incidence. This means that as a vehicle moves along a road past the building, the effective attenuation of the facade will alter. In addition, most of the sound will be at grazing incidence at the upper floors of high-rise buildings (Fig. 11.3).

There is an international standard for measuring the airborne sound insulation of facade elements and facades (ISO 140 1978), but it is difficult to carry out such measurements in practice. The international standard rating method (ISO 717/3 1982) uses the same implied noise spectrum as that used for interior partitions, but this may be criticized because there is usually much more low-frequency noise energy in transportation and from many types of industrial sources than from typical sources inside buildings. If possible, the actual spectrum of the relevant external noise source(s) should be determined,

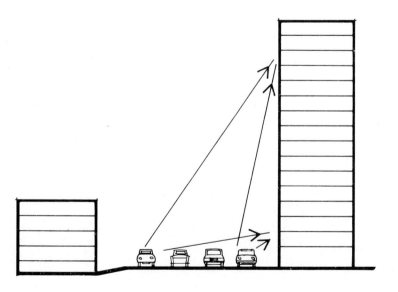

FIGURE 11.3.
Traffic noise incident at grazing angles for high-rise buildings.

and the effective attenuation should be calculated using one-third octave band data for the elements chosen.

Usually there will be more than one type of element used, even for a building facade, and naturally, the roof will be of different construction from the walls. It is necessary to calculate the composite attenuation, which cannot be done on an arithmetic basis, because of the logarithmic nature of the decibel. The "sound transmission coefficient," t, for each element at each frequency band of concern (usually the one-third octave bands from 100 or 125 to 4,000 or 5,000 Hz) must be determined, multiplied by the relevant area of the component, and the result averaged and converted back into sound transmission loss. For example, if, at 500 Hz a masonry wall has a sound transmission loss (STL) of 45 dB and a net area of 50 m², and there is also a window with a sound transmission loss of 28 dB and a net area of 20 m², the average sound transmission loss of the facade will *not* be

$$\frac{(45 \times 50) + (28 \times 20)}{50 + 20} \neq 40 \text{ dB} \qquad \textbf{(11.1)}$$

but it will be determined from the following:

$$\frac{(3.16 \times 10^{-5} \times 50) + (1.6 \times 10^{-3} \times 20)}{50 + 20} = 33 \text{ dB} \qquad \textbf{(11.2)}$$

Natural Ventilation and Sound Transmission Through Building Envelopes

The decision whether to use natural ventilation or air-conditioning plays a vital role in a building's acoustic environment. If there is a free air path, airborne sound will be transmitted with little or no attenuation. As shown earlier, a simple area-averaging technique is incorrect, and it severely underestimates the importance of even small air gaps and leaks. Using the typical brick wall as an example, with an STL of 45 dB at 500 Hz, if it should include an air vent with an effective area of 10 percent of the total wall, the average STL will be only 10 dB (Fig. 11.4). Even a gap with an effective area of only 1 percent of the total will reduce the average transmission loss to about 20 dB. It is common to find air gaps of such dimensions around openable windows and doors unless they are provided with special acoustic seals.

It is extremely important that building designers understand that there is no point in choosing building-envelope components with high sound transmission loss values if there are to be windows or other openings for ventilation purposes. To obtain the expected performance of the facade, it is essential that it be *sealed* against air leaks. If natural ventilation is required, then the openings must be located in acoustically protected areas. One possible solution is to locate ventilation openings facing a courtyard shielded from external noise sources; however, there is then the possibility that noise will travel out through one room's windows, reverberate around the courtyard, and then be transferred to other rooms through their open windows (Fig. 11.5).

If the decision is made to air-condition the building, the external envelope may be designed to have good sound attenuating properties. However, care must be taken with the air-conditioning system so that it does not itself introduce unwanted sound originating from the plant and airflow and that it does not allow noise to be transmitted from one space to another through the ductwork. Generally, the air velocity should be kept as low as possible, which means, for a given air quantity, that ducts must be large; the fans also should be large and relatively slow moving (although it is necessary for them to be chosen to operate at their maximum design efficiency if they are to be quiet). It is essential that the air-conditioning engineer be involved in the early design stages so that sufficient space is allowed not only for the ductwork, but also for the plant rooms.

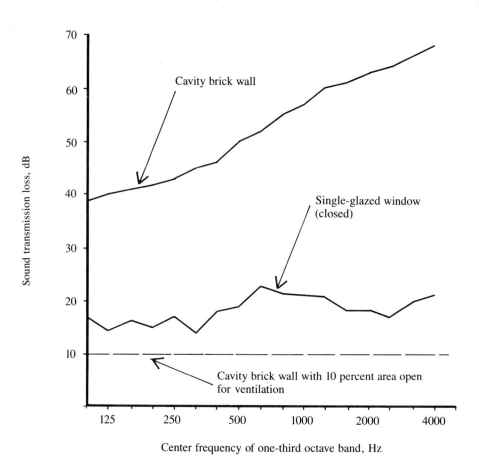

FIGURE 11.4.
Effect of ventilation openings on facade attenuation.

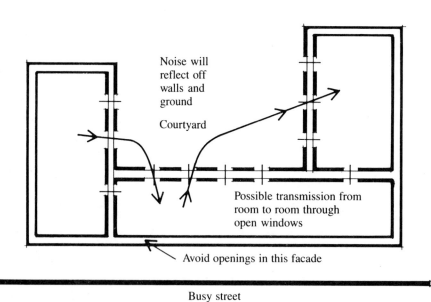

FIGURE 11.5.
Advantages and problems using courtyards for ventilation.

Double Glazing

It is sometimes thought that factory-sealed double-glazed windows also will have good acoustic performance. However, if transportation noise is the problem, with significant low-frequency components, this type of double glazing has little or no advantage over single glazing of the same overall pane thickness. In order to achieve good low-frequency sound attenuation, it is necessary to provide a wide air gap between the panes, preferably on the order of 200 mm. Two separate frames should be used, sealed around their perimeters and

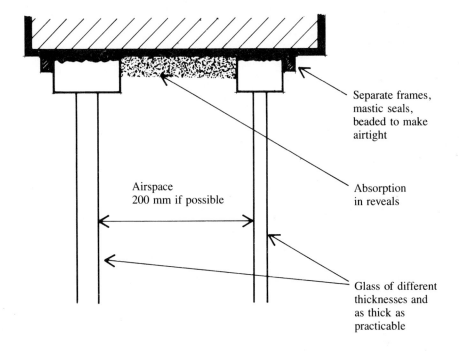

Separate frames,
mastic seals,
beaded to make
airtight

Absorption
in reveals

Airspace
200 mm if possible

Glass of different
thicknesses and
as thick as
practicable

FIGURE 11.6.
Principles of double glazing for good sound
attenuation.

constructed with resilient materials to prevent structure-borne flanking transmission. The provision of absorbent linings in the reveal space is also useful (Fig. 11.6). An acoustically effective type of window, which also should have good thermal properties, consists of one leaf of single glazing and another, spaced at least 75 mm away, of thermal double glazing.

Doors

Doors can be a source of flanking sound transmission; by their very nature, they must be openable. If they are to provide reasonable noise reduction, they must be as heavy as possible and also equipped with resilient seals around head and jambs. In addition, there must be some form of sealing at the bottom. One solution is to use a spring-loaded seal that lifts as the door opens and drops as the door is closed. Trailing seals soon wear out, or if they are effective in reducing noise, they will be difficult to operate and will probably be removed by the building's occupants, thus defeating their purpose.

Roofs

It is obvious when aircraft noise is the main source that the roof must be considered as an important sound-transmission path. The roof also may be important for road traffic and other ground-level sources for low-rise buildings. For example, there is an effective sound-transmission path through the eaves, roof tiles, and ceiling for typical tile-roofed dwellings (the same naturally applies for shingled roofs). Field measurements have shown that the overall sound transmission into a building differs little between timber, brick-veneer, and double-brick facades if the building has a conventional framed roof (Lawrence and Burgess 1986). Double-glazing windows in such buildings has little or no effect in reducing noise intrusion.

Massive roofs, such as those constructed out of reinforced concrete, are required for good sound attenuation. Such roofs are too expensive in most cases for low-rise residential buildings, although they are common for commercial buildings and for high-rise dwellings, hotels, and so on.

Some research has been carried out on the design of improved lightweight roofing systems, such as multiple layers of dense sheet materials and incorporating thermal insulation. Great care must be taken with metal sheet roofing systems because it is extremely difficult to avoid the transmission of rain noise through such roofs. Much more research is needed in these areas.

SUMMARY

The building envelope must be assessed as an overall system when considering sound transmission through it. It is very easy to waste money by choosing elements with good sound-attenuating properties for some parts of the facade and acoustically transparent elements for other parts. For example, the good sound reduction that may be provided by a heavy masonry facade with wide-spaced double glazing may be flanked by a path through an overhanging lightweight floor or eaves construction or by a ventilation opening.

Not only must the original design be carefully carried out, with the acoustic requirements forming part of the brief, along with the thermal and visual criteria, but informed detailing of the construction is essential. This, together with frequent supervision of the construction, is necessary to ensure that small errors that can be of great significance in sound transmission are avoided. It is also helpful if building foremen and tradesmen are briefed regarding the importance of the faithful execution of details if the desired result is to be achieved.

REFERENCES

Burgess, M. A. 1977. Noise prediction for urban traffic conditions: Related to measurements in the Sydney metropolitan area. *Appl. Acoust.* 10: 1-7.

International Standards Organization. ISO 140: *Acoustics—Measurement of Sound Insulation in Buildings and of Building Elements,* Part 5: *Field Measurements of Airborne Sound Insulation of Facade Elements and Facades.*

International Standards Organization. ISO 717/3: *Acoustics—Rating of Sound Insulation in Buildings and of Building Elements,* Part 3: *Airborne Sound Insulation of Facade Elements and Facades.*

Lawrence, A., and Burgess, M. 1986. Road Traffic Noise Reduction of Domestic Facades. In *Proceedings of the Community Noise Conference, Toowoomba.* Australian Acoustical Society, Sydney. 349-356.

Standards Association of Australia. 1985. SAA 2021: *Acoustics—Aircraft Noise Intrusion: Building Siting and Construction.*

U.K. Department of the Environment. 1976. *Calculation of Road Traffic Noise.* Department of the Environment, United Kingdom.

P A R T IV

Prediction and Analysis of Building Energy Performance

It has been shown that it is possible to manipulate building form, structure, and materials to significantly reduce energy loads without adversely affecting indoor environmental quality. One of the most important factors in energy load reduction is the building perimeter. The design of the building envelope is critical for directing light and heat where they are needed. The climate, choice of materials and shading controls, and the thermal mass of the facade are all variables that need analyzing in terms of their interactions with each other.

Measured data from buildings could provide the necessary information for estimating actual energy savings, but the existing data base is small. Therefore, Part IV is principally concerned with the introduction of tools to analyze such interactions, the results from analyses of prototypical models using conventional glazing materials as a baseline, and an evaluation of the potential of advanced glazing materials.

In Chapter 12 on tools for evaluating building energy performance, Nancy Ruck describes simple methods for preliminary assessment of energy savings and the use of daylighting, larger mainframe programs currently used for research purposes, and the newer microcomputer versions now being used in many professional offices. In addition, field studies that have been carried out in recent years are described. These demonstrate some of the problems associated with monitoring a building's energy performance.

In Chapter 13, on the evaluation of glazing performance, Stephen Selkowitz looks at the newer window films and coatings and comments on the latest research results on switchable glazing materials and technological limitations in this area.

Nancy Ruck concludes that human comfort and performance are inextricably woven into the fabric of building design and that emphasis needs to be placed on a more human-oriented approach to environmental controls. Any negative impact produced by the impairment of human performance can reduce energy savings and increase costs. Efforts in the field of energy balance have stagnated on the promotion of more accurate methods to predict static conditions denying the building occupant any stimulus from change. The solution lies in an integrated consideration of other than energy related criteria.

Evaluation of Building Energy Performance

12

Nancy Ruck

THE NEED FOR PREDICTION METHODS

There are many design strategies to reduce energy consumption in buildings, and some have been discussed in detail in Part III. However, to estimate a building's energy performance, the interaction between thermal and lighting loads must be taken into consideration. Although effective daylighting design strategies can reduce electric lighting requirements, the use of daylight in buildings to save energy can present problems resulting from its interaction with the thermal load. The problem of lighting controls in the interior to enable the dimming or turning off of electric lights also needs to be satisfactorily resolved.

Measured performance data from daylighted buildings could provide the necessary information for estimating the real energy savings, but the existing performance data base is relatively small. Only a small number of buildings have been monitored (Warren et al. 1986, Boyer 1986).

If experience with existing buildings cannot provide sufficient guidance for successful solutions, the designer has several options. There are many prediction and evaluation techniques that help a designer understand the energy- and comfort-related implications of a proposed design. These technologies vary widely in their accuracy, comprehensiveness, applicability, and utility. Some are intended to be used early in the design process when general concepts are being evaluated; others are best suited for performance evaluations of detailed designs. Some of the tools address the performance of specific building components, e.g., windows or H.V.A.C. systems, while others address overall building energy performance.

The overriding concern in selecting appropriate prediction and evaluation techniques is to choose a tool that is appropriate for the intended purpose. In many cases, this is difficult because tool developers do not always clearly indicate the range of applicability of their tools. The usefulness of a tool is also linked to the skill of the user. In the hands of an experienced user, some tools can be used successfully for purposes for which they were not nominally intended.

In the sections that follow several of the options now in use or under development for determining daylighting opportunities in buildings and examining the overall energy impacts are examined. These include both simple procedures and more complex mainframe computer models. With continued evolution in the power and performance/cost ratio of microcomputers, it is clear that future tool development will be increasingly microcomputer-based.

SIMPLE PREDICTION METHODS

Energy Analysis Using the Nomograph

Nomographs and other simple graphic methods are available to evaluate the role of daylighting as an energy-saving strategy. In some applications, the use of

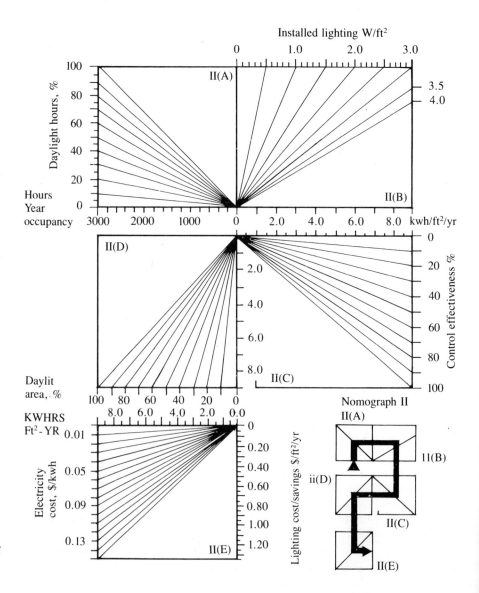

FIGURE 12.1.
Nomograph II: Annual energy use and cost of daylight. (After Selkowitz and Gabel 1984.)

daylight may not be appropriate; in others, its use as an energy strategy may not be cost-effective. It is therefore useful to evaluate the energy-savings potential of daylight early in the design process. Nomographs that serve as a graphic means of obtaining information can be useful for making quick decisions. Although they may appear simple in terms of presentation and use, they may contain results from very powerful simulation tools. Figures 12.1 and 12.2 show two in a series of four nomogaphs derived from analyses using the energy-simulation program DOE-2.1B and developed by Lawrence Berkeley Laboratory (LBL) specifically for daylighting applications (Selkowitz and Gabel 1984). Other daylighting nomographs are incorporated in a broader nomograph approach to overall building analysis (BHKRA Associates 1985).

The LBL nomographs are intended to assist designers in making initial decisions concerning daylight's potential as a strategy for energy conservation and load management in commercial buildings. They do not provide detailed design decisions, nor do they guarantee that workable solutions are possible. They address annual electric lighting energy savings and peak-load savings, but they exclude the thermal effects. However, they offer a quick estimate of the magnitude of potential savings on the basis of which more detailed studies might be pursued.

Nomograph II, shown in Fig. 12.1, can be used initially to determine annual energy consumption (kWh/ft^2-yr) and associated costs ($/ft^2-yr) for an electrically lighted building and then to estimate annual energy and cost savings that will accrue if daylight strategies are utilized. Nomograph IV, as shown in Fig. 12.2, determines the justifiable economic investment in daylight-

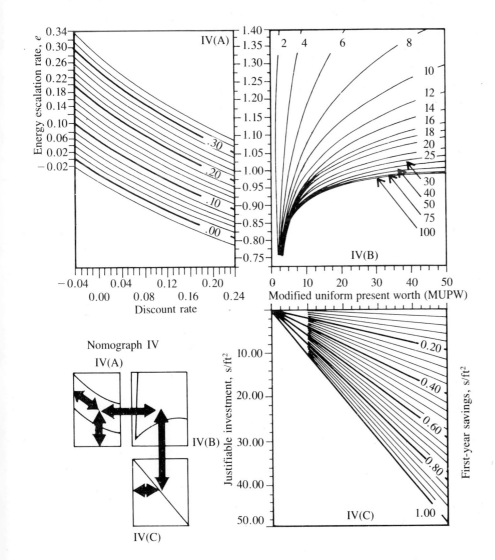

FIGURE 12.2.
Nomograph IV: Justifiable investment in daylighting. (After Selkowitz and Gabel 1984.)

ing (lighting controls, sun controls, and other costs directly associated with daylighting) based on the energy savings determined with the second and third nomograph (impact of daylighting on peak electrical loads). Other simple graphic methods (Robbins et al. 1985, Robbins 1986) have been used to estimate energy savings in buildings.

Skylight Graphs

Another estimation technique based on graphs and derived from extensive computer simulations relates to optimization of the basic physical parameters of a skylight system, that is, skylight size and shape, glazing material, and light-well size and shape, in order to maximize lighting energy and hence cost savings (AAMA 1987). The technical bases of the graphs are described in Arasteh et al. (1985), Johnson (1983), Sullivan et al. (1983), and Winkelmann and Selkowitz (1984).

The information presented in the graphs is valid for most commercial skylighting applications and where the following conditions apply:

1. Climatic conditions similar to those in the representative cities of Seattle, Wash., Madison, Wis., Washington, D.C., Lake Charles, Iowa, Los Angeles, Calif., Fresno, Calif., and Phoenix, Ariz.
2. Occupancy schedules similar to office occupancies (i.e., daytime operation, normal internal heat gains, and lighting requirements).
3. Uniform lighting conditions where there are no major differences in task illuminance.
4. Flat or low-slope roof systems (4:12 pitch), with approximately even skylight distribution across roof and skylight spacing less than or equal to 1.5 times ceiling height, with skylight glazings at least partially diffusing.
5. Daylight controls for lighting.

Buildings that differ greatly from these conditions may require a detailed analysis to accurately estimate their skylighting performance.

A series of worksheets containing simple analysis procedures is used for optimizing the basic parameters for the skylight system. The graphs are used to find the range of effective apertures to target in your design, the fraction of lighting energy saved, total electricity consumption on an annual basis, peak electrical demand, and heating energy consumption. Figures 12.3 to 12.6 show the graphs of these parameters for Phoenix, Arizona.

The graphs for the range of effective apertures and lighting energy savings are given for three different average illuminances of 300 lx (30 fc), 500 lx (50 fc), and 700 lx (70 fc). For total energy consumption, peak electrical demand, and heating energy consumption, the graphs have been developed only for 500 lx (50 fc) with continuous dimming. Judgment should be used in applying these results to other situations.

The maximum and minimum effective aperture values from Figs. 12.3 and 12.4 can be used to determine the skylight-to-floor ratio and hence the skylight area. The fraction of lighting energy saved from Fig. 12.4 is used to determine the total number of kilowatt hours of electricity saved each year from the daylighting system. This is achieved by multiplying the lighting power density (Fig. 12.3) by the gross floor area, the number of full-load lighting hours, and the fraction of lighting energy saved. Division by 1,000 converts to kilowatthours per year.

Total energy consumption includes lighting, air-conditioning, fans, and other uses in the building. Figure 12.5 shows a graph of electricity consumption on an annual per-square-foot basis. There are two bands, one for 1.2 W/ft^2 and the other 2.2 W/ft^2. The top of each band (dashed line) represents a low skylight efficacy of 0.50, and the bottom (solid line) corresponds to a high skylight efficacy of 1.00.

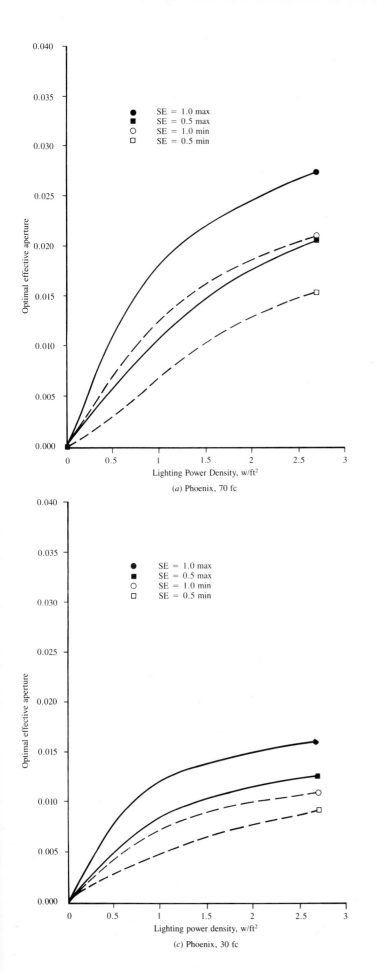

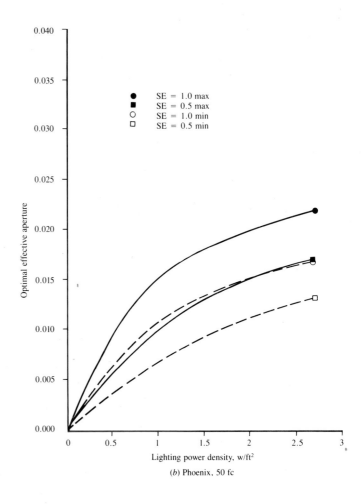

FIGURE 12.3.
Optimal effective aperture as a function of lighting power density for three average illuminances in Phoenix, Arizona. SE is skylight efficiency.

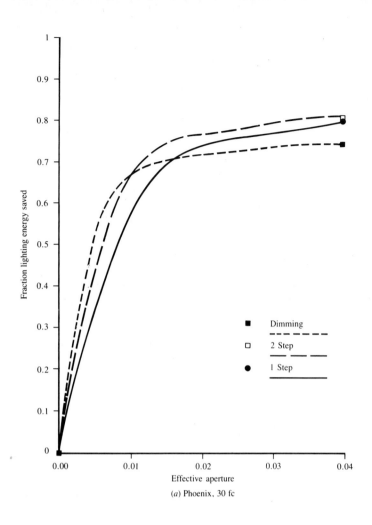

(a) Phoenix, 30 fc

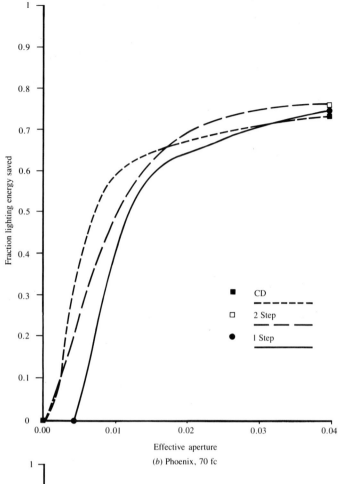

(b) Phoenix, 70 fc

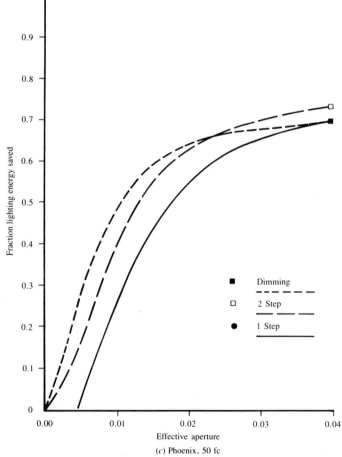

(c) Phoenix, 50 fc

FIGURE 12.4.
Fraction of lighting energy saved as a function of
effective aperture for three average illuminances
in Phoenix, Arizona.

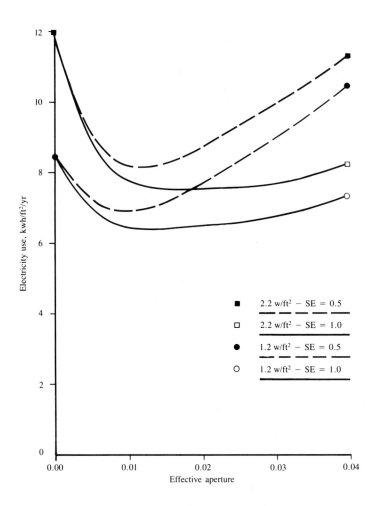

FIGURE 12.5.
Total energy consumption as a function of effective aperture for Phoenix, Arizona.

Peak electrical demand (Fig. 12.6) is a measure of the maximum demand in kilowatts that a building will require during a typical year, and this can have a strong influence on electricity cost. Therefore, this graph can warn of potential peak-demand problems.

Figure 12.7 shows heating energy consumption for the heating system as a function of skylight-to-floor ratio, single glazing *(SG)* or double glazing *(DG)*, and shading coefficient *(SC)*. The opaque portion of the roof has a U-value equal to that prescribed by building codes and in the analysis has been kept constant as the skylight-to-floor-ratio increases, so the heating energy consumption reflects the increased heat loss through the skylights.

Skylight graphs provide an easy way to address the complex problem of controlling energy costs and reducing energy consumption. There are, of course, other considerations, such as occupant comfort and building performance, and these issues can be studied with other daylighting design tools.

Prediction Using Regression Coefficients

The prediction of net energy consumption (lighting, cooling, and heating loads) for various climatic zones can be carried out using regression equations derived from an extensive series of computer runs using a prototypical building model (Sullivan et al. 1983). With data generated by simulating building modules representative of perimeter, core, and rooftop zones of a commercial building, the following regression expression can be used for a quantitative and qualitative analysis of individual components of a building's energy use.

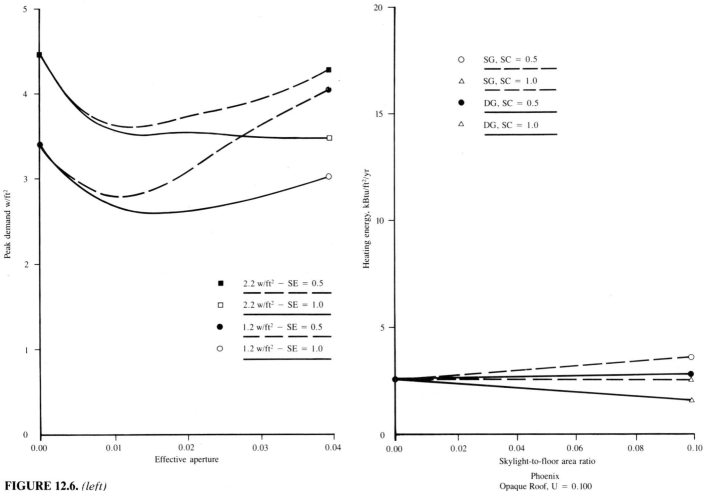

FIGURE 12.6. *(left)*
Peak electrical demand as a function of effective aperture for Phoenix, Arizona.

FIGURE 12.7. *(right)*
Heating fuel consumption as a function of skylight-to-floor ratio.

The equation includes effects arising from building conductance, solar radiation, internal heat gains, and infiltration, as well as correction factors for overhangs and daylighting:

$$E_i = b_1 U_o A_T + b_2 k_o A_g SC + b_3 k_d A_f L + b_4 A_f \qquad (12.1)$$

where E_i = energy required for heating or cooling
b_1, b_2, b_3, b_4 = regression coefficients
U_o = exterior envelope overall U-value (W/m^2/°C)
A_T = exterior wall or roof area (m^2)
A_g = window or skylight area (m^2)
SC = shading coefficient
k_o = correction factor due to overhangs
A_f = floor or roof area (m^2)
L = lighting wattage (W/m^2)
k_d = correction factor due to daylighting

This method of prediction focuses on the relationship between fenestration parameters and thermal loads based on a standard building module with a specific HVAC and plant system. There are many other parameters which, when included in a detailed simulation, might alter the absolute energy requirements; therefore, this prediction method can be regarded only as a means to determine the radiation effects of various window variables in different climatic zones.

A full description is given in the reference (Sullivan et al. 1983). The regression coefficients given in the publication for five locations in the United States can be used in the equation to determine heating and cooling loads and total energy consumption for those localities. Ongoing work has extended this technique to include peak electric demand as well as indicators of thermal and visual comfort. However, it should be borne in mind that many other parameters in actual buildings may influence energy use, such as building shape, pattern of use, shading devices, and equipment types.

BUILDING ENERGY SIMULATION PROGRAMS

Mainframe Programs

An energy-analysis program has several conflicting requirements. On the one hand, it must be as versatile and flexible as possible to allow modeling of a wide variety of architectural design solutions and the performance of these solutions under a range of exterior sun and sky conditions. Accuracy is a desirable attribute, but it must be balanced by the need for adequate speed in order to reduce runtime and costs. Adding complexity to the daylight calculation model not only adds costs in terms of computational time, but also in terms of the input requirements for the program, as well as the educational investment necessary on the part of the user to effectively operate the program. Energy programs currently available include DOE-2 (Building Energy Simulation Group 1984), ESP (Clark and McLean 1985), and BLAST (U.S. Army 1979).

A number of approaches exist for calculating daylight for inclusion in energy programs.

Approaches to Daylight Modeling

The need to quantify the effects of daylighting on building energy use has prompted the development of daylight prediction techniques for use in large building energy-simulation programs. These techniques differ in their level of sophistication in daylight modeling and in the degree to which the daylight model is integrated into the thermal and solar gain models of the main program. One approach has been to mathematically model simple room shapes, converting existing hourly insolation data from weather tapes using luminous efficacy factors and generally to adopt the daylight factor method with two basic sky models, clear and overcast skies. This approach was developed at the National Bureau of Standards and was intended to be used as a self-contained module that could be linked to building simulation modules (Gillette and Kusuda 1980).

Another approach uses hourly daylight profiles derived from small-scale model tests and driven by exterior solar radiation values from a weather tape or by calculated room solar gains (Matthews and Barnaby 1980). A statistical daylight model (SDM) was developed from this approach for use in a modified version of BLAST. Another model-based approach is the ENSAR Daylighting Analysis Program, which is an interactive design tool based on physical models in conjunction with a quick, responsive analysis tool in the form of a computer and data-acquisition system for modeling the distribution of natural light. However, any experimental approach using data from scale models is time-consuming and can require a large investment in a photometric system. Results may be limited to the specific sky conditions (real or simulated) under which the scale model was used.

On the other hand, computional techniques in the form of mathematical modeling are also limited in accuracy because of assumptions incorporated into the mathematical models relating to sky luminance distribution, window systems that incorporate complex shading devices, and room geometry and surface characteristics. Research is currently overcoming these problems.

The determination of daylight illuminance in rooms having complex window systems is being tackled by using photometric equipment to determine the bidirectional transmittance of window components and systems and inputting bidirectional transmission coefficients in the form of equations into computer programs such as SUPERLITE (Kim et al. 1986).

The internal illuminance is, in some of these mathematical modeling techniques, expressed in terms of a ratio, a modified daylight factor (see Chap. 10). In traditional usage, the daylight factor is only valid for overcast skies. However, in several of the newer approaches, a modified daylight factor is defined for each of several different sky conditions. To account for orientation effects, it is possible to normalize to an exterior vertical illuminance under clear skies rather than the traditional horizontal illuminance. These approaches (Robbins 1986, Kim et al. 1986) become similar to the lumen method (Griffith 1976), which has been the basis of North American daylight design practice over the last 30 years.

The method is modified by introducing an extensive series of coefficients of utilization relating to variable design parameters, making the method more broadly applicable (Kim et al. 1986). Direct sun, Commission Internationale de l'Eclairage (CIE) standard clear and overcast skies, and ground-reflected light are included. A further refinement would be to use a daylight availability model with sky luminance distributions that incorporate the effects of orientation and partly cloudy skies. Aydinli (1983), Aydinli and Seidl (1986), and Littlefair (1984) have produced methods based on the average sky related to their particular countries.

These average-sky models can be easily used to determine total lighting use for a whole working year. However, for some applications, such as the modeling of lighting as a casual gain in large computer programs, a more dynamic approach is needed; that is, the state of lighting for each hour and day of the year is required. Gillette (1983) and Winkelmann and Selkowitz (1985) have developed models on this basis in which the momentary sky luminance distribution is assumed to be a linear combination of clear and overcast skies, depending on cloud cover.

Another approach has been adopted by Kittler (1986) whereby a series of representational homogeneous skies (including the clear and overcast skies) is based on actual turbidity values. With this latter approach, absolute values of internal illuminance can be obtained without the utilization of a daylight factor (Trudgian, Kittler, and Ruck 1988). A more detailed description of some of the daylighting modules to energy programs together with comments follows.

DOE-2 Daylighting Model

DOE-2 is unique among most building energy simulation models in that a daylighting model is integrated into the overall hourly thermal and lighting calculation allowing for occupant use of shading systems in response to solar gain and glare. The daylighting model for DOE-2 is based on a compromise between computing requirements for maximizing accuracy, minimizing computational time and cost, minimizing input requirements, and maximizing versatility. The daylighting simulation determines the hourly, monthly, and yearly illuminance levels and glare indices, the impact of daylighting on electrical energy consumption and peak electrical demand, and the daylighting impact on cooling and heating requirements and on annual energy costs.

The analysis for the whole building is based on separate analyses of each of the identified thermal or daylighted building zones. Zones must be chosen and daylight sensors must be specified in a manner that allows for the interior daylight distribution. The program accounts for daylight availability, site conditions, and window management in response to sun control and glare, and it incorporates various lighting design and lighting control strategies.

The DOE-2 daylighting subroutine has been designed for flexibility and expansion. The current program, DOE-2.1C, can calculate interior illuminances from conventional window designs assuming clear, overcast, and cloudy sky conditions. The program also includes sun-control systems that are ideal diffusers. The thermal interaction of daylight strategies is automatically accounted for within the DOE-2 program.

The primary daylighting calculation occurs in a preprocessor to the main program. For each daylighting zone, the daylighting module calculates all daylight factors on Commission Internationale de l'Eclairage (CIE) standard overcast and clear sky conditions as well as direct sun conditions. Glare indices are also calculated and stored for later use. For a given hour, the sun position and sky type are used to obtain the appropriate precalculated values.

This model works well for simple design solutions. The program now also has new functional keyword features that allow an investigation of the performance of advanced glazing technologies. The new function is able to replace a constant value such as light transmittance with a function that can be dependent on many other factors; for example, light transmittance can be a variable function of solar intensity and temperature.

In future versions, a new coefficient of utilization model will be added. This model should increase computational speed as compared with the daylight-factor method and should allow more flexibility in the range of designs that can be accommodated. This model utilizes a series of precalculated or premeasured coefficients that relate the illuminance at a task location to a normalized source of illuminance at the window location. Multiplying the appropriate coefficients by the window area, window transmittance, and exterior illuminance at the window provides the interior horizontal illuminance at the task location.

The illuminance results are fitted to regression equations that become coefficient of utilization functions. For more complex architectural designs, the coefficients will be developed by systematic scale-model tests using a sky simulator and will be stored in the library. Minimizing data-manipulation time by generating appropriate indices is a development that will reduce runtime.

The model provides three pathways for the hourly illuminance calculation, as shown schematically in Fig. 12.8. If the room and window design is simple, the preprocessor will be able to calculate the coefficient of utilization equations directly. If the architectural design is more complex (venetian blinds or light shelves), the coefficient of utilization equations will be precalculated or measured for the devices and stored in the library of the program. If the user desires to model a unique building design for which precalculated values are not available, instructions will be provided to develop the coefficients from the user's own model tests.

Daylight Illumination Calculation Programs

As building design solutions have become more sophisticated, computer models have become increasingly complex. The advent of hour-by-hour energy-analysis programs has suggested the need for daylighting models that can calculate interior illuminance in greater detail. Although greater accuracy is one goal, the most important reason to move to more powerful daylight calculation models is to be able to model the realistic (and complex) designs in common use today.

Very powerful and complex daylight illuminance models, such as the SUPERLITE program, provide many of the capabilities desired in an energy-analysis program. However, modeling is computationally too complex to be

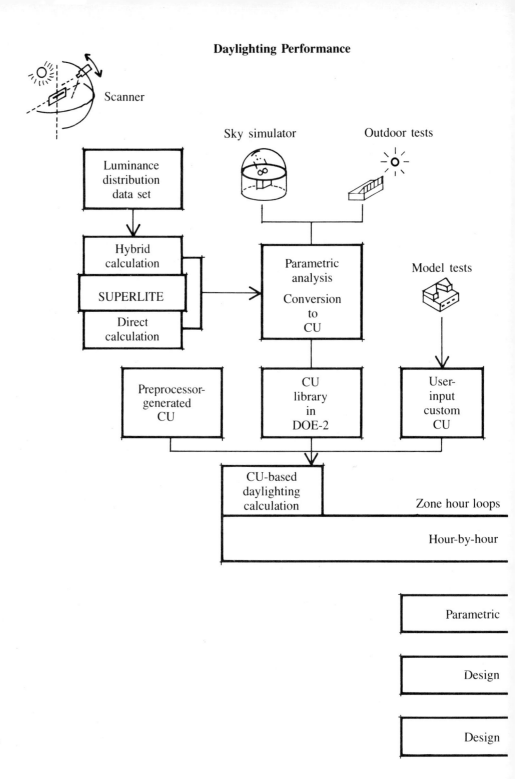

FIGURE 12.8.
Diagram of DOE-2.1C modeling capabilities.
(After Selkowitz and Winkelman 1983.)

utilized in an hour-by-hour program. SUPERLITE is a large mainframe computer model that predicts the spatial distribution of daylight illuminance in a building zone based on exterior sun and sky conditions, site obstructions, details of fenestration and shading devices, and interior room properties. The program can also include the measured luminance properties of shading devices, which can be used as a driving function for illuminance predictions.

The mathematical basis of the SUPERLITE algorithms has been described by Modest (1982). The program can model a uniform sky and Commission Internationale de l'Eclairage standard overcast and clear skies with or without direct sun. In principle, other sky distributions, such as those mentioned early in this chapter, also could be used. From the luminance distributions of these skies, the program can calculate the luminances of exterior and interior sur-

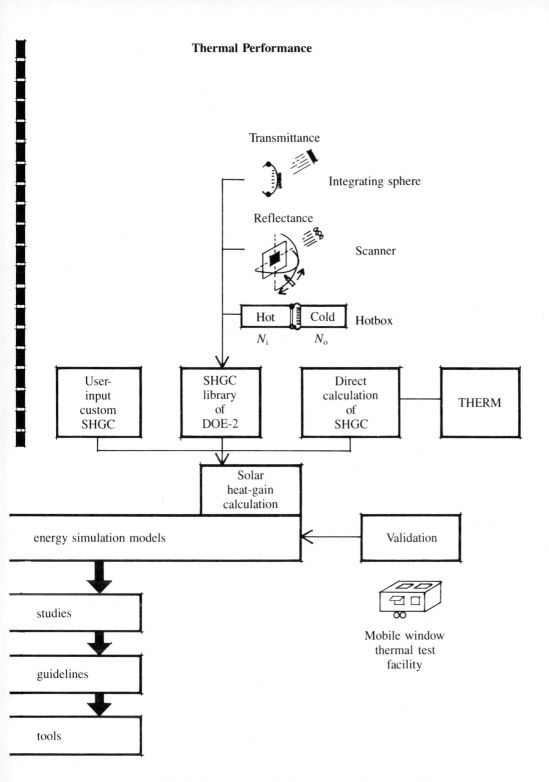

faces over a grid of points. The number of subdivided surfaces can be increased if the luminances of the surfaces vary significantly. From these calculations, the workplane illuminances can be determined by integrating the surface luminances over the appropriate solid angles.

A major advantage of SUPERLITE is its ability to model nonrectangular surfaces and other complex geometries (Modest 1981). Various types of diffusing curtains and overhangs with opaque, translucent, or transmitting materials also can be taken into account. The program also can model electric lighting systems combined with daylighting strategies.

Both SUPERLITE and DOE-2 are large computer models that require a substantial investment in training. They are being used mostly for research, and they provide the technical basis for the derivation of coefficients and regression equations from which are developed simpler and more accessible tools for building design.

Lumen III was originally developed as an electric lighting model, but it also can serve as a daylight program, calculating room illuminances from clear and overcast skies and for clear and diffuse glazing, overhangs, and specified shading devices such as venetian blinds. External obstructions, including adjacent sunlit surfaces, are also included in the calculation procedures. It uses flux transfer algorithms by DiLaura and Hauser (1978).

Microcomputer Programs

The following microcomputer programs were developed for use in offices. Since the microcomputer market is changing rapidly, some of these may no longer exist and other versions are likely to be developed.

MICROLITE 1-2-3 (Bryan and Fergie 1986)

This program is updated from the MICROLITE I.0 to receive input data by transfer from a LOTUS 1-2-3 spreadsheet or an AutoCAD file. The program analyzes daylight distribution within a room. It also includes several processors to aid the user in comparing how different design alternatives affect daylight distribution within a space.

CONTROLLITE (Lighting Systems Research Group 1985)

This program can compute lighting-energy reductions that result from using daylight, scheduling, and other control strategies. Daylight illumination under clear or overcast sky conditions can be calculated. The program is based on algorithms developed by Bryan et al. (1981).

DAYLIT (Ander et al. 1986)

This program uses the IES method (i.e., the lumen method) in its daylighting calculations. Clear and overcast skies are combined in a weighted average based on the ratio of clear and overcast skies experienced to duplicate actual weather patterns. Thermal loads also are calculated to produce an output of energy consumption and energy costs.

DAYLITE

This program calculates daylight factors in a room using the flux-transfer method for clear and overcast sky conditions and direct sunlight. It also can accommodate skylights and calculates annual lighting energy consumption.

FENESTRA (Cuttle and Baird 1986)

In this program, which is currently under development, the convention of determining daylight illumination on the basis of the standard overcast sky or average sky has been rejected. Instead, the principal assumptions are that the sky component is due to a uniform sky and that the direct component is due to a point source, the sky being divided into 10-degree zones. Data are generated from hourly recorded values of solar irradiance.

BUNLIT 1.0 (Trudgian, Kittler, and Ruck 1988)

This is a daylighting program that calculates electrical energy consumption in integrated systems. BUNLIT can be used independently or it can be interfaced with the energy simulation program BUNYIP. This program differs from other daylighting programs in that it uses the homogeneous sky model which encompasses all types of real skies with a simplicity and accuracy that makes it ideal for microcomputer application. The sky model is based on homogeneous sky theories by Kittler (1986) and requires only irradiance data for interpolation.

Other microprograms include PC-DOE, a version of DOE-2, TAS, a modified version of SERI-RES by Littler, and BEACON from Japan.

LIMITATIONS OF EXISTING TOOLS

It may appear that existing design tools address all the design-related activities. In addition to the above-mentioned programs, tools for economic analysis, site planning, structural analysis, and energy analysis are also available. However, the use of these tools as independent entities requires time-consuming specialized processes either for formatting the same data in different ways or converting the output format for one tool to serve as input for another. Even if currently available tools were combined or implemented in a combined package, there still would be major limitations. When evaluating building design criteria that affect human comfort (e.g., glare potential and view access), besides their interactions one with another, there is also a lack of appropriate variables or performance indices for effective communication and evaluation. Appropriate indices would not only indicate performance in these areas, but also minimize data-manipulation time.

It is evident that while there has been a rapid increase in the number of computer-based tools, notably microcomputers, and while these tools can greatly increase the speed of solving specific problems, they have not yet attained the required capabilities. An ideal tool for building design could be conceived as one in which all building and occupancy types could be accommodated and all critical design issues could be addressed, from human comfort and energy requirements to aesthetics, one that is interactive, one that provides a means for evaluating nonquantifiable design aspects, and one that is cost-effective. The magnitude of the information required by an ideal design tool introduces the issue of efficient data handling, especially to support visual information in the form of images. It is suggested that computing processes in the future will take advantage of the continual advances in computer graphics and imaging technologies to generate, store, retrieve, and manipulate images. Large data bases of images can cover a wide range of information.

It is difficult to effectively evaluate the qualitative aspects of the environment, e.g., the quality of the luminous environment, view access, and aesthetic appeal, without the use of realistic images. During the past few years, researchers have been working on computer models to generate realistic images of environments under different lighting conditions (Modest 1982, Ward 1986). Researchers also have been exploring the potential of such technologies for generating, storing, and manipulating such images to increase overall efficiency.

Over the last 5 years, the concept of an advanced tool for the design of buildings has been emerging that is oriented toward an understanding of the luminous and thermal performance of the building envelope with respect to building performance and the health and well-being of occupants. This knowledge is now being used to generate a new interactive tool to evaluate building performance.

VALIDATION AND EVALUATION BY MEASUREMENT

Many methods have been employed to validate computer energy simulation programs from comparison with actual measurements to comparison of separate results from several analysis methods. In this section, two methods are described which were applied to new buildings as well as a method used to retrofit an existing building.

Measured data describing interior light distribution, lighting control system performance, and electrical energy consumption for lighting were obtained from a monitoring program over a year on a building in San Francisco that incorporated daylight features specifically designed to displace electrical energy consumption (Warren et al. 1986). The daylighting features included a central atrium, a large floor-to-floor dimension, sloping ceilings, and reflective light shelves to provide solar shading and enhance light penetration. Various glazing materials were selected as integral components of the daylighting design. The atrium used clear glass on vertical north-facing surfaces and diffusing glass on sloped south-facing surfaces. Clear glass was used above the light shelves on both north and south elevations, and tinted glass was used below the light shelves to reduce light transmission. A reflective coating was added to the tinted glass. Sun-control devices on the south fenestration consisted of an external translucent roller shade above the light shelf on the south elevation and an internal movable blind below the light shelf. It was intended that ambient illumination for casual tasks and circulation be provided by daylight supplemented by an indirect fluorescent lighting system. Desk lamps were provided for task illuminance.

The design process began with two alternative solutions, energy forecasts being made by hand calculation and computer software. An investigation was made of state-of-the-art methods to evaluate daylighting. The Lumen II program was evaluated, but as with hand calculations, the limitations with regard to room shape and window design restricted its usefulness in terms of calculations for innovative devices such as light shelves.

A comprehensive daylight measurement program was therefore carried out in conjunction with the building design and involved ⅛- and ⅜-inch scale models and a full-scale mockup to test electric lighting alternatives. The model testing was carried out for both clear and overcast sky conditions. Following an evaluation of these model tests and the integration of daylight with other building design issues, daylighting techniques were selected and final model/simulation studies were carried out to forecast the annual contribution of daylight to possible energy savings. Several different prediction techniques were used. When and where supplemental electric lighting was required was estimated by graphic methods. Lighting control strategies were then analyzed.

The DOE-2 program was used to interpolate changes in the building's total annual energy usage with the adjustment of ambient lighting load schedules in relation to test results from the physical scale-model testing program. The results indicated that 70 percent of the electrical energy required to maintain the required ambient illuminance of 300 to 350 lx (30 to 35 fc) would be displaced by daylighting. An estimate of the proposed building's annual energy consumption, excluding energy used by the computer equipment, was 19,000 BTU/ft²/yr, less than half the level forecast for the original conventional building design.

After consideration of occupancy, the building was subsequently monitored in typical zones. The monitoring program included battery-operated data loggers to poll illuminance levels in the specified zones, temperature, and lighting power. Illuminance profiles were obtained across the building from a series of ambient illuminance measurements at partition height. The contribution of the electric lighting system was determined from nighttime measurements. The building monitoring program produced data that in combination with field observations showed that the architectural daylighting features were performing well and that the differentiation of task and ambient lighting systems was an effective energy-conserving strategy. However, the combination of very low transmittance glazing (17 percent) and shading by the external light shelf caused the south exterior zone under the light shelf to be one of the dimmest areas in the building, and although it had access to daylight, it required continuous supplemental lighting. The central atrium also was less effective at providing light deep into adjacent interior spaces on the lower floors.

Analysis of data also brought out the fact that additional electrical energy savings and peak electrical demand reduction could be generated by modification and adjustment of the lighting control system. There was widespread variation in the performance of the lighting control system which resulted in excess lighting power being used. Electric lighting use correlated poorly with daylight levels despite the use of dimming controls. Analysis of monitored results revealed that most problems were caused by poor placement of the photocell controls or inadequate calibration. Plans have been developed for correcting these problems. With proper control, it was estimated that potential energy reductions of about 60 percent could be realized, as predicted by the model/simulation studies. This would provide energy savings, assuming 3,750 hours of daytime occupancy during a year, of 22.4 kWh/m^2 (2.08 kW/ft^2). With electricity charges of \$0.08 per kilowatthour, the annual cost savings, excluding benefits from reduced peak demand, would be \$1.79/m^2 (\$0.166/ft^2). The simple payback for a properly operating control system would be 2.6 years (Warren et al. 1986).

Several independent analysis techniques, including graphic and computer approaches, supported by modeling studies have been employed and compared to determine energy savings due to daylighting for an earth-covered office building in Sacramento, Calif. After the project was completed, onsite illuminance measurements were recorded for validation purposes and to gain further insight into the expected annual performance of this building.

The potential energy savings for the project are considered to be in the 40 percent range due solely to the displacement of electric lighting energy. A computerized version of the IES daylighting analysis method (Fitzgerald et al. 1982) using a three-step lighting control operating under clear and overcast skies was used to estimate savings. This study predicted 23 percent lighting energy savings for the total project. A graphic-analysis approach also was used, giving results of 83 percent savings based on the daylighted areas or 41.5 percent based on the gross project area (Robbins et al. 1985). This method also includes both clear and overcast skies, but direct sun is again not included. Comparisons between these independent analyses showed that the graphic method produces the most realistic results. Although the comparative findings from the various methods used are not within close tolerances owing to differences in assumptions, the expected performance levels seem to be well established when more than one technique is employed simultaneously.

Other evaluation studies have been carried out for daylighting retrofits (Thomas et al. 1986). A series of innovative designs was carried out, and their contribution to daylighting was monitored by photometric sensors. The relative merits of each daylighting retrofit was analyzed in terms of three sky conditions: cloudy dull, cloudy bright, and sunny clear, together with two ground reflectivity conditions: bare ground and snow-covered ground. In this case, no computer simulations were carried out. It was found through the full-scale model studies that a multi-element double-reflector system encased between panes of insulating glazing provided the most even interior light distributions of all window systems examined.

SUMMARY

A review of current tools used to simulate building performance indicates that numerous tools are available, including nomographs, regression procedures, and programs for micro-, mini-, and mainframe computers. Although these prediction and evaluation techniques can help a designer understand the energy- and comfort-related implications of a proposed design, it is necessary to choose the tool that is appropriate for the intended purpose. It is possible that in the future, as researchers in environmental science develop a better understanding of the interactions between the environment, buildings, and building users, a tool may be made available to cover a range of environmental quality issues in addition to the quantitative aspects, thus eliminating the current necessity of having to switch from one tool to another.

Although computer-based tools allow the testing and analysis of design alternatives under a wide variety of conditions, the information available to the designer has not fundamentally changed. The performance of alternative solutions with respect to all design criteria cannot as yet be predicted. It is obvious that unless such tools include integrated consideration of other than energy-related criteria, their usefulness to designers is limited.

ACKNOWLEDGMENT

The reproduction of Figures 12.3 to 12.7 from the AAMA *Skylight Handbook* is gratefully acknowledged.

REFERENCES

American Architectural Manufacturers Association (AAMA). Des Plaines, Ill. Windows and Daylighting Group, LBL and Charles Eley Associates. 1987. *Skylight Handbook Energy Design Guidelines.*

Ander, G. D., Milne, M., and Schiler, M. 1986. Fenestration Design Tool: A Microcomputer Program for Designers. In *Proceedings of the 2nd International Daylighting Conference, Long Beach, California.* 187-193.

Arasteh, D., Johnson, R., Selkowitz, S., and Sullivan, R. 1985. The Effects of Skylight Parameters on Daylighting Energy Savings. LBL Report 17456, Berkeley, Calif.

Arasteh, D., Johnson, R., Selkowitz, S., and Sullivan, R. 1985. Energy performance and savings potentials with skylights. *ASHRAE Trans.,* Vol. 91, Part 1, 154-179.

Aydinli, S. 1983. Daylight in interiors in consideration of average sky conditions. *Proceedings of the 20th Session of the Commission Internationale de l'Eclairage. E11/1-2.*

Aydinli, S., and Seidl, M. 1986. Determination of the Economic Benefits of Daylight in Interiors Concerned with the Fulfillment of Visual Tasks. In *Proceedings of the 2nd International Daylighting Conference, Long Beach, Calif.* 145-151.

Boyer, L. L. 1986. Multiple Validation of Annual Energy Savings Analysis Techniques for Preliminary Daylight Design. In *Proceedings of the 2nd International Daylighting Conference, Long Beach, Calif.* 125-130.

Burt, Hill, Kosar, Rittelmann Associates (BHKRA). 1985. *Energy Nomographs: A Graphic Calculation Technique for the Design of Energy Efficient Buildings.* Tennessee Valley Authority. Chattanooga, Tennessee.

Brown, J. P. 1983. *SOLITE 1 Computer Program.* San Diego, Calif.

Bryan, H. J., and Fergie, R. J. 1986. A Coherent Microcomputer Environment for Daylighting Design. In *Proceedings of the 2nd International Daylighting Conference, Long Beach, Calif.* 173-177.

Bryan, H. J., Clear, R. D., Rosen, J., and Selkowitz, S. 1981. QUICKLITE 1: New procedure for daylighting design. *Solar Age.* 6: 37-47.

Building Energy Simulation Group, Lawrence Berkeley Laboratory and Group Q-11, Solar Energy Group, Los Alamos National Laboratory. 1984. *DOE-2 Engineers Manual,* Version 2.1B. National Technical Information Service, Springfield, Va.

Choi, U. S., Johnson, R., and Selkowitz, S. 1984. The impact of daylighting on peak electric demand. *Energy Build.* 6: 387-399.

Clark, J., and McLean, D. 1986. *ESP: A Building and Plant Energy Simulation System.* ABACUS, Strathclyde University, Glasgow.

Cuttle, K., and Baird, G. 1986. FENESTRA: A Personal Computer Aid for Window Design. In *Proceedings of the 2nd International Daylighting Conference, Long Beach, Calif.* 182-186.

Di Laura, D.L., and Hauser, G.A. 1978. On calculating the effects of daylighting on interior spaces, *J. Illum Eng. Soc.* Vol.1, 2-14.

Fitzgerald, D. K., Boyer, L. L., and Grondzik, W. T. 1982. Energy Analysis Process for Daylighting Utilization in Office Buildings. In A. Bowen and R. Vagner (Eds.), *Passive and Low Energy Alternatives I.* Pergamon Press, New York. 6-1 to 6-7.

Gillette, G. 1983. *A Daylighting Model for Building Energy Simulation.* Building Science Series 152. NBS, Washington.

Gillette, G., and Kusada, T. 1980. *A Daylighting Computational Procedure for Use in DOE-2 and Other Dynamic Building Energy Analyses.* National Bureau of Standards, Washington.

Griffith, J. W. 1976. *Predicting Daylight as Interior Illumination.* Libby-Owens-Ford Glass Co., Toledo, Ohio.

Johnson, R. 1983. Building Envelope Thermal and Daylighting Analysis in Support of Recommendations to Upgrade ASHRAE/IES Standard 90. LBL Report 16770, Berkeley, Calif.

Kim, J. J., Papamichael, K. M., Spitzglas, M. and Selkowitz, S. 1986. Determining Daylight Illuminance in Rooms Having Complex Fenestration Systems. In *Proceedings of the 2nd International Daylighting Conference, Long Beach, Calif.* 204-208.

Kittler, R. 1986. Luminance Models of Homogeneous Skies for Design and Energy Performance Predictions. In *Proceedings of the 2nd International Daylighting Conference, Long Beach, Calif.* 18-22.

Lighting Systems Research Group. 1985. CONTROLITE 1.0 lighting control systems and daylighting analysis program: user's manual. Lawrence Berkeley Laboratory, Berkeley, Calif. Report LBL-1744Rev.

Littlefair, P. J., 1984. Daylighting Availability for Lighting Controls. In *Proceedings of the CIBSE National Lighting Conference, Cambridge, England.*

Matthews, S., and Barnaby, C. S. 1980. Research into the Empirical Correlation of Insolation Measurements with Daylighting Performance. In *Proceedings of the Annual DOE Passive and Hybrid Solar Energy Program Update Meeting.*

Modest, M. F. 1981. Daylighting Calculations for Nonrectangular Interior Spaces with Shading Devices. Presented at Illuminating Engineering Society Annual Technical Conference, Toronto, Canada; and LBL Report 12599, Berkeley, Calif.

Modest, M. F. 1982. A general model for the calculating of daylight in interior spaces. *Energy Build.* Vol. 4:4 (August 1982).

Robbins, C. L. 1986. *Daylighting Design and Analysis,* Van Nostrand Reinhold, New York.

Robbins, C. L., Hunt, K. C., and Buhl, M. 1985. *A Graphical Method of Predicting Savings Attributed to Daylighting.* SERI/TR-254. Solar Energy Research Institute, Golden, Colo.

Selkowitz, S., and Gabel, M. 1984. LBL Daylighting Nomographs. LBL Report 13534, Berkeley, Calif.

Selkowitz, S., Kim, J. J., and Spitzglas, M. 1983. A New Coefficient of Utilization Model for Daylighting Calculation in Energy Analysis Programs. In *Proceedings of the 1983 International Daylighting Conference, Phoenix, Arizona.* 219-221.

SolarSoft Inc., Box 124, Snowmass, CO, 81654. Attn: William Ashton.

Spitzglas, M. 1983. A "Transmission Function" Approach to Daylight-Introducing Systems in the Built Environment. In *Proceedings of the 1983 International Daylighting Conference, Phoenix, Arizona.* 223-225.

Spitzglas, M., and Selkowitz, S. 1986. New Approaches to the Photometry of Fenestration Systems and Their Optical Components. In *Proceedings of the 2nd International Daylighting Conference, Long Beach, Calif.* 120-122.

Sullivan, R., Nozaki, S., Johnson, R., and Selkowitz, S. 1983. Commercial Building Energy Performance Analysis Using Multiple Regression Procedures. LBL Report 16645, Berkeley, Calif.

Thomas, G. P., Manwell, S. P., and Kinney, L. F. 1986. Development of a Monitoring System and Evaluation Method for a Daylighting Retrofit. In *Proceedings of the 2nd International Daylighting Conference, Long Beach, Calif.* 222-234.

Trudgian, R., Kittler, R., and Ruck, N.C. 1988. *The Personal Computer Daylight Design Aid.* Research Report, Faculty of Architecture, University of New South Wales, Kensington.

U. S. Army Construction Engineering Research Laboratory. 1979. *BLAST, The Building Loads Analysis and Thermodynamics Program Users Manual,* Vol. 1 (ReportE-153).

Ward, G. 1986. Advanced lighting design using ray tracing. Internal Report, Lawrence Berkeley Laboratory, Berkeley, Calif.

Warren, M., Benton, C., Verderber, O., Morse, S., Selkowitz, S., and Jewell, J. 1986. Evaluation of Integrated Lighting System Performance in a Large Daylighted Office Building. LBL Report 21466, Berkeley, Calif.

Winklemann, F. C., and Selkowitz, S. 1985. Daylighting simulation in the DOE-2 building energy analysis program. *Energy Build.* 8: 271-286.

Winkelmann, F. C., and Selkowitz, S. 1984. Daylighting Simulation in the DOE-2 Building Energy Analysis Program. LBL Report 18508, Berkeley, Calif.

Evaluation of Advanced Glazing Technologies

13

Stephen Selkowitz

DESIGN VERSUS NEW TECHNOLOGY

The building envelope has been dictated by the resources and technological options available and by the functions of the building. For example, the medieval fortress was a logical design within its historical context. A glass-walled automobile showroom is an equally appropriate design in a different context. Today, glass plays a significant role in the design of almost all building envelopes. Since its emergence during the last century as a major building material, it has evolved into a versatile design element performing functions today that would have been unthought of only a few years ago.

Windows and other glazed openings can solve complex needs that require variable properties and tradeoffs between conflicting conditions. For example, the glazing that provides a view must also provide visual privacy, and glazing that admits daylight and hence reduces electric lighting needs also can create discomfort and disability glare. The sun that provides desirable warmth in the winter also contributes to thermal discomfort and cooling requirements in the summer.

The problem that confronts building designers results from the highly variable climatic influences that impinge on building exteriors. Many options are available to extend the degree of thermal and solar control that glass can provide. The glass selection can influence total transmittance, spectral properties, and directional properties, which will satisfy some demands but not necessarily all the performance requirements. In addition, it has been suggested in Chap.

10 that there is a growing concern for the recognition of the value of the variable and natural qualities of daylight for human well-being. It also was suggested that the local climate should be more closely attuned to the building's indoor thermal conditions and that this can be accomplished by an "active" building envelope with devices in the envelope itself to control heat and light admission.

With the new advanced glazing technologies, the transparent components of a building envelope are becoming more flexible in their response to heat and light. In this chapter, conventional glazing materials are studied from the energy-conservation viewpoint. The effects of variations in glazing parameters such as window size, transmittance, and shading coefficient are systematically examined. It is shown that the energy performance of windows can be enhanced by using advanced optical materials that increase the efficacy of daylighting by selectively controlling daylight and solar transmission.

CONCEPTS USED TO EVALUATE GLAZING PERFORMANCE

Effective Aperture

To simplify interpretation of results in the evaluation of advanced glazing technologies by computer simulation, a lumped parameter can be used which is the product of the ratio of glass area to floor-to-ceiling wall area times the visible transmittance of the glazing. This parameter is called the "effective aperture." Its use allows a convenient comparison of the energy performance of dissimilar fenestration designs. Plotting annual energy use in a perimeter zone as a function of effective aperture allows a quick assessment of the potential impact of windows. Effects of mullions and other opaque elements can be accounted for with the wall-to-window ratio term, while a dirt depreciation factor can be incorporated into the visible transmittance term. This parameter is also useful when analyzing daylighted conditions because various combinations of window-to-wall ratios and transmittances that yield the same effective aperture will generally have the same influence on component and total building energy performance.

Spectral Selectivity

Window systems of the future may use new coatings deposited directly on glass or plastic to provide the same sophisticated solar control that now requires external or internal shading devices. The performance requirements for such ideal glazings are complex because of the multiple functions they must serve. Two of the most important functions for new glazing coatings or materials are reducing adverse cooling impacts and increasing beneficial impacts of daylight. Glazing materials having selective spectral transmission over different wavelength ranges can increase light transmittance and at the same time reduce infrared transmittance.

Blue-green glass and some metallic coatings have spectral sensitivities that allow higher transmittance in the visible portion of the spectrum. New low-E coated glazings have increased light transmittance and reduced heat transmittance and therefore exert more control on heating and cooling loads. Continuing improvements in coatings should further increase spectral control. A comparison of the spectral transmittance of typical conventional and low-E glazings, shown in Fig. 13.1, illustrates the selective transmittance effect of a low-E coating (see Chap. 10).

Glazing Luminous Efficacy Function

The thermal impact of fenestration can vary significantly with glazing type. Different tinted or coated glazings can have different visible transmittances for

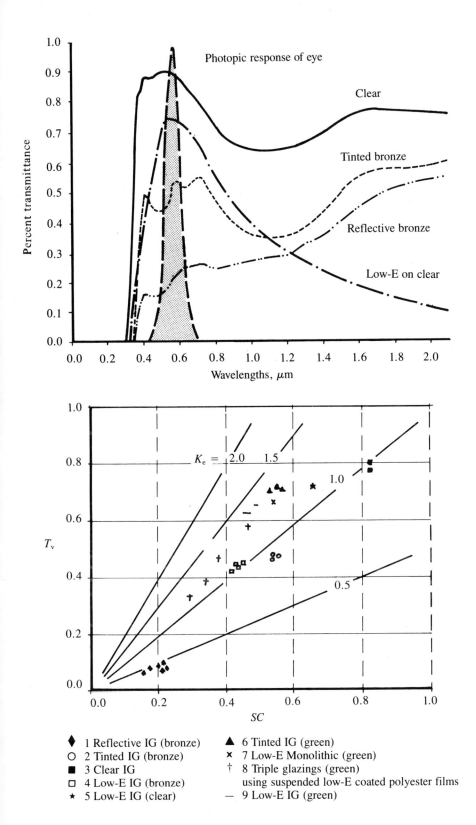

FIGURE 13.1.
Spectral transmittance of various glazing types compared with the photopic response of the eye.

Legend for Figure 13.2:

◆ 1 Reflective IG (bronze)   ▲ 6 Tinted IG (green)
○ 2 Tinted IG (bronze)   × 7 Low-E Monolithic (green)
■ 3 Clear IG   † 8 Triple glazings (green)
□ 4 Low-E IG (bronze)   using suspended low-E coated polyester films
★ 5 Low-E IG (clear)   — 9 Low-E IG (green)

FIGURE 13.2.
Visible transmittance versus shading coefficient for conventional and low-E glazing types.

the same shading coefficient or, alternatively, different shading coefficients for the same transmittance. A luminous efficacy constant (K_e), that is, the ratio of the glazing visible transmittance to the shading coefficient (the total solar heat gains associated with a given fenestration system), can usefully characterize the thermal performance of the glass. Figure 13.2 gives the visible transmittance as a function of the shading coefficient for representative low-emittance and conventional glazings. For these commercially available glazings K_e ranges from .3 to 1.4.

IMPACT OF DAYLIGHTING ON ENERGY CONSUMPTION

Effective Aperture versus Lighting Energy Using Controls

General trends of electric lighting requirements with daylighting or, conversely, electric lighting savings are shown in Fig. 13.3 (Johnson et al. 1986). Electric lighting control strategies are modeled with both on-off switching and continuous dimming in response to the daylight illuminance levels at a single control point. It can be seen that electric lighting requirements first drop off substantially as the effective aperture increases and then asymptotically approach a minimum lighting energy fraction. This minimum is not zero, because some building occupancy occurs during hours when daylight is not available. At large apertures, the dimming controls will require more lighting energy than step switching because the continuous dimming system is modeled with 10 percent power consumption at zero light output.

In all cases, the electric lighting savings approach the knee of the curve at a fairly small effective aperture. For example, simple one-step switching pro-

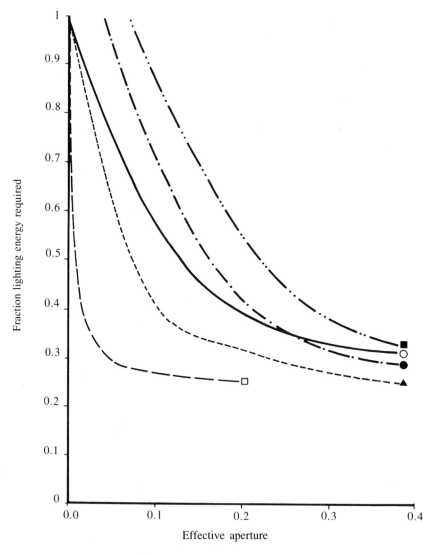

FIGURE 13.3.
Lighting energy as a function of effective aperture for a south zone (Northern Hemisphere) using daylighting.

■ 1-step switching, 538 lx (50 fc)
● 2-step switching, 538 lx (50 fc)
○ Continuous dimming, 538 lx (50 fc)
▲ Continuous dimming, 323 lx (30 fc)
□ Skylights—continuous dimming, 538 lx (50 fc)

vides no savings until the effective aperture reaches approximately 0.07 (e.g., a window occupying 30 percent of the window wall and with approximately 25 percent transmittance). This is so because the electric lights switch off only when daylight provides the full design illuminance of 538 lx (50 fc). However, the continuous dimming system begins to provide savings as soon as daylight is introduced. Results are similar in the Southern Hemisphere.

It can be concluded that at very large effective apertures, when the daylight contribution is maximized, the on-off system outperforms the continuous-dimming system because of its 10 percent power consumption at zero light output. The single lighting control zone used in most of these simulations is modeled with a single control sensor located 3 m (10 ft) from the window in a 5-m-deep (16-ft) office. Further electric light savings are possible by subdividing the space into multiple control zones.

Although minimal electric energy fractions are similar for vertical fenestration and skylights, electric lighting requirements with skylights drop off much more quickly as a function of effective aperture, and the knee of the curve occurs at a very small effective aperture, approximately 0.04. With evenly distributed skylights, properly spaced relative to ceiling heights, the daylight distribution in the space is substantially more uniform than in the sidelighted space with vertical fenestration.

IMPACT OF DAYLIGHTING ON PEAK ELECTRICAL DEMAND

Studies also have been carried out to show how daylighting strategies reduce peak electric demand by reducing electric lighting use (Choi, Johnson, and Selkowitz 1984). However, the glazing that admits daylight also introduces cooling loads, and these add to peak electric requirements. Maximal peak demand reductions will be obtained with a daylighting design that optimizes the balance between electric lighting savings and cooling penalties.

Figure 13.4 shows a component breakdown of peak demand for an office module with and without daylighting controls. It can be seen that the peak-

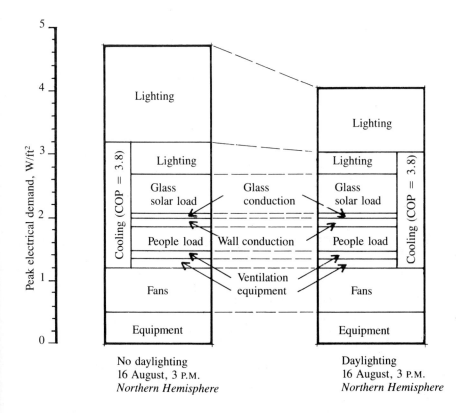

FIGURE 13.4.
Peak electrical demand. Breakdown by components of an office module with and without daylighting: lighting, 1.7 W/ft^2; WWR = 0.5; SC = 0.60; T_v = 0.40.

demand component associated with removing the solar gain from windows is about the same size as the electric lighting savings.

Figure 13.5 shows the peak electrical demand for a typical office building design as a function of effective aperture. Electric lighting power density is 18.3 W/m^2 (1.7 W/ft^2), and electric lights are dimmed in response to daylight only in the perimeter zones.

It can be seen that as the effective aperture increases from zero, daylighting reduces peak demand to a minimum at an effective aperture of about 0.1 with savings of 8.3 percent as compared to a windowless building and 11.6 percent as compared to a building with the same window system but without electric lighting dimming controls. In this example, the perimeter zones comprise 37.5 percent of the total floor area. As the ratio of the perimeter office zone to total floor area increases, the relative peak-demand savings will increase. At larger effective apertures, the additional savings in electric lighting become smaller,

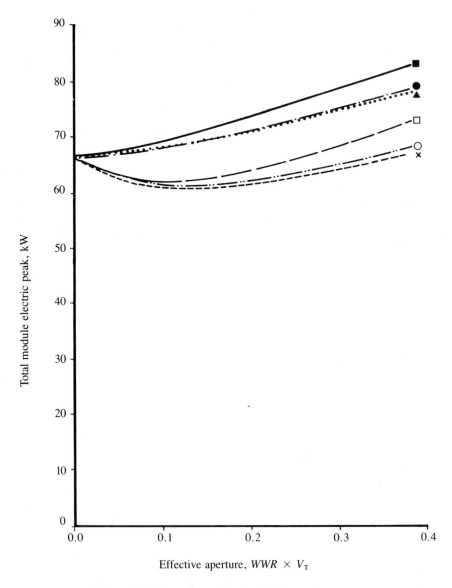

■ No window management, no daylighting
□ No window management, continuous dimming
● Interior shades, no daylighting
○ Interior shades, continuous dimming
▲ Exterior shades, no daylighting
× Exterior shades, continuous dimming

FIGURE 13.5.
Peak electrical demand as a function of effective aperture for a typical office building with and without daylighting.

while cooling requirements continue to increase, so the total peak demand rises from the minimum.

It is evident that a daylighted building with moderately sized glazing area will normally have a lower peak demand than a building without glazing, but further increases in window area and/or transmittance to increase daylight admittance will reach a point, depending on climate and orientation, beyond which peak electrical demand will increase. Control of solar gain is necessary if daylighting is to maximize reductions in peak loads. Existing glazing materials and window shading systems offer many good options; others will emerge as a result of ongoing research.

Impact of Daylighting on Total Energy Consumption

In recent studies, the influence of daylighting on net annual energy performance has been characterized in general terms (Sanchez and Rudoy 1981, Arasteh et al. 1985, Selkowitz, Arasteh, and Johnson 1984). Figure 13.6 is an example of these results. The curves show net annual energy consumption as a function of effective aperture for a south zone in Lake Charles, La. (representative of a hot humid climate in the Northern Hemisphere) with and without daylighting. In the nondaylighted case (*solid line*) energy consumption increases with effective aperture. This is attributable to solar gain-induced cooling loads. With day-

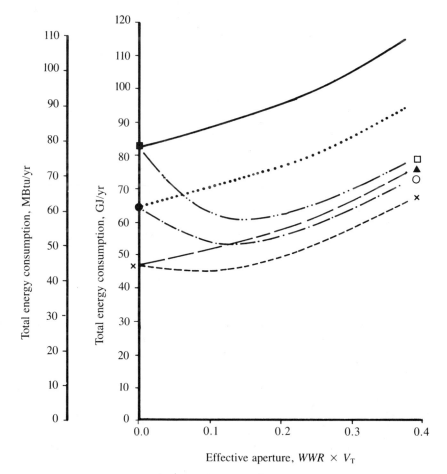

■ 29.1 W/m² (2.7 W/ft²), no daylighting
□ 29.1 W/m² (2.7 W/ft²), continuous dimming
● 18.3 W/m² (1.7 W/ft²), no daylighting
○ 18.3 W/m² (1.7 W/ft²), continuous dimming
▲ 7.5 W/m² (0.7 W/ft²), no daylighting
✕ 7.5 W/m² (0.7 W/ft²), continuous dimming

FIGURE 13.16.
Annual electricity cost per square foot of daylighted perimeter zone space as a function of effective aperture for a high rate structure (New York).

lighting (*broken line*), electric lighting consumption is reduced and net annual energy consumption for any effective aperture is substantially reduced.

Beyond a certain critical design point for minimal energy use, the constantly increasing cooling load begins to diminish the net benefits from daylight. Eventually, the initial benefits may be negated, and the daylighted design then requires more energy than a design having little or no glazing and totally dependent on electric lighting. These trends, which are also demonstrated in other climates and orientations, make apparent the need to understand in detail the impacts of daylighting on cooling loads.

THERMAL PERFORMANCE OF ADVANCED GLAZING TECHNOLOGIES

Cooling Loads versus Electric Lighting Requirements

From an energy perspective, window performance is distinctly different from that of insulated walls and roofs. The window's net energy benefits can balance thermal losses against winter solar gain and daylighting benefits. However, although daylighting can greatly reduce electrical energy for lighting requirements, with poor window design, it can lead to unnecessary cooling load increases that are higher than those associated with electric lighting. The annual cooling load impact of daylight will be affected by the glazing and shading system characteristics and the daylight distribution within the space.

The cooling load associated with the lighting can vary tremendously. An indication of the the variance is shown in Fig. 13.7, which shows annual cooling loads increasing as a function of the percentage lighting requirement for four different window systems. A comparison of three electric lighting installations with no allowance for daylighting is also shown. The thermal impact of the luminous efficacies of both daylight and electric light are compared by varying the lighting power density of the electric light source and the K_e of the glazing for the daylighting source. As mentioned previously, the glazing efficacy factor (K_e) is the ratio of net visible transmittance to shading coefficient. For daylighting purposes, a higher K_e means less solar gain is associated with a given quantity of daylight.

Figure 13.7 shows the daily cooling load increase (from a base-case building with no electric lights and no windows) as a function of the daily lighting requirement met by electric lights and by daylight on a clear day in March in the Northern Hemisphere. For electric lights only, the cooling load increases linearly with percent of lighting requirement met. For the daylighting case (without electric lighting), the lighting requirements outside daylight hours cannot be met and therefore never meet 100 percent of the requirements. Increasing K_e is analogous to decreasing the lighting power density of the electric lighting.

The differences in cooling loads resulting from using glazing systems with different K_e values can be seen by comparing three glazing systems, tinted glass ($K_e = 0.67$), blue-green glass ($K_e = 1.1$), and optimum glazing where $K_e = 2.0$ in Fig. 13.7 with varying K_e. The smallest value, 0.67, is typical of conventional gray- or bronze-tinted glass; the intermediate value, 1.1, is typical of blue-green glass and several recently introduced glasses with low-E coatings. The largest value, 2.0, is approximately the theoretical limit at which only visible light is transmitted and all near-infrared is rejected.

In contrast to the electric lighting, as daylight approaches its maximal contribution, the cooling load increases drastically and each next increment of effective aperture adds even more to the cooling load. The two boundary cases shown, in which K_e is equal to 2.0 and in which the daylight flux is uniformly distributed throughout the space, demonstrate the theoretical potential for reducing cooling impacts associated with daylight. The impact of cooling loads is very small, up to a case with about 85 percent of the lighting requirements

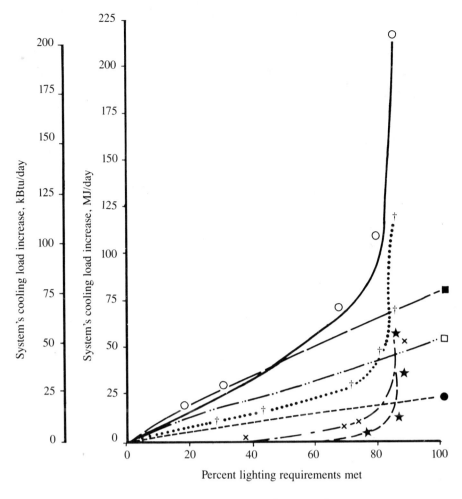

■ Electric lights, 29.1 W/m² (2.7 W/ft²)
□ Electric lights, 18.3 W/m² (1.7 W/ft²)
● Electric lights, 7.5 W/m² (0.7 W/ft²)
○ Tinted glass, $K_f = 0.67$
† Blue-green glass, $K_e = 1.1$
× Optimal glazing, $K_e = 2.0$
★ Optimal glazing, equal flux

FIGURE 13.7.
Increase in cooling requirements as a function of lighting requirements met using various electric light sources and glazing materials in south zone (Northern Hemisphere) on a clear day.

met. Similar results occur if the analysis is performed over the entire season rather than for the single day shown here.

In a sidelit space, the illumination levels are often three to four times higher near the window than in the back of the room. In a skylighted space, however, the average annual, maximal, and minimal workplane illuminations differ by about 50 percent. The annual cooling loads resulting from a glazing system such as skylights instead of vertical windows are therefore much lower, as shown in Fig. 13.8. This figure shows cooling energy per unit floor area of daylighted space as a function of electric lighting savings for Lake Charles, La. It also shows cooling energy increasing with effective aperture for the most efficient electric lighting but decreasing (up to 60 percent savings) for the less efficient lighting systems. Skylights produce less cooling load than windows for equivalent lighting energy savings.

As discussed previously, the maximal daylight contribution is less than 100 percent because the lighting schedule requires light during nondaylight hours and the dimming system uses 10 percent residual power at zero light output. In all cases, there is a steep rise in cooling as the daylighting contribution approaches maximum. The more uniform lighting distribution of skylights provides the same lighting savings with smaller effective apertures than windows and less total transmitted daylight and thus less solar gain.

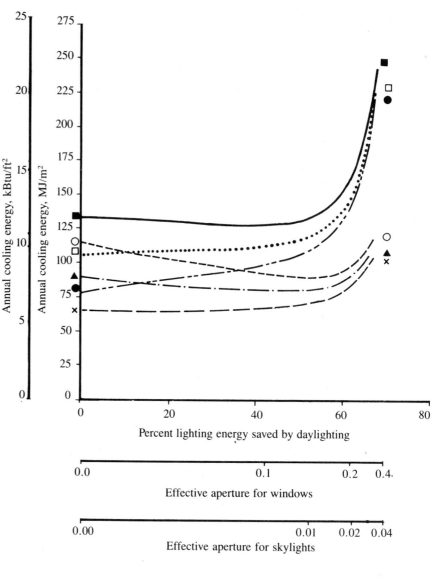

FIGURE 13.8.
Annual cooling load as a function of electric lighting energy saved in a south zone (Northern Hemisphere) and in a skylighted building.

■ South glazing, 29.1 W/m² (2.7 W/ft²)
□ South glazing, 18.3 W/m² (1.7 W/ft²)
● South glazing, 7.5 W/m² (0.7 W/ft²)
○ Skylights, 29.1 W/m² (2.7 W/ft²)
▲ Skylights, 18.3 W/m² (1.7 W/ft²)
✕ Skylights, 7.5 W/m² (0.7 W/ft²)

Because of the varying intensity of transmitted radiation, interior daylight illuminances vary depending on sky conditions, time of day, and season. Moderate and large apertures provide daylight savings on cloudy, overcast days, while under other sky conditions, such as clear sky with sunlight, these apertures provide excess daylight and therefore higher solar gain.

Effect on HVAC System Sizing

The influence of daylighting on chiller size is similar to its influence on cooling load. Figure 13.9 shows chiller size as a function of effective aperture for a typical office floor with the vertical fenestration having a K_e value of 0.67 and electric lighting power density of 18.3 W/m² (1.7 W/ft²). This configuration represents many typical office building designs.

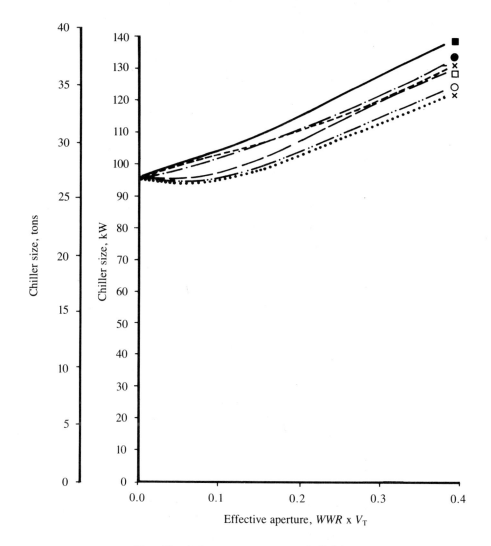

FIGURE 13.9.
Chiller size as a function of effective aperture with and without daylighting for vertical windows.

■ No window management, no daylighting
□ No window management, continuous dimming
● Interior shades, no daylighting
○ Interior shades, continuous dimming
▲ Exterior shades, no daylighting
× Exterior shades, continuous dimming

Without daylighting, fenestration simply adds a cooling load and chiller size is a monotonically increasing function. With the use of daylighting, the chiller size remains approximately the same as for an opaque wall up to an effective aperture of about 0.15. At larger effective apertures, chiller size is always smaller than the nondaylighted case but increases with effective aperture at about the same rate as the comparable nondaylighted case.

Window effects on chiller sizing are shown in Fig. 13.10 for Lake Charles, La. (a hot, humid climate in the Northern Hemisphere). Increasing K_e values from 0.5 to 1.0 produces the largest reduction of chiller size, but substantial additional reductions are obtained with higher K_e values for higher values of effective aperture. At $2000 per ton of installed cooling equipment, large initial cost savings can accrue through the use of daylighting.

In contrast to the results with vertical windows, in a space with skylights chiller size immediately decreases as daylighting is used at small effective apertures. Figure 13.11 gives chiller size as a function of skylight effective aperture for a building in San Francisco. In other U.S. cities, chiller size is also

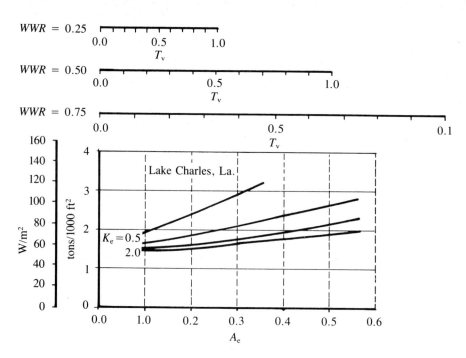

FIGURE 13.10.
Chiller size as a function of effective aperture in Lake Charles, La. (a hot, humid climate in the Northern Hemisphere).

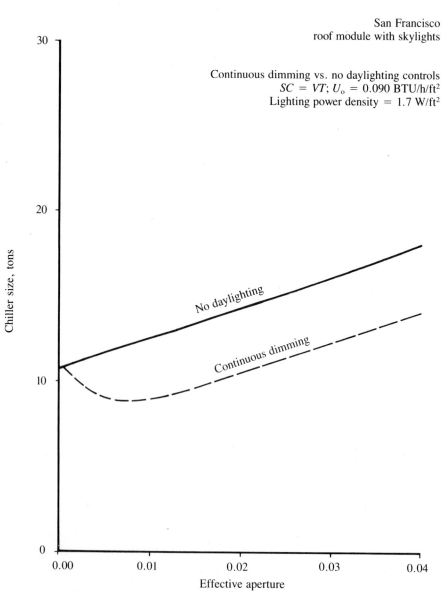

San Francisco
roof module with skylights

Continuous dimming vs. no daylighting controls
$SC = VT$; $U_o = 0.090$ BTU/h/ft^2
Lighting power density = 1.7 W/ft^2

FIGURE 13.11.
Chiller size as a function of effective aperture for skylights.

minimized with an effective aperture of approximately 0.01. Spaces without skylights ($A_e = 0$) require larger chillers than those of spaces with skylights that use daylight until the effective aperture exceeds approximately 0.02 as shown in Fig. 13.11. In most cases, annual energy consumption and peak electrical demand are also minimized within these effective aperture ranges.

NEW GLAZING SYSTEMS

Control of Solar Gain and Light

Window systems of the future may use new coatings deposited directly on glass or plastic to provide the same sophisticated solar control that today requires mechanical devices. The performance requirements for such ideal glazing systems are complex because of the multiple functions they must serve.

Two of the most important functions for new glazing coatings or materials are reducing adverse cooling impacts and increasing daylight transmittance. Several approaches can be taken to accomplish these functions. To reduce cooling impacts, optical switching films can be used to actively modulate the transmittance or spectral and angle selective transmittance can be employed to passively reject solar gain. To increase the beneficial impacts of daylight, selective coatings and switching films can be used for glare control. Daylight also can be collected and distributed to the internal core of a building by concentrators and light guide systems.

Ideally, one would like to control the intensity of the transmitted radiation, its spectral content, and perhaps its spatial distribution in a room. New glazings with dynamic and responsive control of sunlight will serve these functions in the future, and a vast array of window accessories is now available to alter the intrinsic glazing properties in response to changing external conditions and internal needs (see Chap. 10).

Technology Status and Prospects

From a performance perspective, actively controlled coatings are desirable because the glazing can then be continuously adjusted to meet the changing needs of the building and its occupants. The complexity of these needs for both residential and nonresidential buildings is shown in Fig. 13.12.

Research on advanced glazing materials is being carried out internationally on light-sensitive (photochromic), heat-sensitive (thermochromic), and electrically activated (electrochromic) coatings that respond differently to different climatic elements (Lampert 1981). These materials have different effects on lighting and cooling energy requirements, as shown schematically in Fig. 13.13.

No single type of thermochromic or photochromic coating is likely to provide the full range of required response, although an electrochromic window with an appropriate controller has that potential. Electrochromic coatings are multilayer coatings that are more complex than the current generation of low-E coatings now on the market. However, they are still in the experimental stage, and it may be some time before such coatings are used in buildings, although they are likely to be first used primarily in automobiles and as display devices.

This type of technology is compatible with the trend toward more sophisticated building management-control devices and systems and the trend to integrate electronics and sensors into the machines we depend on in the workplace. An example of the variability over a 24-hour period of the solar heat gain through several fenestration options in a west zone (Northern Hemisphere) is given in Fig. 13.14. The concurrent daylight levels at the control point are shown in Fig. 13.15. A similar range occurs in the south zone (Northern Hemisphere), but with seasonal variation. The fenestration systems modeled,

Functions and applications

Key □ = Primary
■ = Secondary

Capability or control function	Residential			Nonresidential					
	Windows	Skylights	Sunspaces	View windows	Daylight windows	Envelope systems	Skylights	Atria	Interiors
Glare (visual performance)				□	■	■	■	■	□
Glare (visual comfort)	□	□	■	□	■	■	□	■	□
Privacy	■		■	□					■
Control of interior fading	■		■	□	□	□			□
Daylight control				□	■	■	■	■	□
Thermal comfort	■	□	■	□	□	■	□	□	
Solar gain control (cooling)	■	■	■	■	■	■	■	■	
Chiller/HVAC size	■			□	■	■	■	■	
Peak demand control	■	■	■		■	■	■	□	
Winter solar gain control	□		□			□		□	
Control responsiveness					■	■	■	■	
Control reliability					■	■	■	■	

FIGURE 13.12.
Functions of the building envelope and their application.

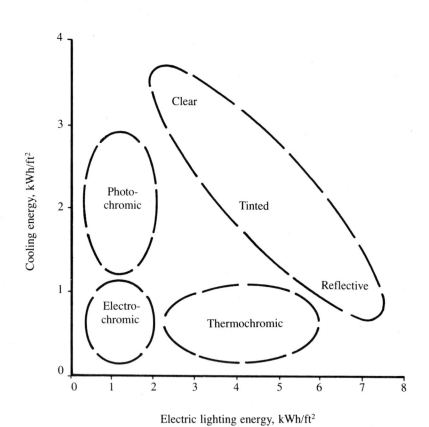

FIGURE 13.13.
Projected cooling and lighting energy requirements for conventional glazings and three types of optical switching materials.

all with an effective aperture of 0.3, include three conventional types of glazing and two that have advanced materials with optical switching properties. They can be identified as follows:

PR: Photochromic glass, responsive to solar radiation. Shading coefficient varies linearly from 0.8 to 0.2 as total solar radiation incident on glass varies from 31.5 W/m² (10 Btu/ft²/h) to 315 W/m² (100 Btu/ft²/h), $K_e = 1$.

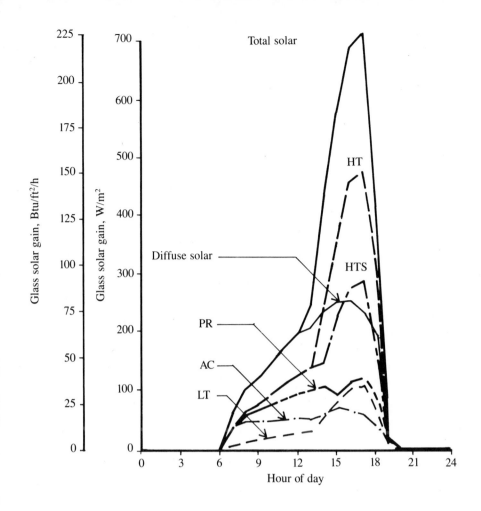

FIGURE 13.14.
Hourly glass solar gain for a west orientation on a clear July day (Northern Hemisphere).

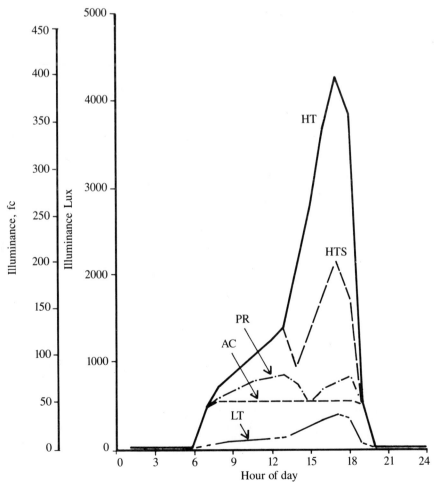

FIGURE 13.15.
Hourly daylight illuminance level at the control point used in Fig. 13-14 on a clear July day (Northern Hemisphere).

255

AC: Electrochromic glass with the glass transmittance controlled to hold daylight levels at 538 lx (50 fc) at control point. Maximal light transmittance is 0.8 and $K_e = 1.0$.

HT: Conventional high-transmission glazing with shading coefficient of 0.8 and light transmittance of 0.78.

HTS: High-transmission glazing with window shade that reduces solar gain by 40 percent and light transmittance by 65 percent.

LT: Conventional low-transmission glazing with shading coefficient of 0.18 and light transmittance of 0.07.

With high-transmission glass, daylight from a diffuse sky provides design lighting requirements most of the morning hours. During afternoon hours, with direct sunlight as a source, even with shading, the daylight level is about twice the required level, imposing additional cooling loads with no additional daylighting benefits. Low-transmission glass minimizes solar heat gain but provides little usable daylight.

A fenestration control option that continuously modulates the transmission of the daylighting aperture so that the daylight level at the control point never exceeds the design level is represented by AC. A similarly controlled mechanical device such as external venetian blinds might be expected to perform similarly. This level of control, while allowing maximal lighting benefits with daylighting, is required to mitigate the afternoon solar gain while retaining required daylight levels in the morning, as shown in Figs. 13.14 and 13.15.

Effect of Glazing Technology on Comfort and Cost

Comparing the relative cost-effectiveness of daylighting designs and more efficient electric lighting systems is difficult because it requires accounting for nonenergy design benefits such as the well-being and productivity of occupants. The traditional importance of windows in satisfying human needs and the value of the variable and natural qualities of daylight have been emphasized.

In addition, the quality of the visual environment depends on providing an acceptable size and shape of window to maintain contact with the outside world and limit glare. As outlined in Chap. 3, daylight admitted through windows is also considered essential for photobiological processes, such as controlling the biorhythms of the body and stimulating metabolic functions.

The avoidance of discomfort glare from the sky or sun seen through the windows or skylights is an important consideration in daylighting design. The reduction of sky luminance has in the past been achieved by using tinted and reflective glasses that also reduce daylight transmittance. Other options involve use of sun control devices or window design that separate the view function and light admitting functions. As discussed in Chap. 10, many glazing options are now available that incorporate solar control to improve thermal comfort. While tinted and reflective glasses reduce solar heat gain, this benefit must be measured against reduced light transmission. The newer window films and coatings that selectively reflect short-wave infrared radiation while still transmitting most visible light and maintaining good levels of daylight are to be commended.

To ensure thermal and visual comfort, controls are important. The architect can exercise substantial control over the energy performance of windows by selecting appropriate sizes and locations. Dynamic control of window thermal and optical properties provides the best solution to maintaining comfort and energy performance over a wide range of environmental conditions. A variety of other techniques such as the introduction of comfort algorithms into computer programs are now under development or available to further improve building design and modify window performance.

Daylight is frequently viewed as an added first cost (e.g., increased glazing and lighting control costs). However, reductions in chiller size can provide a first-cost savings that can offset the costs of electric lighting controls, solar control devices, and improved glazing products.

Annual operating costs are influenced by the effectiveness of the daylighting design in relation to the electric lighting system, the climate, and local utility rates. Annual electricity costs of daylighted space are shown in Fig. 13.16. The utility rate structure for New York City is used to calculate annual costs. The daylighted case with continuous dimming saves about $6/m^2 in annual costs compared to the nondaylighted case. Both show rapidly rising costs at effective aperture rises above .2. The lowest curve with electrochromic glazings shows continuously decreasing annual operating cost, independent of window size. Changes in lighting power density and electric rates would change actual savings, but the general trends illustrated here will be broadly applicable.

The integrated design of fenestration and lighting systems in which solar gain is controlled, daylight is admitted, and electric lights are dimmed in response to daylight levels will reduce net annual energy consumption and peak electric demand. These reductions lower operating costs over the life of the building. The magnitude of the savings will depend on the specifics of building design, climate, heating fuel costs, and the utility rate structure. In the United States, lighting control systems presently cost from $6 to $12 per square meter of floor area and more sophisticated glazing and shading systems may add additional costs. These first costs may, however, be offset by reductions in chiller and cooling equipment costs. In Madison, Wis., with 18.3 W/m^2 lighting power density, an effective aperture of 0.2, and managed shades, daylighting

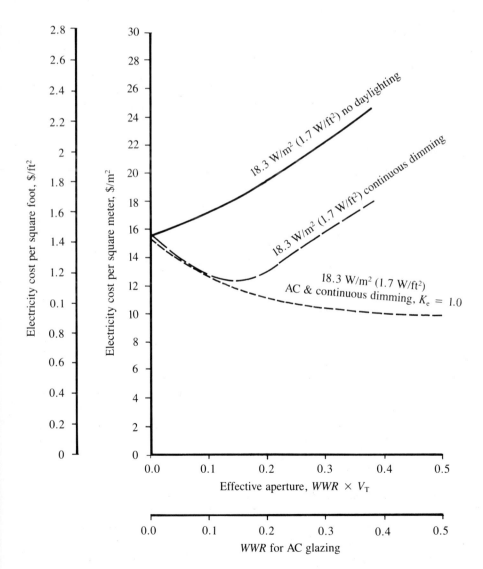

FIGURE 13.16.
Annual electricity cost per square foot of daylighted perimeter zone space as a function of effective aperture for a high rate structure (New York).

reduces chiller requirements by about .0055 tons/m^2 in the perimeter zone. This is a cost reduction of $11 per square meter of floor area, which is about equal to the cost of the lighting control system.

SUMMARY

The comparative performance of conventional and low-E glazings has been simulated using the DOE-2.1C program. It has been found that a significant fraction of electric lighting and some cooling energy can be saved by dimming electric lights in response to available daylight in perimeter areas. This strategy, if properly implemented, can reduce total electricity consumption for a typical office building to less than that of a building with no windows or daylighting. The degree to which the use of daylighting can reduce lighting loads depends primarily on the window-to-wall ratio and the visible transmittance of the glazing.

An assessment of the energy effects of conventional glazings for a range of window-to-wall ratios in a daylighted office building in both hot and cold climates has shown that low-E glazings are a "cooler" source of daylight than their conventional counterparts because for a given visible transmittance they reflect a much larger fraction of incident solar infrared radiation (Sweitzer, Arasteh, and Selkowitz 1986).

Results from computer simulations examining the effects of daylighting on electric lighting requirements, cooling loads, and peak electrical demand have demonstrated that energy savings from daylighting can be compromised or negated by improper design. Unfavorable solar optical properties of glazing materials, inadequate solar control, nonuniform daylight flux distribution, and excessively large effective apertures can impose cooling-load penalties greater than electric light savings.

Nevertheless, in the context of conventional building design using glazings with static properties, daylighting can reduce energy requirements and operating costs. A building with daylight-responsive lighting controls will normally be more energy efficient than on identical building without such controls. Substantially greater savings are possible with more sophisticated architectural designs having carefully sized and placed apertures and using advanced glazing technologies, solar control devices, and lighting control hardware.

Energy savings can be affected by the spectrally selective solar optical properties of the glazing material as characterized by K_e and by an active solar heat gain control system with a dynamic range and responsiveness that operates in response to the variations in incident solar flux density.

The direct economic benefit of reduced operating costs from energy and peak-demand savings are a function of utility rates. However, in order to complement and validate computer simulation results, measured data from real buildings are needed. At present few well documented case studies are available.

The most important benefit from the application of daylighting to buildings with daytime work-use patterns is not energy consumption, but increased occupant satisfaction and hence productivity. The need for a pleasant, comfortable environment without distraction and glare is paramount. New advanced glazing technologies with increased control of daylight and solar heat gain can improve visual and thermal comfort. In addition to improved technology, better design tools are needed. The introduction of new thermal and visual comfort algorithms to energy-simulation programs will be important contributions toward providing the tools that designers need to create an environment that is both thermally and visually comfortable as well as energy efficient.

ACKNOWLEDGMENT

Part of this work was supported by the Assistant Secretary for Conservation and Renewable Energy, Office of Building Energy Research and Development, Building Systems Division of the U.S. Department of Energy, U.S.A.

REFERENCES

Arasteh, D., Johnson, R., Selkowitz, S., and Sullivan, R. 1985. Energy performance and savings potentials with skylights. *ASHRAE Trans.,* Vol. 91, 154-179.

Choi, U. S., Johnson, R., and Selkowitz, S. 1984. The impact of daylighting on peak electric demand. *Energy Build.* 6: 387-399.

Johnson, R., Arasteh, D., Connell, D., and Selkowitz, S. 1986. The Effect of Daylighting Strategies on Building Cooling Loads and Overall Energy Performance. *Proceedings of the ASHRAE OSE Conference, Thermal Performance of Exterior Envelopes of Buildings III,* Clearwater Beach, Fl.

Lampert, C. M. 1981. Heat Mirror Coatings for Energy Conserving Windows. LBL Report 11446, Berkeley, Calif.

Sanchez, N., and Rudoy, W. 1981. Energy impact of the use of daylighting in offices. *ASHRAE Trans.* Vol. 87, pp. 245-260.

Selkowitz, S., Arasteh, D., and Johnson, R. 1984. Peak Demand Savings from Daylighting in Commercial Buildings. In *Proceedings of the 1984 ACEEE Summer Study.* Santa Cruz, Calif.

Sweitzer, G., Arasteh, D., and Selkowitz, S. 1986. Effects of Low-Emissivity Glazings on Energy Use Patterns in Nonresidential Daylighted Buildings. *ASHRAE Trans.,* Vol. 93, Pt. 1, pp. 1553-1566.

Conclusion: Human Comfort, Energy Savings, and Cost

Nancy Ruck

It is evident that human comfort is inextricably woven into building design and the fabric and servicing of buildings. It is also evident that the physical effects of heat, light, and sound cannot be considered as isolated elements in building design and that their interactions must be given priority according to the building's function. Greater user attention is also necessary owing to the advent of new technologies such as screen-based equipment.

The choice between human well-being, energy savings, and cost is a dilemma often confronted by building designers. For example, utilizing daylight in buildings to promote human well-being and to conserve energy can minimize the use of electricity for lighting, but because of possible increased costs for cooling and heating, the overall costs of building management may be higher.

In view of the arguments put forward in this book, it is suggested that design objectives should aim at a solution whereby basic human needs have first priority. Generally, energy savings relating to such measures as the use of daylight to reduce electric lighting in commercial and industrial buildings are relatively small compared with occupancy costs. In commercial and industrial buildings particularly, improving occupant satisfaction and productivity deserves as much or more attention than energy conservation.

Buildings create environments. They provide the temperature, humidity, lighting, and ventilation necessary for people to live and work productively, and since productivity is invariably the largest component of a building's operating costs, it can be severely affected by human stress. Any negative impact such as the impairment of visual performance by glare can reduce the value of any energy savings.

Today, owing to rising energy costs, the age-old principles of environmental control by the filtering of natural climatic elements through the building envelope are now being resurrected. The concept of a climate-adapted building from both the human and energy viewpoints and the development of new dynamic technologies have increased the potential for active rather than passive envelope design to control the indoor environment and hence the placing of less emphasis on mechanical equipment requiring nonrenewable energy sources.

In addition, building interiors have been provided with such sophisticated electronic devices as computers and video display terminals. Evidence of dissatisfaction with environments accommodating such automated equipment has raised the question of whether a building can still provide, in this technological age, visual, thermal, and acoustic satisfaction. Such equipment relies heavily, at present, on human adaptation to the heat, noise, and vibration generated, and these electronic devices also have created new visual, work, and attention stresses. There is no doubt that such environments demand a greater degree of consideration for the user.

There are many similarities and contradictions in the requirements for thermal, visual, and acoustic satisfaction in buildings. A summary of their interactions, the limited ranges of building occupants' thermal, visual, and acoustic capabilities, the influence of context, air-conditioning, and screen-based equipment on human comfort and future perspectives as regards building design and performance are discussed in this conclusion.

THE HUMAN COMFORT RANGES

It is evident from Part I that there are limited ranges of heat, light, and sound in which building occupants can experience thermal, visual, and acoustic comfort and that these limits are mainly determined by the capabilities of our visual and hearing systems and our ability to make physiological and behavioral adjustments to balance energy exchanges between the body and the environment. Above these ranges, excessive heat, glare, and noise can disable human response. External influences, both cultural and locational, also play a part.

A thermal environment can be said to be comfortable when it imposes no constraint on human well-being; a visual environment is satisfying when an eye task such as reading a book or searching for an object can be quickly and easily performed without distraction and stress. Similarly, in the human perception of sound, which follows closely that of the human perception of light, there is a limited range of frequencies and intensities to which people respond favorably.

The comfortable limits of human sensation have been determined to some extent by the human adaptation mechanisms. For example, the human eye can operate over a very large range of luminances from starlight to bright sunlight by adjusting through a photochemical process to the prevailing average conditions of luminance. The presence of excessively high luminances in the field of view, beyond an adaptation range, can cause discomfort or visual disability. As discussed in Chap. 6, luminance control in interiors can be exercised by several methods, including external and internal shading devices, shielding devices, and the use of new advanced glazing materials.

Adaptation to heat is not so positively defined. Human beings require continuous physiological and behavioral adaptation to balance energy exchanges between the body and the environment. Excessive heat is a major constraint to human well-being, for example, in the low latitudes near the equator, which emphasizes the importance of microclimate control.

At the other end of the spectrum, however, so-called thermally comfortable environments created from the constancy of temperature concept and taken for granted in most urban environments are now being viewed with some concern. Such artificially contrived environments are not only energy-source-reliant, but they also create a monotony that can produce occupant dissatisfaction.

In the case of the acoustic environment, people can readily adapt to noises that are predictable and not excessively high. They tend to become used to a noise over time, if it is a steady-state noise that remains fairly constant in level. However, annoyance reactions to noise interfering with communication are common. A great deal depends on the general acoustic environment and the context in which the noise is heard. Most electronic equipment generates noise, and like heat, the noise is cumulative. Unfortunately, distraction caused by noise is not always directly related to volume. Its frequency, rhythm, and tone all contribute to noise pollution. Therefore, while some electronic devices such as computer terminals are quieter than mechanical equipment such as typewriters, their noise can be more irritating to the human ear.

INTERIOR DESIGN AND THE IMPACT OF SCREEN-BASED EQUIPMENT

Interior design can encourage or discourage occupant activities, and the interior environment is determined partly by influences transmitted by the envelope and partly by influences generated within the building or by its control systems. Today, particularly in the commercial and industrial fields, buildings not only control such systems as heating, lighting, and air-conditioning, but they also control communication, which requires the inclusion of electronic equipment such as screen-based equipment. The provision of a comfortable and stimulating environment for occupants under these circumstances is becoming an important facet of building design.

It has been suggested that the computer and the visual display terminal can negate the benefits of building and workspace automation. All electric and electronic devices generate heat. Hence the amount of heat they produce can substantially increase a building's cooling load. The ratio of one person to a visual display terminal can effectively double the cooling load for which a building is designed. The heat generated by the equipment also tends to be local and unpredictable, depending on the monitors running at the time. This can lead to occupant discomfort.

Evidence is growing that electronic equipment can pose a potential health hazard to building occupants by generating noise and vibration, body strain, and visual stress problems. In environments containing word processors, for example, space requirements and lighting need to be reconsidered, since the conventional overhead, uniformly located lighting installation is no longer satisfactory. General lighting schemes with uniform ceiling lighting can create problems, such as reflections on display screens, and disallow individual modification of lighting levels. Paperwork requires higher illuminance levels than screen-based tasks, although both kinds of work are generally carried out adjacent to each other and at the same time. In addition, light from windows and electric lights can cause glare and reduce visibility. It is evident that a great deal of flexibility is needed in building planning, and this flexibility must include floor configurations and ceiling heights, HVAC systems, and in particular, lighting.

THE INFLUENCE OF CONTEXT

Human responses also will vary with context. Observations of human sensations by controlled laboratory studies have not taken into account this adaptation phenomenon. In actual environments, some form of adaptation occurs beyond simple adjustment to ambient conditions. For example, higher sound levels can be tolerated if there is satisfaction with the environment or the general character of the neighborhood. Acceptability of noise is also dependent on the character of the noise. Thermal response is influenced by both the location and habituation and by the activity or function of the space. Variability in thermal perception is also related to nonthermal dimensions such as humidity, color of surroundings, physical arrangement of rooms, and personality. Similarly, in lighting terms, excessive contrast or glare can be tolerated in some environments but not in others depending on the space's function. No single condition can be judged as being equally comfortable by all. Both hearing and visual sensitivity are reduced with age, although there is no clear evidence that there are differences in thermal preference with age.

However, meeting visual or noise criteria such as adequate illuminance or an acceptable sound level may result in a failure to meet other requirements, such as those associated with thermal comfort. It is the task of the designer to consider the interactions of one physical component with another when designing for comfortable and productive environments.

THE CONSTANCY HYPOTHESIS

The validity of a universal optimum in ambient warmth was questioned in Chap. 1. There is some evidence that uniform conditions are not as desirable as an environment that mimics the natural cycles of climatic change. It was noted in Chap. 1 that, despite the advent of the HVAC system and the so-called thermally comfortable indoor environments to which we are generally accustomed, our present dependence on equable and mild indoor warmth in the form of a constant set temperature is being queried as a result of considerable dissatisfaction within both homes and office buildings. If people's thermal perceptions are in accord with naturally occurring levels of warmth outdoors, there can be no universally applicable standard to set the quantity of heating and cooling required. At most, standards can apply to specific climatocultural zones as determined by the preference of the local inhabitants.

A new approach to indoor climate management that abandons thermostatic control as such and is based on a controlled variability would enable increased human adjustment by adaptation. One method to achieve atmospheric variability indoors which corresponds closely to the outdoors without loss in thermal satisfaction is to alter the thermostat function so that it becomes mobile by either programming or a designated chip to allow additional inputs. As mentioned in Chap. 5, such thermobiles could be incorporated in all new building designs.

This concept of controlled variability also can be introduced into the operating of mechanical equipment by extending the "dead band," the time when neither heating nor cooling occurs, so that heating is initiated only when the temperature falls below a certain value and cooling is initiated when the temperature rises above a specific value. This would result in a tendency for indoor temperatures to follow outdoor temperatures (see Chap. 9).

A control system can therefore be made more flexible and responsive to occupant needs. In addition, being able to maintain some individual control over the environment is also an important psychological attribute of a well-designed building, since occupants are now demanding control over many aspects of the work environment from task lighting to enjoying daylight and the view and to maintaining acoustic privacy.

To extract the greatest benefit from an air-conditioning design, it also must be considered in conjunction with the design of the building envelope, for even under the best circumstances, a HVAC system can only provide partial human comfort within a building. The physiologic state of an individual may place that person in thermal stress even though measurements in the building indicate that the person should be comfortable. Cold air movements, uncontrolled hot or cold interior surfaces, equipment radiating heat, mean radiant temperature conditions created by large glass areas, and human activities can cause stress among building occupants.

THE DYNAMIC BUILDING ENVELOPE

As mentioned in Chap. 6, this viewpoint of variable thermal conditions is compatible with a growing concern for the recognition of the value of variable and natural qualities of daylight for human well-being. It has been suggested that a building's indoor thermal conditions should be more closely attuned to the local climate. It also was demonstrated in Chap. 8 that it is possible to use external climatic energy sources by filtering and distributing them to occupied space via the envelope for end uses such as lighting, cooling, heating, and ventilation. Standard environmental control procedures that reject climatic influences are now being modified in some buildings by climate-adapting procedures and the use of advanced glazing technologies, in which the positive and negative influences such as heat loss and gain are selectively filtered and balanced by these new materials. The use of existing static glazing materials such as reflective coatings on glass may limit solar gain but have a blinding effect externally, creating both a safety problem and discomfort to motorists and pedestrians. Electrochromism is one of the new technologies that offers dynamic (visible or near-infrared) control using low voltages. A new advanced crystal technology also has made economical, large-area, electrically controlled windows a commercial reality.

However, although the envelope is constantly exposed to varying demands, its opaque components are mainly composed of elements having static properties. This has resulted in a static material logic in building design that cannot lead to optimal solutions in terms of energy savings as long as the properties of the envelope are not fully variable. The philosophies behind efforts in the field of energy balance in buildings, for example, have stagnated on the improvement and sophistication of accurate methods for the prediction of static performance, and this has tended to obscure the deeper viewpoint of the actual cyclic variation of the physical environment.

It is evident that attention should be given to a more human-oriented approach to temperature and light control in buildings, for with the new advanced glazing technologies, the transparent components of the building envelope are becoming more flexible in their response to heat and light. A great deal more research is required to integrate the design of the building envelope with the external environment and the building services to achieve greater human satisfaction in building interiors.

The new developments in mechanical environmental control systems, as discussed in Chap. 9, also provide new design possibilities for introducing variability in interior environmental conditions. However, the high cost and energy consumption of these innovations could restrict their usefulness to projects with large budgets or special requirements. With a climate-adapted building and an "active" building envelope, together with devices in the envelope itself to control heat and light admission, energy consumption can be reduced and heating, ventilating, and air-conditioning equipment can be used for interior fine-tuning only. In this case, the cost of the plant will be reduced and the satisfaction of the occupants will be increased.

There are other contradictions and therefore difficulties in obtaining an optimal response to varying climate demands, e.g., the need to have high thermal resistance to solar heat gains during the hot season but preferably high admittance to solar radiation (while avoiding heat losses) during the cold periods. Similar analogies may be put forward for the acoustic environment, since it is sometimes necessary to use the building envelope as a filter between noisy surroundings and the building's occupants. This entails sealing all openings against air leaks and, when using double glazing, providing a wide air gap between the two panes to prevent the ingress of low-frequency traffic noise. Therefore, ventilation is in conflict with excluding noise, and allowing daylight and view makes thermal insulation more difficult. Some of the requirements of the building envelope are therefore in conflict. What is expected varies from one building to another, and priorities should be established at the beginning of the design process.

It is therefore evident that the building envelope components can no longer be considered in isolation, and it is necessary to examine at an early stage in the building design the implications and interactions of, for example, daylight design decisions with other design criteria such as electric lighting, energy consumption, heat loss and heat gain, sound transmission loss, and cost. The need to balance daylighting and visual requirements with some of the consequential disadvantages of glass in the building envelope is self-evident.

It is possible that in the future, in energy terms, the building envelope can be designed to act as a gate system through which a continuous and variable flow of energy takes place. The objective of the building envelope then becomes a building system whereby the direction and rate of flow can be controlled for optimal utilization of energy. It is possible to allow the energy to flow on alternative paths—a direct flow inward and outward with a diversion path whereby energy can be collected and stored for later use. The building envelope then becomes a dynamic system providing variable routines that take outdoor conditions and indoor environmental requirements and develop and utilize them to maintain an optimal energy-conserving regime.

AN ADVANCED COMPUTER-BASED ENVELOPE DESIGN TOOL

It has been demonstrated in Chap. 13 that in order to maximize energy benefits and minimize costs, it is necessary to understand building energy performance in sufficient detail to assess the component impacts. With the introduction of energy simulation computer programs and other appropriate design tools, it is now possible to predict energy use and peak electric demand in various climates as functions of fenestration parameters and include the effects of climate, orientation, glazing area, U-value, shading coefficient, visible transmittance, lighting power density, and lighting control strategies.

Results indicate that some compromise is necessary, particularly when utilizing daylight to offset electric lighting loads. However, increasing window area and/or transmittance to increase daylight savings reaches a point, depending on climate and orientation, beyond which total energy consumption increases as a result of greater cooling loads. It should be noted that the analysis of such advanced glazing technologies as low-E glazings has demonstrated that additional electric lighting and cooling energy savings can be achieved with such glazing materials in both hot and cold climates and that their use also can create more comfortable interior environments and provide greater design freedom without adverse energy consequences. Although the first cost for such windows is higher than for those using conventional glazings, low-E glazings now have an established reliability and can offer multifunctional control of window energy performance.

Computer-based tools allow the testing and analyzing of design alternatives under a wider variety of conditions than has been previously practical with slower techniques, but they have not fundamentally changed the information available to the design professional. These tools are currently applicable to certain limited functions of the design process, such as the analysis procedures demonstrated in Chap. 13. They do not address the varying information needs of the designer at various points in the design process. The memory and speed of the computer have not yet been applied to the conceptual and creative part of the design process to assist in increasing efficiency and to reduce uncertainty.

Unless computer tools include integrated considerations of other than energy-related criteria, their usefulness to designers is limited. Most designers are not interested in considering one design criterion at a time, and the interest in energy conservation is only one of the many considerations necessary in building design and performance. Concurrent effects such as luminous or thermal comfort, view, and economics need to be included in the analysis process, and alternative fenestration systems must be examined on the basis of all these parameters.

There are also at present few means for effectively evaluating the quality of the luminous environment, view access, and aesthetic appeal without the use of realistic images. Architectural design is a visually oriented process, and the generation of appropriate images for a predicted interior environment is necessary for both communication and evaluation.

Although increasingly powerful computer graphics packages are now becoming available for workstations, the development of microcomputer technology and high-resolution imaging systems offers the most promise because such technological advances are able to overcome the limitations of computer graphic solutions. By using imaging systems, hardware and software can be used to capture, store, transmit, and present detailed realistic images from sources such as real environments or from direct computation. This concept of an advanced tool for the design of buildings is currently emerging and will result in a greater understanding of the thermal and luminous performance of the building envelope with respect to building performance and the occupants' health and comfort.

THE FUTURE OF BUILDING DESIGN

Technological advancement and changing human needs will determine the future of building design. Not only is the use of screen-based equipment changing working conditions and imposing new requirements for human comfort, their introduction is also bringing new concepts in the design of air-conditioning and lighting installations. For example, the use of general ambient lighting in conjunction with local task lighting to allow for multiple tasks is one answer to needed changes in interior luminance patterns, and with ambient lighting comes the potential opportunity to utilize daylight and to relate more closely to outdoor sky conditions.

It is envisioned that the future will bring greater technological control of the indoor environment, since with new and improved electronic devices it is possible to employ control algorithms that allow a control system to be adjusted to local conditions. This presupposes a more sophisticated integration of daylight with electric light and mobile thermostat functions that can be programmed according to appropriate temperature gradients. The requirements of human variability and energy savings can therefore in some instances be made compatible.

While the interest in energy savings will be maintained, we can anticipate that there will be additional interest in the less quantifiable aesthetics and amenities associated with climate-adapted interior space. The new technologies to serve the needs of the building designer will include not only selective coatings for glazing to reduce heat loss and solar gain, alter the spectral qualities of transmitted light, and create a wide range of reflected appearances

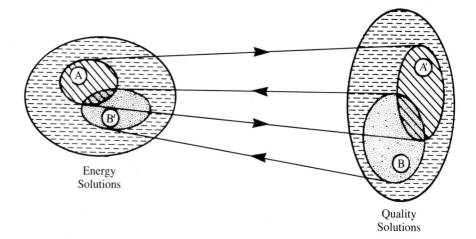

Energy
Solutions

Quality
Solutions

FIGURE A.
Schematic representation of the relationship between lighting quality and energy use. (After Selkowitz and Griffith 1986.)

on the indoor and outdoor surfaces, but also coatings that alter the distribution of transmitted light and dynamically control intensity.

In addition performance criteria will be examined in the context of total building performance, for it is anticipated that design criteria will evaluate both the technical aspects of performance and issues of comfort and amenity that are at present less quantifiable. However, there are at present few design criteria that address all the preceding issues against which performance can be compared.

An approach and possible framework for an evaluation procedure for a high-quality lighting design that maximizes visual comfort and minimizes energy consumption is shown schematically in Fig. A (Selkowitz and Griffith 1986). The framework is an attempt to relate the qualitative aspects of all lighting design solutions. On the left-hand side of the figure is conceptualized the range of all possible lighting-quality solutions that meet the basic requirements of the lighting problem. Within the range of all possible solutions represented by points within the space, there is a smaller "design space" boundary A containing those solutions which meet the minimal criteria for "good lighting quality." Each of the design solutions has an annual energy use associated with it, and each also provides differing levels of visual performance. On the right-hand side of the figure is shown the solution space for energy performance. Within the energy space, those solutions with the best energy performance are outlined as in boundary B. If the best visual-quality solutions are mapped to the energy boundary space A′, some of the points also fall within the best energy space B while others lie outside. The same is true in reverse.

To decide between these alternatives, some form of implicit economic judgment could be exercised. However, the cost of lighting quality and visual performance is not necessarily the same for each energy solution. An extension of the analysis described to include visual performance could be used to identify and quantify the tradeoffs, as shown in Fig. B.

The value of increased or reduced visual quality or performance could be inferred from changes in rental value, vacancy rates, and other side effects that result from the lighting-quality feature. In the set of envelopes at the right of the figure, the costs or benefits of energy, lighting quality, and visual performance are translated into an economic value that allows tradeoffs to be made and solutions optimized with the constraints of the design criteria governing the solution. Such a procedure is not simple, but it is an indication of an approach to initially reconcile the wide range of design factors that require consideration.

Many advances in the efficiency of building design, construction, and operation over the past decade have produced significant reductions in energy consumption and hence energy costs. From the techniques used, several issues have emerged, in particular the need for more pleasant and productive work environments. There is no inherent conflict between attaining success in the

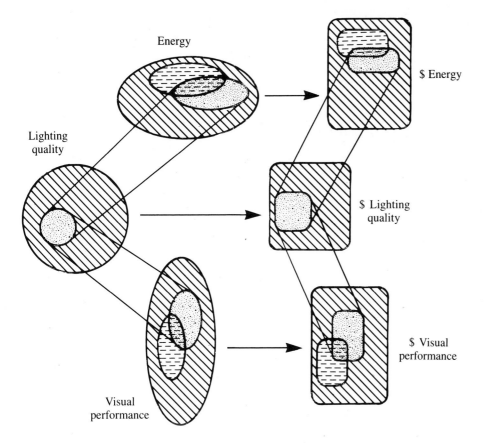

Energy

Lighting quality

Visual performance

$ Energy

$ Lighting quality

$ Visual performance

FIGURE B.
Schematic representation of the relationships between lighting quality, visual performance, and energy and the costs or benefits of each. (After Selkowitz and Griffith 1986.)

latter issues and success in the reduction of energy costs, but there is the danger that environmental quality and human well-being will be overlooked in our endeavors to design cost-effective buildings without regard for their impact on human occupants.

SUMMARY

It is evident from the ideas discussed that in building design, human requirements, energy savings, and cost-effectiveness should be examined as factors for optimal integration, and the minimization of energy savings must be consistent with the functional objectives of the building and the needs of the occupants. In housing, these may include comfort and health; in offices, they also include productivity.

It is also evident that unless prediction tools include integrated considerations of other than energy related criteria their usefulness to designers is limited. Appropriate visual, thermal, and acoustic response to the built environment together with view and economics needs to be included in the analysis process.

Overall, the design objective should be to seek a solution with regard for human well-being in which minimal overall costs are obtained with the greatest possible energy savings.

REFERENCE

Selkowitz, S., and Griffith, G. W. 1986. Effective daylighting in buildings revisited. *Light. Design Appl.* March 1986. 34-47.

Glossary

The following definitions have been developed with the object of conveying a basic understanding of the concepts involved. For more precise definitions, reference should be made to an appropriate source.

Absolute humidity: amount of water vapor in air.

Absorption: transformation of radiant energy to a different form of energy by the intervention of matter.

Acclimatization: physiological and psychological changes experienced with changing environmental warmth.

Accommodation: the spontaneous adjustment of the eye for the purpose of looking at an object situated at a given distance.

Adaptation (of the body): psycho-physiological and cultural adjustment to environmental stimulus over time.

Adaptation (of the eye): the process taking place as the eye becomes accustomed to the luminance or the color of the field of view.

Air-to-air transmittance: the rate at which a wall or roof transmits heat energy from the air on one side of it to the air on the other side. It involves not only the resistance of the construction itself, but also the surface resistances on both sides.

Algorithm: repetitive computational procedure.

Ambient temperature: temperature of immediate vicinity.

ASHRAE: American Society of Heating, Refrigerating and Air-Conditioning Engineers.

ASHRAE scale: verbalized scale of thermal sensations commonly employed in laboratory studies.

Bedford scale: verbalized scale of thermal comfort commonly employed in field studies.

Candela (cd): the unit of luminous intensity.

Candela per square meter (cd/m²): unit of luminance recommended by the C.I.E. Non-metric unit: footlambert (ft-L). 1 ft-L = 3cd/m²

Chi-squared: elementary statistical test of association.

Circadian cycle: physiological responses due to the diurnal cycle.

Chromaticity: mathematical/graphical means of describing color.

Chromaticity co-ordinates: two numbers that fix the position of a point on a chromaticity diagram, representing the proportions of primary colors that make up a particular color.

Color appearance (of a lamp): determined by the spectral emission of the lamp and quantified by its chromaticity co-ordinates.

Color rendering: a general expression for the degree to which a lamp faithfully shows the colors of objects.

Color rendering index: a measure of the degree to which the colors of objects illuminated by a given lamp conform to those of the same objects under an ideal light source such as daylight or a tungsten filament lamp of similar color appearance.

Color temperature (of a light source): a measure of the "yellowness" (or warmth) to "blueness" (or coolness) in the color appearance of a light source. The higher the temperature of an incandescent source, the less yellow (or the more blue) it appears. Unit: Kelvin (K).

Comfort zone: range of temperatures, or specific index values, over which an individual or the majority of a group do not experience discomfort.

Conduction: the transfer of heat through a stationary medium. The heat energy is transferred from molecule to molecule.

Cones (of the eye): retinal receptor elements which are primarily concerned with the perception of light and color by the light adapted eye (*see* photopic vision).

Constancy hypothesis: assumption that climatic, geographic, and cultural differences do not introduce variability in thermal sensation, expectation, or preference.

Contrast sensitivity: reciprocal of the minimum relative luminance difference perceptible.

Convection: heat transfer by means of a fluid medium. The heat energy is transferred mainly by the movement of the fluid.

Corrected effective temperature: effective temperature as modified by mean radiant temperature.

Correlated color temperature (of a lamp): the temperature of an ideal tungsten filament light source that emits light of a color appearance closest to that of the lamp being considered. Unit: Kelvin (K).

Daylight: visible global radiation—the sum of sunlight and skylight.

Daylight factor: the ratio of the skylight illuminance at a point on a given plane due to the light received directly or indirectly from a sky of assumed or known luminance distribution to the simultaneous illuminance on a horizontal plane exposed to an unobstructed hemisphere of this sky. Direct sunlight is excluded from both values of illuminance.

°C: degrees Celsius (used to describe the temperature of an object).

Design temperature: idealized indoor temperature for which a building should be designed.

Differential attributes scale: perception scale in which properties of an environment are judged for appropriateness of description in steps ranging from "very accurate" to "very inaccurate."

Diffuse reflection: reflection where incident flux is redirected and scattered in many directions.

Diffuse transmission: transmission in which the light is scattered in many directions.

Diffusion: the spatial distribution of a beam of light which after reflection at a surface or transmittance through a surface travels on in numerous directions.

Disability glare: glare which impairs the vision of objects without necessarily causing discomfort.

Discharge lamp: a lamp which depends on an electric discharge through a gas or a metal vapor or a mixture of several gases or vapors.

Discomfort glare: glare which causes discomfort without necessarily impairing the vision of objects.

Effective aperture: the ratio of window to wall area multiplied by the transmittance of the glazing.

Effective temperature: thermal index combining effects of air temperature, humidity, and air movement for two conditions of clothing.

Electrochromic (material): optical switching material consisting of a film deposited by electrochemical deposition on doped tin oxide-coated glass. The film is capable of dynamically controlling the incoming solar radiation by switching from an uncolored state to a colored one as a result of an applied current.

Emissivity: the ratio of radiation from unit area of a surface to the radiation from unit area of a full emitter (black body) at the same temperature.

Endocrine (gland): secretes directly into blood or lymph—ductless.

Entropy: randomized energy state.

Equal interval attitude scale: attitude test based on standardized weighting of responses to specific questions.

Equatorial comfort index: thermal index combining effects of air temperature, humidity, and air movement for use with heat acclimatized people.

Equivalent sphere illuminance (ESI): the level of sphere illuminance that would produce a visibility equivalent to that produced by a specific lighting environment.

Equivalent warmth: thermal index combining effects of air temperature, humidity, and mean radiant temperature for use in moderately warm environments.

Externally reflected component (of daylight factor): that part of the illuminance at a point which is received directly from external reflecting surfaces illuminated by a sky of assumed or known illuminance distribution.

Factor analysis: statistical procedure for grouping associated variables.

Flicker: impression of fluctuating brightness or color.

Fluorescent lamp: a discharge lamp in which most of the light is emitted by a layer of fluorescent material excited by the ultraviolet radiation from the discharge.

Gaseous contaminant: an impurity in the air which is a gas or vapor other than those considered to constitute normal air (e.g. carbon monoxide, formaldehyde).

Glare: the discomfort and/or impairment of vision experienced when parts of the field of view are excessively bright in relation to the general surroundings.

Glazing efficacy factor: the ratio of the glazing transmittance to the shading coefficient of the glazing.

Heat strain index: thermal index estimating the ratio of evaporative cooling required to the maximum amount of cooling possible.

Heating degree-days: a measure of the amount of heating required in a given climate. For every day of the year when the average temperature falls below a certain value (usually $18°C$), the number of degrees below that value is added up.

Homeotherm: an animal that maintains core temperature constancy.

Homogeneous (sky): a sky with turbidity, pollution, or cloudiness uniformly dispersed over the entire sky hemisphere. Fully overcast skies are examples of real homogeneous skies as are uniformly polluted skies.

Illuminance: the luminous flux falling on a surface divided by the area of the illuminated surface. It is expressed in lux (1 lumen per meter^2) or in footcandles.

Illumination: a general expression for the amount of light falling on a surface. The physical measure of illumination is illuminance.

Incandescent (electric) lamp: a lamp in which the light is produced by means of a body (filament) heated to incandescence by the passage of an electric current.

Index of thermal stress: thermal index indicating cooling rate produced by sweating to maintain thermal balance.

Index of warmth: a composite measure of all or some parameters involved in thermal balance.

Infrared radiation: any radiation whose components lie within the wavelength range of 780 to 10^5 nanometers.

Internally reflected component (of daylight factor): that part of the illuminance at a point received directly from internal reflecting surfaces which are illuminated directly or indirectly by a sky of known or assumed luminance distribution.

Isohyet: line on map joining equal precipitation values.

Isopleth: line on map joining equal values.

Isotherm: line on map joining locations with the same mean temperature.

K: Kelvin (or degrees Kelvin). Temperature *differences* in Kelvin or Celsius degrees have the same value, but temperature *readings* on the two scales are different because the Kelvin scale starts at Absolute Zero, whereas the Celsius scale has its zero at the freezing point of water (approx. 273K).

Linear regression: statistical technique that predicts the most likely association between variables assuming a linear relationship.

Lumen: the luminous flux emitted within a unit solid angle (one steradian) by a point source having a uniform intensity of one candela.

Lumen per square meter (lux): unit of illuminance recommended by the C.I.E. Non-metric unit: footcandle, lumen per square foot. $1 \ \text{lm/ft}^2 = 10.76$ lux.

Luminance: the physical measure of the brightness of a surface such as a lamp, luminaire, sky, or reflecting material, in a specified direction. It is the luminous intensity emitted by an area of the surface divided by the area. Unit: Candela per square meter (cd/m^2) or apostilb. Non-metric unit: foot-Lambert. $1 \ \text{ft-L} = 3 \text{cd/m}^2$.

Luminous efficacy (of a lamp): the ratio of the luminous flux emitted by a lamp to the electrical power which it consumes. Unit: lumen per Watt (lm/W).

Luminous flux: the amount of light emitted by a lamp, luminaire, or the sky irrespective of the directions in which it is distributed and expressed in terms of its capacity to produce a luminous sensation. Unit: lumen (lm).

Luminous intensity (in a given direction): the concentration of luminous flux emitted in a specified direction. It can also be expressed as the luminous flux emitted by a source in an infinitesimal cone containing the given direction by the solid angle of that cone.

Mean radiant temperature: a measure of the effect of the radiant energy from all sources that radiate toward a given point. It is the temperature of a black body which, if located at that point, would have a zero net radiant energy exchange with its surroundings.

Mesopic vision: vision in conditions intermediate between those of photopic and scotopic vision.

Metabolic rate: amounts of metabolism varying according to activity. Usually expressed as Watts per square meter of body area.

Metabolism: liberation of chemical energy within the body.

Multiple correlation coefficient: composite statistical association between more than two variables.

Nanometer: one thousand millionth of a meter.

Neutrality: level of warmth incurring minimum thermoregulatory activity, temperatures at which no thermal sensation is experienced.

New effective temperature: thermal index combining effects of air temperature and humidity for use with lightly clothed people.

Normalized certainty scale: perception scale in which both comfort and discomfort are judged in steps ranging from "completely certain" to "uncertain."

Operative temperature: thermal index combining effects of air temperature and mean radiant temperature.

Optical radiation: that part of the electromagnetic spectrum which lies within the wavelength range of 100 nm and 1 mm.

Particulate contaminant: an impurity in the air which is formed of small particles of solid or liquid material (e.g. dust, mist).

Passive solar building: a building designed to maintain reasonably comfortable conditions for its occupants, relying on solar energy and the design and materials of the building.

Photobiological: biological effects from light.

Photochromic (material): optical switching material consisting of a film deposited on glass which is automatically responsive to light intensity.

Photon: an elementary quantity of radiant energy (quantum). Its value is equal to the product of Planck's constant and the frequency of the electromagnetic radiation.

Photopic vision: vision mediated essentially by the cones and associated with adaptation to a luminance level of several cd/m^2.

PMV (or predicted mean vote): thermal index predicting average sensation on the ASHRAE scale by combining all heat balance parameters.

Point of neutrality: thermal condition under which no warmth stimulus is perceived.

Precipitable vapor pressure: pressure exerted in air by amount of water that could be precipitated from a vertical column at a given moment.

Predicted 4-hour sweat rate: thermal index estimating amount of sweating produced by combined effects of air temperature, humidity, air movement, mean radiant temperature, metabolic rate, and amount of clothing worn.

Principal components analysis: statistical technique that groups main associated variables.

Probit transformation: statistical procedure for determining boundaries of comfort votes by linearizing cumulative percentage frequencies of vote categories.

Pupil: variable aperture in the iris, through which the rays producing the image enter the eye.

Rachitis: rickets, disease with softening of bones.

Radiation: emission or transfer of energy in the form of electromagnetic waves or particles or heat transfer through a vacuum or a transparent medium. The medium is not involved in the transfer process.

Radon: a radioactive gas occurring in nature usually emanating from some soils and ground water.

Reflectance: the ratio of the luminous flux reflected from a surface to the incident luminous flux. Reflectance is usually expressed as a decimal in the range of 0 to 1.

Reflected glare: glare produced by specular reflections of luminous objects appearing on or near the object viewed.

Regression coefficients: algorithms established by regression analysis.

Relative humidity: percentage saturation of air with water vapor.

Relative spectral power distribution: description of the spectral character of radiation (of a light source), i.e., the way in which the relative spectral concentration of a radiant quantity varies throughout the spectrum.

Relative strain index: thermal index developed from the heat strain index combining effects of metabolic rate, air temperature, mean radiant temperature, clothing insulation, clothing wettedness, and vapor pressure of the air.

Resultant temperature: thermal index combining effects of air temperature, humidity, and air movement for use in moderate climates.

Retina: a membrane at the back of the eye which is sensitive to light and composed of photoreceptors (cones and rods) and nerve cells that transmit the stimulation of the receptors to the optic nerve.

Rods (of the eye): retinal receptor elements which are primarily concerned with the perception of light by the dark adapted eye (see scotopic vision).

Scotopic vision: vision mediated essentially by the rods and associated with adaptation to a luminance level below some hundredths of a cd/m^2.

Shading coefficient (SC): the ratio of the solar heat gain through any window to the solar heat gain factor for standard glass under the same conditions.

Skylight: visible diffuse radiation caused by sunlight being diffused and scattered in the atmosphere.

Sodium (vapor) lamp: a discharge lamp in which the light is mainly due to sodium radiation.

Sunlight: visible direct beam radiation from the sun.

Thermal discomfort: subjective responses occurring outside the comfort zone.

Thermal inertia: the heat capacity of a building or of a building element. A building with a high value of thermal inertia responds slowly to heat gains or losses.

Thermal shock: adverse bodily response to transition between places with a large temperature gradient.

Thermal stress: response to hot or cold stimulus.

Thermobile (thermomobile): electronic microclimate control system to maintain preferred levels of warmth varying with outdoor temperature change.

Thermochromic (material): an optical switching material consisting of a film deposited on glass which is automatically responsive to temperature.

Thermopreferendum: preferred and actively sought thermal environment.

Thermoregulation: physiological and behavioral means by which a homeotherm maintains internal temperature constancy.

Translucent (material): a material that transmits light principally by diffuse transmission. Objects are not seen distinctly through such material.

Transmission: Passage of radiation through a medium without change of frequency of the radiation components.

T-test: statistical test for significance in differences between means of samples.

Turbidity Factor (according to Linke): ratio of the vertical optical thickness of a turbid atmosphere to the vertical optical thickness of a clean dry atmosphere as related to the whole solar spectrum.

Ultraviolet radiation: any radiation whose components lie within the wavelength range of 10 to 380 nanometers.

Urban atmospheric dome: air over urban areas as restricted by pollution inversions, surface features, and active energy use.

Visibility level (VL): a measure of how effective real lighting can be in making a particular task or object visible to a specified person. It can be expressed as the ratio of the luminance contrast of a task under reference lighting to the luminance contrast of the task when it can only just be seen.

Visible radiation: any radiation capable of causing a visual sensation directly. The wavelength range of such radiation can be considered to lie between 380 and 760 nanometers.

Visual acuity: the ability to see fine details or the angle subtended at the eye by the detail that can be discriminated.

Visual field: all points of space at which an object can be perceived when the head and eye are kept fixed.

Visual performance: the quantitative assessment of the performance of a visual task taking into consideration speed and accuracy.

$W/m^2 K$: Watts per square meter Kelvin. The rate of heat transmittance through a unit area of surface, caused by a unit temperature difference.

Wavelength: the distance between two successive points of a periodic wave in the direction of propagation, in which the oscillation has the same phase.

Index